VOLUME NINETY THREE

ADVANCES IN BOTANICAL RESEARCH

Latex, Laticifers and Their Molecular Components: From Functions to Possible Applications

ADVANCES IN BOTANICAL RESEARCH

Series Editor

Jean-Pierre Jacquot
Professor, Membre de l'Institut Universitaire de France, Unité Mixte de Recherche INRA, UHP 1136 "Interaction Arbres Microorganismes", Université de Lorraine, Faculté des Sciences, Vandoeuvre, France

VOLUME NINETY THREE

ADVANCES IN
BOTANICAL RESEARCH

Latex, Laticifers and Their Molecular Components: From Functions to Possible Applications

Edited by

ROBERT NAWROT

Department of Molecular Virology,
Institute of Experimental Biology, Faculty of Biology,
Adam Mickiewicz University in Poznań,
Poznań, Poland

ELSEVIER

ACADEMIC PRESS

An imprint of Elsevier

Academic Press is an imprint of Elsevier
125 London Wall, London EC2Y 5AS, United Kingdom
525 B Street, Suite 1650, San Diego, CA 92101, United States
50 Hampshire Street, 5th Floor, Cambridge, MA 02139, United States
The Boulevard, Langford Lane, Kidlington, Oxford OX5 1GB, United Kingdom

First edition 2020

ISBN: 978-0-08-102995-4
ISSN: 0065-2296

For information on all Academic Press publications visit our website at https://www.elsevier.com/books-and-journals

Publisher: Zoe Kruze
Acquisition Editor: Sam Mahfoudh
Editorial Project Manager: Peter Llewellyn
Production Project Manager: Divya Krishna Kumar
Cover Designer: Alan Studholme

Typeset by TNQ Technologies

Contents

Contributors

Sophia Bałdysz
Department of Molecular Virology, Institute of Experimental Biology, Faculty of Biology, Adam Mickiewicz University in Poznań, Poznań, Poland

Céline Bottier
CIRAD (French Agricultural Research Center for International Development), UMR IATE (Agropolymer Engineering and Emerging Technologies Research Unit), Montpellier, France; IATE, Univ Montpellier, CIRAD, INRA, Montpellier SupAgro, Montpellier, France

Xia Cai
Key Laboratory of Resource Biology and Biotechnology in Western China, Northwest University, Ministry of Education, Xi'an, China

Diego Demarco
Universidade de São Paulo (USP), Departamento de Botânica, São Paulo, São Paulo, Brazil

Cléverson Diniz Teixeira Freitas
Universidade Federal do Ceará (UFC), Departamento de Bioquímica e Biologia Molecular, Fortaleza, Ceará, Brazil

Anna Goździcka-Józefiak
Department of Molecular Virology, Institute of Experimental Biology, Faculty of Biology, Adam Mickiewicz University in Poznań, Poznań, Poland

Peng Huang
Hunan Key Laboratory of Traditional Chinese Veterinary Medicine, Hunan Agricultural University, Changsha, Hunan, China; College of Horticulture and Landscape, Hunan Agricultural University, Changsha, Hunan, China

Michalina Krakowiak
Department of Molecular Virology, Institute of Experimental Biology, Faculty of Biology, Adam Mickiewicz University in Poznań, Poznań, Poland

Flávia Maria Leme
Instituto de Biociências, Universidade Federal de Mato Grosso do Sul (UFMS), Cidade Universitária, C.P. 549, Campo Grande, Mato Grosso do Sul, Brazil

Cristina Ribeiro Marinho
Departamento de Ciências Farmacêuticas, Faculdade de Ciências Farmacêuticas de Ribeirão Preto, Universidade de São Paulo (USP), São Paulo, Brazil

Maria Camila Medina
Universidade de São Paulo (USP), Departamento de Botânica, São Paulo, São Paulo, Brazil

Francimauro Sousa Morais
Instituto Federal de Educação Ciência e Tecnologia do Amazonas (IFAM), Manaus, Amazonas, Brazil

Oskar Musidlak
Department of Molecular Virology, Institute of Experimental Biology, Faculty of Biology, Adam Mickiewicz University in Poznań, Poznań, Poland

Robert Nawrot
Department of Molecular Virology, Institute of Experimental Biology, Faculty of Biology, Adam Mickiewicz University in Poznań, Poznań, Poland

Erika Prado
Universidade de São Paulo (USP), Departamento de Botânica, São Paulo, São Paulo, Brazil

Márcio Viana Ramos
Universidade Federal do Ceará (UFC), Departamento de Bioquímica e Biologia Molecular, Fortaleza, Ceará, Brazil

Mengshan Sun
Hunan Key Laboratory of Traditional Chinese Veterinary Medicine, Hunan Agricultural University, Changsha, Hunan, China; College of Horticulture and Landscape, Hunan Agricultural University, Changsha, Hunan, China

Simone Pádua Teixeira
Departamento de Ciências Farmacêuticas, Faculdade de Ciências Farmacêuticas de Ribeirão Preto, Universidade de São Paulo (USP), São Paulo, Brazil

Alicja Warowicka
Department of Animal Physiology and Developmental Biology, Institute of Experimental Biology, Faculty of Biology, Adam Mickiewicz University, Poznań, Poland; NanoBioMedical Centre, Adam Mickiewicz University, Poznań, Poland

Jianguo Zeng
Hunan Key Laboratory of Traditional Chinese Veterinary Medicine, Hunan Agricultural University, Changsha, Hunan, China; College of Horticulture and Landscape, Hunan Agricultural University, Changsha, Hunan, China; National and Local Union Engineering Research Center of Veterinary Herbal Medicine Resource and Initiative, Hunan Agricultural University, Changsha, Hunan, China

Xueyan Zhao
Key Laboratory of Resource Biology and Biotechnology in Western China, Northwest University, Ministry of Education, Xi'an, China; Xi'an Botanical Garden of Shaanxi Province, Xi'an, China

Pierre Gadal Obituary

Pierre Gadal, 1938–2019

Pierre Gadal

After a PhD on galls and the metabolism of tannins in plants, Pierre Gadal became assistant professor at Université Paul Sabatier in Toulouse. He then became a full professor at Université de Nancy I in Lorraine (1973–1981) and finally at Université de Paris-Sud Orsay until the end of his career in 2003. He headed a CNRS-associated lab (Physiologie Végétale Moléculaire) in Nancy and later in Orsay to finally become the initiator and the first director of Institut de Biotechnologie des Plantes (CNRS-Université Paris-Sud), an institute which played a very important role in the structuration of plant biology around Paris. Pierre Gadal had a large number of pedagogic and administrative responsibilities, in particular he supervised the training of PhD students at Paris-Sud during many years. Pierre Gadal was very influential in the domain of plant physiology in France and was awarded a prize from the French Academy of Sciences in 1998 as a crowning of his scientific career. His scientific contributions have gained international recognition with more than 200 articles published in peer-reviewed journals. More specifically his major scientific contributions were linked to the elucidation of the regulation of CO_2 fixation in plants through the study of PEP carboxylase and NADP-malate dehydrogenase, but also to the pathways allowing nitrogen assimilation in plants (mostly glutamine synthetase and GOGAT). His most-recent interests concerned the carbon-nitrogen relations in plants involving isocitrate dehydrogenase. Pierre Gadal and his group have been leaders in plant biochemistry, introducing in the late 80's the use of genetic engineering and structural biology. Pierre Gadal was been a series editor for *Advances in Botanical Research,* a serial book series by Elsevier, supervising volumes 61 to 86. He died in October 2019 after fighting a difficult disease. Reactions to his departure have been unanimous: he was a key figure in plant biology who will be missed. All the students and colleagues who had the chance to work with him were impressed by his charisma and enthusiasm. He was always stimulating us to go forward and at the same time he paid extreme attention to his colleagues helping them as much as he could, irrespective of their social positions. He will be thoroughly missed by all of us.

Preface

Latex-bearing plants are extremely diverse, with more than 20,000 species from over 40 different families of angiosperm plants, like Asteraceae, Euphorbiaceae, Apocynaceae or Papaveraceae. The occurrence of latex represents a highly convergent trait, which has evolved numerous times independently in phylogenetically distant plant families and orders. The probable evolutionary advantages are connected to the synthesis and storage of huge amounts of highly concentrated toxic, defense-related substances against herbivores and pathogens. Moreover, latescent plants frequently occur in tropical regions, where interactions between plant and herbivorous insects are more intense than in temperate regions. Plant latex, named also a milky sap, is a complex emulsion of different molecular components, such as terpenoids, alkaloids, proteins, starches, sugars, oils, tannins, gums, etc. It exudes from the plant immediately after mechanical injury and in most cases coagulates quickly upon air exposure. It enables the immediate transport of defense substances to the precise point of the pathogen attack. Thanks to preformed defense substances ready to be used at the moment of attack it makes the system much more advantageous than non-laticiferous inducible defense systems.

This book includes contributions of experts working in different areas of plant latex research across the globe, covering broad information on plant latex, the structure of laticifers, proteins, and low-molecular compounds of the latex, their possible functions and applications in pharmacology, industry and medicine. The opening Chapter 1 by Marcio Ramos *et al.* gives novel insights into plant latex-borne defense and the latex itself, introduces its components, which play a role in plant defense.

Latex is an exudate of specialized cells called laticifers, which are best defined by their origins, development, and overall anatomy. The excellent chapter on the "Structural diversity and distribution of laticifers" was written by Simone Pádua Teixeira *et al.* (Chapter 2), which deals with the structure, ultrastructure, and development of laticifers in plants. The very valuable part shows the extensive bibliographic survey on taxonomic distribution of laticifers, what will be especially useful for researchers and students of the subject.

Latex-bearing plants and their products have been utilized by humanity for centuries for medicinal or industrial purposes. One of such plants is

Chelidonium majus L., perennial latex-bearing plant belonging to the family *Papaveraceae*, which has been used for centuries in folk medicine, as early as the Greek botanist Dioscorides in the 1st century A.D. wrote "De Materia Medica". The plant restored the sight to birds progeny (Greek "*chelidon*" = swallow bird), also mentioned in the French Furetière dictionary (17th century). Chinese medicine, as well as South Asia Ayurveda (ancient Indian medical system) knew about a broad variety of benefits of this plant. My interest in *Chelidonium majus* originated from the traditional use of its exuding yellow latex against warts and condylomas, which are the effect of human papilloma virus (HPV) infection, hence presenting antiviral activity. Research conducted to reveal this activity has led to the description of *C. majus* latex proteins, which could possibly act by facilitating the transport of low-molecular compounds. The plant latex proteins are described in full in the chapter by Oskar Musidlak *et al.* entitled "Plant latex proteins and their functions" (Chapter 3), with detailed information regarding *C. majus* latex proteins. The plant is also described in the chapter by Alicja Warowicka *et al.* "Pharmacologically active compounds from latex-bearing plants" in the context of diverse low-molecular components produced by its latex (Chapter 5).

As mentioned earlier, plant latex is rich in diverse molecular components, functions of which are still not very well understood. It can be considered as a treasury of bioactive specialized metabolites, like alkaloids, terpenoids, lignans, cannabinoids, cardiac glycosides, or tannins. They show diverse biological activities, like anticancer, anti-inflammatory, antiproliferative, antiviral, antimicrobial, antioxidant, antiparasitic or insecticidal. They can be used as bioproducts important for human use, like medicines, toxins, pharmaceuticals, pesticides immune allergens, and industrial products.

The idea of this book was to summarize the current state of knowledge on two kinds of molecular components of the latex: macromolecular, which comprises proteins, and low-molecular, which mainly concerns alkaloids, terpenoids, starches, sugars, oils, tannins, and many others. Hence, the chapters from 3 to 5 deal in details with these molecules. In Chapter 3, written by my group, we tried to deal with different families of proteins in the latex and with their possible functions. Special emphasis was put into contribution of proteins to latex-borne defense, as such function of the latex comes out of the research of many groups. We also mark the antiviral activity of the latex of different plants in terms of proteins engaged in such activity. Chapter 4 by the group of Jianguo Zeng describes low-molecular compounds of latex-bearing plants *Macleaya cordata* and *Cannabis sativa*, and their biosynthesis, with the special emphasis on the synthesis of sanguinarine in *M. cordata* and cannabidiol in *C. sativa*. On the other hand, Chapter 5 by Alicja Warowicka *et al.*, deals with the subject of pharmacologically active compounds from latex-bearing plants, many of them with established medicinal use. It summarizes different groups of latex molecules, from alkaloids, terpenoids and phenolics to glycosides, saponins, tannins and acetogenins. It also describes the allergenic potential of the latex pathogenesis-related proteins and ends with the very interesting potential use of latex compounds in nanotechology.

The last two chapters present the latest knowledge on the useful bioactive compounds from *Euphorbia* species, the largest genus of Euphorbiaceae, and from *Hevea brasiliensis*, which is a source of commercially used rubber. Plants from the family Euphorbiaceae are known for the production of diverse terpenoids, and were used in traditional folk medicine in different areas of the world. Chapter 7 by Xueyan Zao and Xia Cai elaborates on latex proteins and enzymes involved in terpenoid biosynthesis of selected species: *Euphorbia kansui* Liou and *Euphorbia helioscopia*, showing the complexity of the pathways leading to the synthesis of latex compounds.

One of the most economically important species from Euphorbiaceae family is a tropical plant *Hevea brasiliensis*. It grows in a form of a tree and presents a main source of renewable latex to produce commercial natural rubber, which is used in the industry to produce tires, gloves, medicinal tubing, and others. Celine Bottier in Chapter 8 provides excellent insights into other biochemical components of *H. brasiliensis* latex by describing protein, lipid, carbohydrate, and mineral contents, which also contribute to latex important industrial as well as medicinal properties.

To conclude, many latex plants contain valuable low-molecular compounds or secondary metabolites, which are widely used in industry, pharmacology, medicine and other areas. However, broad research on different aspects of latex-bearing plants is still lacking. New generation genome sequencing techniques and the reduction of costs, are expected to provide new opportunities for the research on the latex-bearing plants. I hope that this book, which gathers the present knowledge on plant latex and its molecular components, will push the researchers into further investigation. I am aware of the fact, that not all areas and potential useful plants were covered in the book, as the latex world is extremely intricate and diverse, with limited number of scientists working on the subject. However, we tried to give the compendium of the current state-of-the-art, which will hopefully stimulate the reader for further search on plant latex, its components, and potential applications in different areas of human lives.

I would like to acknowledge for the kind invitation of Professor Jean-Pierre Jacquot to coordinate this project, which was a great and unique adventure. I would like to also thank to all authors for contributing excellent reviews to this volume, creating a unique survey on multiple aspects of plant latex and laticifers. I hope that it will reach a broad visibility among students and scientists, who wish to receive a pronounced and detailed monographic overview on different aspects of latescent plants. My dream is that the new generation of scientists, not only from the plant sciences area, but also pharmacy, medicine and industry, will discover this treasury of substances with the benefit to us all.

Welcome to the fascinating world of plant latex.

Robert Nawrot

December 8th, 2019

CHAPTER ONE

Plant latex and latex-borne defense

Márcio Viana Ramos[a,*], Cléverson Diniz Teixeira Freitas[a], Francimauro Sousa Morais[b], Erika Prado[c], Maria Camila Medina[c] and Diego Demarco[c]

[a]Universidade Federal do Ceará (UFC), Departamento de Bioquímica e Biologia Molecular, Fortaleza, Ceará, Brazil
[b]Instituto Federal de Educação Ciência e Tecnologia do Amazonas (IFAM), Manaus, Amazonas, Brazil
[c]Universidade de São Paulo (USP), Departamento de Botânica, São Paulo, São Paulo, Brazil
*Corresponding author: E-mail: vramos@ufc.br

Contents

Abstract

Laticifers are present in plants belonging to a broad range of taxonomic groups, and originated independently during the evolution of species. Although laticifers are cells with specialized metabolism, their ultrastructural aspects are closely similar among different species. They appear early after seed germination and elongate throughout the plant body. Recent evidence suggests that independent clusters of laticifers may arise in different organs and tissues of the plant and then exhibit distinct chemical and biochemical contents. Many proteins and secondary metabolites are synthesized and stored in laticifers. Despite similarities in metabolism, proteomes and the secondary metabolites existing in laticifers are distinct among species even when they are closely

Advances in Botanical Research, Volume 93
ISSN 0065-2296
https://doi.org/10.1016/bs.abr.2019.09.002

related taxonomically. Studies of molecular composition, structural aspects and activities of latex molecules suggest that laticifers and latex are strongly involved in plant defense.

1. Introduction

An impressive number of plant species distributed in diverse taxonomic groups share a common and intriguing property. These plants exude a milky fluid when they undergo a mechanical injury caused by various biological aggressors (Konno, 2011). This fluid is frequently white and has a coagulation-like effect after release, sealing the injury and then becoming sticky. This milky fluid is called latex and is the product of the metabolic activity of specialized cells called laticifers. The laticifers can be organized as a unique row of cells growing continuously, accompanying the plant's growth and being distributed throughout the plant body. Laticifers ramify and can also anastomose in such a way that a network resembling a web is formed, advancing through plant tissues. Their tube-like architecture covers all tissues and is particularly intense in leaves (Hagel, Yeung, & Facchini, 2008). Accordingly, latex flows faster from leaves and petioles (green parts), but can also leak from fruits, roots or trunk. Adjacent cells composing the laticifers can have their walls totally or partially dissolved. This observation and the branching and anastomosing events serve as the basis to classify different types of mature laticifers and to study their evolutionary aspects (Hagel et al., 2008).

All typical subcellular structures of plant cells, such as mitochondria, plastids, ribosomes and vacuoles, are all present in laticifers, so the ordinary molecular machinery involved in basal metabolism is present (Agrawal & Konno, 2009). In addition, these cells produce diverse metabolites originated from secondary pathways, such as phenylpropanoids, phenolic compounds, and terpenoids (Agrawal, Petschenka, Bingham, Weber, & Rasmann, 2012). Many of these compounds have been shown to be deleterious to insects and microorganisms. However, there is still limited information about these secondary compounds in terms of seasonal synthesis, storage and other physiological aspects. Laticifers are also interesting in terms of protein content. Pathogenesis-Related proteins (PR-proteins) are plentiful in different latex fluids, a fact that conceptually supports a role in plant defense by laticifers (Ramos, Demarco, da Costa Souza, & Freitas, 2019). This miscellany of latex-borne compounds and their implication in plant protection is still not fully understood.

2. General aspects and definition

Latex can differ in composition and properties depending on the plant group considered due to its multiple emergence during evolution. Some studies indicate that the protection conferred by this secretion ensured the biological success of latescent plants in relation to non-latescent ones (Agrawal & Konno, 2009).

Latex can be characterized as a suspension of small particles dispersed in a liquid with different refractive indices, giving it an emulsion appearance (Fig. 1A). However, latex is much more than that. It corresponds to the whole protoplast of the laticifers (Demarco, Kinoshita, & Castro, 2006). This means that the latex is not only composed of primary and secondary metabolites, but it also contains different types of organelles, in addition to various nuclei present in the protoplast (Fig. 1B). Therefore, when anatomical cuts are performed without due care for the preservation of the latex, only the laticifer walls are observed.

Despite the presence of membranes, organelles, nuclei and several components that are not secretory products, latex is recognized and studied for the primary and secondary metabolites produced by laticifers, which are stored basically within their large vacuoles (Fig. 1C–O). The proportion of these metabolites is variable, which may alter the color of the latex. Thus, not all latex is white; it can also be yellowish, greenish, orange, reddish and even colorless (Mahlberg, 1993).

Many latex metabolites act as chemical barriers to herbivory or infection due to the presence of toxic and dissuasive food components (Dussourd, 2017). Also, after release, latex acts as a physical barrier because its compounds coagulate rapidly (Bauer et al., 2014), sealing wounds against the entry of pathogens such as bacteria and viruses (Konno, 2011). In addition, the high viscosity of latex and its rapid coagulation in contact with air can also immobilize insect jaws (Dussourd, 2017).

3. Major molecules present in latex

Although latex of different plant species has mainly terpens, it also contains substances such as rubber, mucilage, phenolic compounds, alkaloids and cardenolides (Konno, 2011). Carbohydrates such as starch grains are also present (Demarco et al., 2015), and proteins such as cysteine proteases are common in the latex of some species (Freitas et al., 2007). Interestingly,

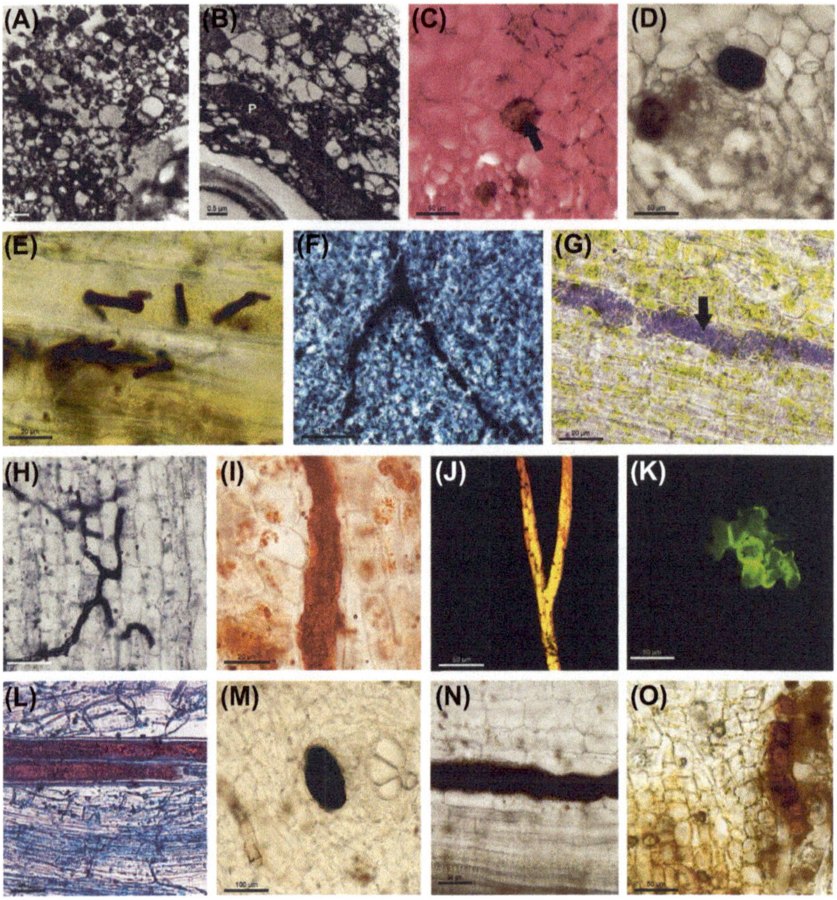

Fig. 1 Latex constituents. **A–B.** Electron micrographs of latex in *Asclepias curassavica* L. [tropical milkweed]. (A) Emulsion aspect of latex containing many small lipophilic particles of different contrast and sizes in a hydrophilic medium. (B) Presence of organelles within latex. Note an elongated plastid (P) in peripheral position. **C–O.** Latex metabolites identified by histochemical tests. (C) Carbohydrates (dark pink) detected by PAS reaction in *Allium cepa* L. [onion]. An arrow indicates latex. (D) Mucilage (black) identified using tannic acid and ferric chloride in *Musa paradisiaca* L. [banana tree]. (E) Starch grains (black) detected by Lugol's reagent in *Hura crepitans* L. [sandbox tree]. (F) Proteins (navy blue) stained with aniline blue black in *Hura crepitans.* (G) Terpenes (purple) identified by NADI reagent in *Urera baccifera* (L.) Gaudich. [scratchbush]. An arrow indicates latex. **H–K.** Lipids observed under bright field (H-I) and blue light (J-K). (H) Lipids stained black using Sudan black (B) *Hippobroma longiflora* (L.) (G) Don [star of Bethlehem]. (I) Lipids stained red with Sudan IV. *Mandevilla splendens* (Hook.) Woodson [shining mandevilla]. (J) Yellow fluorescent lipids detected with the use of Nile blue. *Euphorbia milii* Des Moul. [crown of thorns]. (K) Green fluorescent lipids stained with neutral red. *Musa paradisiaca.* **L.** Neutral lipids (pink) identified using Nile blue in

each class of metabolite acts in the plant's defense (Table 1). However, these compounds can act synergistically, whereby their toxicity is enhanced when acting together (Ramos et al., 2019), as in the case of terpenes associated with phenolic compounds (Wink, 2008) or different proteins exhibiting defense functions against insects or fungi (Ramos et al., 2010; Souza et al., 2011).

3.1 Secondary metabolites

Latex is composed predominantly of lipids (Fig. 1G—M), with terpenes (triterpenes and tetraterpenes) being the major class (Prado & Demarco, 2018). However, other classes of terpenes with higher or lower molecular weight can be found in latex, especially *cis*-1,4 polyisoprene, popularly known as natural rubber (Hagel et al., 2008). Its main function is related to coagulation (Konno, 2011). Although it does not occur universally in all latescent species, rubber can make up 45% of the latex gross weight in *Hevea brasiliensis* (Willd. ex A. Juss.) Müll. Arg. and up to 30% in *Ficus elastica* Roxb. (Mooibroek & Cornish, 2000).

Cardenolides, derived from triterpenoids, have a defense function. Agrawal et al. (2012) showed a direct relation between cardenolide concentration and the toxic effect of latex, due to the high absorption of these nonpolar compounds by the mucous membranes of insects. In *Calotropis procera* (Aiton) R. Br. (Apocynaceae), more than 30 different structurally diverse cardenolides have been reported. Although these cardenolides are highly toxic to many insects through inhibition of the $Na+/K^+$ ATPase pump, *C. procera* serves as food for at least three insect pests (Lepidoptera: *Danaus plexippus*; Hemiptera: *Aphis nerii* and *Oncopeltus fasciatus*) that are insensitive to cardenolides. These species evolved to overcome the toxicity of cardenolides and are capable of ingesting and storing them into their body and thus using latex-borne cardenolides in their own defense against predators (Agrawal et al., 2012). However, many aspects underlying the structural diversity of cardenolides and the ability of insects to store them are unknown.

The phenolic compounds found in latex, such as flavonoids, tannins and diphenols, are also related to the plant's defense. Tannins can inhibit digestive enzymes, thus exerting a negative effect on the growth of herbivores.

Euphorbia milii. **M.** Fatty acids (dark green) detected by copper acetate and rubeanic acid in *Musa paradisiaca*. **N.** Phenolic compounds (dark gray) identified by ferric chloride in *Ipomoea tricolor* Cav. [morning glory]. **O.** Alkaloids (red-brown) detected by Dragendorff's reagent in *Ficus benjamina* L. [weeping fig].

Table 1 Secondary metabolites present in plant latex with defense action.

Family	Species	Secondary metabolites	Class	Activity	Reference
Apocynaceae	*Periploca angustifolia* Labill.	cis-1,4-polyisoprene	Terpenoids	anti-herbivore and phytopathogenic defenses	Dghim, Bouaziz, Mezghani, Boukhris, & Neffati (2015)
Apocynaceae	*Calotropis procera* (Aiton) R. Br.	Calactin, calotoxin, usharin, usharidine, vouscharin.	Cardenolides	Anthelmintic activity	Shivkar & Kumar (2003)
Apocynaceae	*Calotropis procera* (Aiton) R. Br.	Voruscharin, Ushcharidin, Usharin, Calotropagenin, digitoxin.	Cardenolides	Insecticidal activity	Rasmann, Johnson, & Agrawal (2009)
Apocynaceae	*Calotropis procera* (Aiton) R. Br.	Acetate calotropenyl, calotropenyl, 1-(2′,5′-dimethoxyphenyl)-glicerol.	Terpenoids, aryl-propanoid	Inhibited the egg hatching of *Haemonchus contortus*	Cavalcante et al. (2016)
Apocynaceae	*Calotropis procera* (Aiton) R. Br.	Quercetin-3-O-rutinosides, kaempferol-3-O-rutinoside, isorhamnetin-3-O-rutinoside, and 5-hydroxy-3,7-dimethoxyflavone-4 −O-D-glucopyranoside.	Flavonoids	Insecticidal activity against both insects	Nenaah (2013)
Apocynaceae	*Himatanthus sucuuba* (Spruce ex Müll. Arg.) Woodson	Plumericin, α-amyrin cinnamate, α-amyrin acetate, Lupeol Acetate, Lupeol Cinnamate, Lupeol β-Phenylpropionate	Iridoids and Terpenoids	Antifungal and cytotoxic activity	Wood et al. (2001)
Asteraceae	*Taraxacum officinale* F.H. Wigg	Phenolic inositol esters, taraxinic acid b-D-glucopyranosyl ester, β-amyrin, cycloartenol acetate, α-amyrin acetate and Lupeol acetate.	Phenolic compound, Terpenoids lactone glycosides terpenoids	Repellent activity (Latex metabolite extracts strongly repelled *Diabrotica balteata* larvae)	Huber et al. (2015)
Asteraceae	*Lactuca sativa* L.	Lactucopicrin, Lactucin, Lettucenin A	Terpenoids	Antifungal activitity	Sessa, Bennett, Lewis, Mansfield, & Beale (2000)
Euphorbiaceae	*Jatropha neopauciflora* Pax	Catechin and Catechol.	Flavonoid and compound phenolic	Antibacterial activity	Hernandez-Hernandez et al. (2017)

Euphorbiaceae	*Euphorbia palustris* L.	Benzoyl ingenol-laurate, amyrin decadienoate esters, cis-1,4-polyisoprene, and 24-methylenecycloartanol.	Terpenoids	Antifungal activity *Fusarium sporotrichioides*, *Fusarium proliferatum* and *Alternaria alternata*	Krstić et al. (2016)
Euphorbiaceae	*Synadenium grantii* Hook f.	Euphol and citrostadienol	Terpenoids	Cytotoxic activity	Oliveira et al. (2013)
Euphorbiaceae	*Euphorbia resinifera* O. Berg	Euphorol A, Euphorol B, Euphorol C, Euphorol D, Euphorol E, Euphorol F, Euphorol G, Euphorol H, Euphorol I	Terpenoids	Cytotoxic activity	Wang et al. (2016)
Euphorbiaceae	*Euphorbia tirucalli* L.	Phorbol, 12-deoxyphorbol esters and ingenol, b-sitosterol, euphorbol hexacosonate, 12-deoxy-4b-hydroxyphorbol-13-phenyl acetate-20-acetate, 12, 20-dideoxyphorbol-13-isobutyrate, glut-5-en-3-b-ol and euphol, 4-deoxyphorbol di-ester, Cycloeuphordenol, Cyclotirucanenol, Euphol, Tirucalicine, Tri-methyl ellagic acid, Terpenic alcohol, isoeuphorol, taraxasterol and tirucallol, 3,7,12-tri-O-acetyl-8-isovaleryl-ingol	Terpenoids	Antibacterial, antifungal and antimicrobial activity.	Mali & Panchal. (2017)
Euphorbiaceae	*Euphorbia peplus* L.	(2S★,3S★,4R★,5R★,6R★,8R★, 11R★,13S★,14S★,15R★, 16R★)-5,8,15-triacetoxy-3-benzoyloxy-11, 16-dihydroxy-9-oxopepluane. (2R★,3R★,4S★,5R★,7S★,8S★,9S★, 13S★,14S★,15R★)-2,5,9,14-	Terpenoids	Antifeedant activity (*Helicoverpa armigera*) and antifungal activity against three strains of agricultural phytopathogenic fungi, *Rhizoctonia*	Hua et al. (2017)

(Continued)

Table 1 Secondary metabolites present in plant latex with defense action.—cont'd

Family	Species	Secondary metabolites	Class	Activity	Reference
		tetraacetoxy-3-benzoyloxy-8,15-dihydroxy-7-isobutyroyloxyjatropha-6(17),11Ediene, (2R*,3R*,4S*,5R*,7S*,8S*,9S*,13S*,14S*,15R*)-2,5,8,9,14-pentaacetoxy-3-benzoyloxy-15-hydroxy-7-isobutyroyloxyjatropha-6(17),11E-diene, 2R*,3R*,4S*,5R*,7S*,8S*,9S*,13S*,14S*,15R* -2,5,7,8,9,14-hexaacetoxy -3- benzoyloxy- 5-hydroxyjatropha-6(17),11E-diene, (2S*,3S*,4S*,5R*,7R*,9R*,13R*,15R*)-3,5,7,15-tetraacetoxy-9-nicotinoyloxy-14-oxojatropha-6(17),11-diene, (2R*,3R*,4S*,5R*,7S*,8S*,9S*,13S*,14S*,15R*) -2,5,9,14-tetraacetoxy-3-benzoyloxy-8,15-dihydroxy-7-isobutyroyloxyjatropha-6(17),11E-diene, (2R*,3R*,4S*,5R*,7S*,8S*,9S*,13S*,14S*,15R*)- 2,5,14-triacetoxy-3-benzoyloxy-8,15-dihydroxy-7-isobutyroyloxy-9-nicotinoyloxyjatropha-6(17),11E-diene, (2R*,3R*,4S*,5R*,7S*,8S*,		*solani*, *Colletotrichum litchi* and *Fusarium oxysporum* f. sp. *niveum*.	

		9S*,13S*,15R*)-2,5, 8-triacetoxy-3-benzoyloxy-15-hydroxy-7-isobutanoyloxy-9-nicotinoyloxy-14-oxojatropha-6(17), 11E-diene, (2S*,3S*,4R*,5R*,6R*, 8R*,9R*,11R*,13S*, 14S*,15R*,16R*)-5, 8,9,15- tetraacetoxy-3-benzoyloxy-11, 16-dihydroxypepluane, 20-deoxyingenol-5-angelate, Ingenol 20-acetate-3-angelate, 20-deoxyingenol-3-angelate, Peplusol, Pepluanin A.			
Campanulaceae	*Lobelia cardinalis* L.	Lobeline	Alkaloids	Insecticidal activity	Oppel, Dussourd, & Garimella (2009)
Moraceae	*Ficus carica* L.	Caffeic acid, 3.4-Dihydroxybenzoic acid, p—OH—Phenylacetic acid, p-Coumaric acid, Luteolin, N-argenine, Ferulic acid	Phenolic compound and Flavonoids	Antiviral activity	Aref et al. (2011)
Moraceae	*Ficus carica* L.	Lanosta-8, Urs-12-en-24-oic acid, Aristolone, Olean-12-en-3-ol, acetate, Maragenin I acetate, α-Amyrenyl acetate	Coumarin and terpenoids	Antimicrobial activity	Aref, Mars, Fekih, Aouni, & Said (2012)
Moraceae	*Morus australis* Poir.	1,4-dideoxy-1,4-imino-D-arabinitol,1-deoxynojirimycin, and 1,4-dideoxy-1,4-imino-D-ribitol.	Alkaloids	Insecticidal activity	Konno et al. (2006)

The concentration of phenolic compounds in latex varies depending on the group of plants analyzed. For instance, they are present in high concentrations in *Musa paradisiaca* L., *Ipomoea tricolor* Cav (Fig. 1N) and *Ficus benjamina* L. latex.

Although alkaloids are found in approximately 35 families, mostly angiosperms, only 14.2% of genera have these substances (Samanani, Liscombe, & Facchini, 2004). This restricted distribution seems to be related to the type of environment, with a predominance of alkaloids in plants that grow in tropical habitats. These compounds have high toxicity against herbivores and pathogens or may be food deterrents (Dussourd & Denno, 1991). This class of secondary metabolites varies widely in terms of molecular structure, as well as ecological and pharmacological activity, and confers an orange color (Matile, Jans, & Rickenbacher, 1970). The first study of alkaloids was performed with the latex of *Papaver somniferum* L. (Annett, 1920), which has approximately 20 types of alkaloids, including morphine and codeine, potent analgesics used in the pharmaceutical industry (Moussaoui et al., 2001).Other species of the same genus has also shown diversity of alkaloids (Oh et al., 2018). As seen in some families, such as Amaryllidaceae, Apocynaceae and Asteraceae, laticifers can produce a large number of alkaloids (Khatoon, 2017), which can act synergistically with other alkaloids or compounds, thus presenting antimalarial, anesthetic, antibacterial, and antiviral properties (Wittstock & Gershenzon, 2002).

When comparing the total composition of latex by histochemical analysis, it can be noted that regardless the group of plants, coloration or toxicity, the latex always has all or almost all the main chemical classes of secondary metabolites (Table 2), showing that, despite the many variations observed in the chemical analysis, in general all latices are similar and have evolved in a convergent way in the different groups.

3.2 Carbohydrates

Carbohydrates, including polysaccharides and mucilage, are among the substances found in latex (Fig. 1C–D). The presence of starch grains occurs in the Hippomaneae and Euphorbieae (Fig. 1E) (Demarco et al., 2015; Prado & Demarco, 2018). However, in Euphorbieae the grains have various morphologies and are even used in the interpretation of evolutionary and taxonomic aspects. The rounded and spindle morphology of starch grains in laticifers is interpreted as being less specialized. Progressive specialization is associated with larger starch deposition in polarized form, in order to assume an osteoid and occasionally lobular morphology, thus leading to an

Table 2 Histochemical analysis of latex components in different plant families.

	Amaryllidaceae	Musaceae	Euphorbiaceae		Moraceae	Urticaceae	Apocynaceae	Caricaceae	Convolvulaceae	Campanulaceae
Histochemical test	***Allium cepa* L.**	***Musa paradisiaca* L.**	***Euphorbia milii* des moul.**	***Hura crepitans* L.**	***Ficus benjamina* L.**	***Urera baccifera* (L.) gaudich.**	***Mandevilla splendens* (hook.) woodson**	***Carica papaya* L.**	***Ipomoea tricolor* cav.**	***Hippobroma longiflora* (L.) G. Don**
Lipids										
Sudan black B	+	+	+	+	+	+	+	+	+	+
Sudan IV	+	+	+	+	+	+	+	+	+	+
Neutral red	+	+	+	+	+	+	+	+	+	+
Neutral lipids										
Nile blue	+	+	+	+	+	+	+	+	+	+
Terpenoids										
NADI reagent	+	+	+	+	+	+	+	+	+	+
Fatty acids										
Copper acetate and rubeanic acid	+	+	+	+	+	+	+	+	+	+
Phenolic compounds										
Ferric chloride	+	+	+	+	+	+	+	+	+	+
Potassium dichromate	+	+	+	+	+	+	+	+	+	+

(Continued)

Table 2 Histochemical analysis of latex components in different plant families.—cont'd

	Amaryllidaceae	Musaceae	Euphorbiaceae		Moraceae	Urticaceae	Apocynaceae	Caricaceae	Convolvulaceae	Campanulaceae
Histochemical test	*Allium cepa* L.	*Musa paradisiaca* L.	*Euphorbia milii* des moul.	*Hura crepitans* L.	*Ficus benjamina* L.	*Urera baccifera* (L.) gaudich.	*Mandevilla splendens* (hook.) woodson	*Carica papaya* L.	*Ipomoea tricolor* cav.	*Hippobroma longiflora* (L.) G. Don
Alkaloids										
Dragendorff's reagent	+	+	+	-	+	-	+	-	-	-
Wagner's reagent	+	+	+	-	+	-	+	-	-	-
Carbohydrates										
PAS reaction	+	+	+	+	+	+	+	+	+	+
Mucilage										
Tannic acid and ferric chloride	+	+	+	+	+	+	+	-	+	-
Acidic mucilage										
Ruthenium red	+	+	+	+	+	+	+	-	+	-
Alcian blue	+	+	+	+	+	+	+	-	+	-
Starch										
Lugol's reagent	-	-	+	+	-	-	-	-	-	-
Protein										
Aniline blue black	+	+	+	+	+	+	+	+	+	+

interpretation of the specialization pattern of these taxa (Mahlberg, 1975). Although latex production requires substantial energy (Agrawal & Konno, 2009), starch grain mobilization has not been verified as an energy source (Biesboer & Mahlberg, 1978). On the other hand, it has been shown that the carbohydrates present in latex are the main source of carbon used by the laticifers for the synthesis of triterpenes in *Euphorbia lathyris* L. (Groeneveld, Hageman, & Vellenga, 1982).

3.3 Proteins in latex

Laticifers are living cells, so proteins involved in all ordinary activities inherent to cell metabolism are present in latex of all latescent plants. In addition, specific proteins involved in the rubber synthesis and other specific compounds may be present according to the species considered. However, there is recent evidence that latex-borne substances can be produced at specific sites along with the laticifers and that independent network systems may exist in the plant, exhibiting independent proteomes. Although the qualitative and quantitative protein profiles of latex fluids are very diverse, all of them have in common the presence of proteins involved in primary plant metabolism (Kitajima et al., 2012, 2018).

The latex protein profiles of different plants are markedly distinct even when they belong to closely related species (Fig. 2). The protein profiles are also modified dynamically according to different conditions, such as time of development, environmental conditions and specific biotic or abiotic stresses. Despite such obvious differences, many common aspects among latex proteins exist.

The most widely studied latex proteins are those involved in plant defense. Protease inhibitors, chitinases/chitin-binding proteins, osmotin-like proteins, lipases, and typical enzymes participating of the anti-oxidative metabolism have been found in latex. However, lectin, a typical plant protein, has been reported in *H. brasiliensis*, and is called hevein, a well-recognized antifungal protein (Vanparijs, Broekaert, Goldstein, & Peumans, 1991). These proteins have been demonstrated to be constitutive or *de novo* synthesized in reaction to infections or herbivory in different latices. Proteases are the most common protein in latex. Proteases belonging to the different catalytic classes (cysteine, serine, aspartic, and metalloprotease) have been reported in different latex fluids. There is little knowledge about the type of proteases and their functions in latex. The general aspects about the involvement of proteases in plant defense were recently discussed in details (Hou, Jamieson, & He, 2018). Later, a new conceptual interpretation of

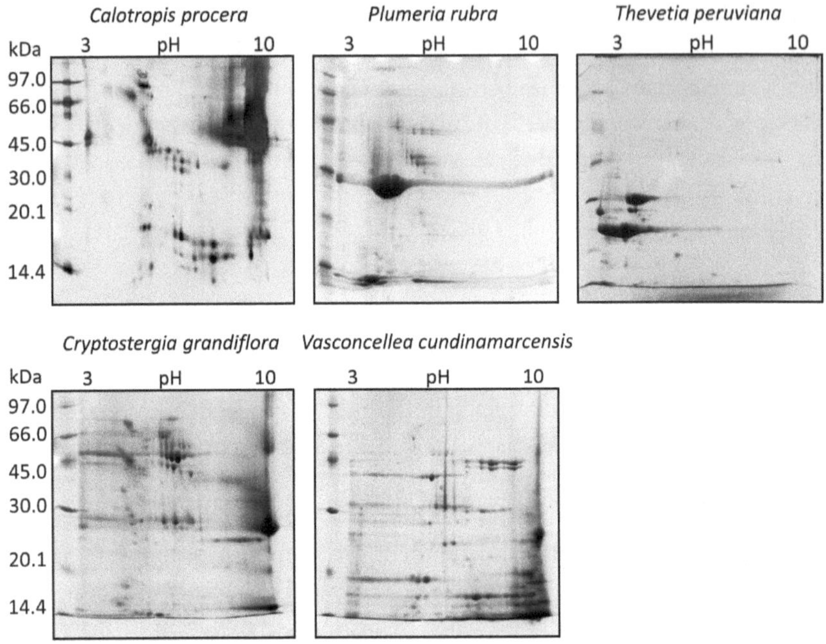

Fig. 2 Protein profiles of latex obtained of different plants species accessed by two-dimensional polyacrylamide gel electrophoreses (12.5%). Latices were collected and processed to obtain the soluble proteins. Regardless the taxonomic position, closer related species or distanced ones exhibit remarkable differences in terms of protein profile.

the attributions of latex proteases was proposed (Ramos et al., 2019). In this review, a pro-active role of latex proteases in plant defense was proposed for the first time.

For instance, up to 30% of all *Carica papaya* L. latex content is formed by proteolytic enzymes (Moussaoui et al., 2001), including papain, which is the best known and has a high commercial value, because of its use in the food industry. In general, latex proteases are synthetized and stored as zymogens and become active only after some injury to the laticifers occurs. This mechanism protects the plant against autolysis (Azarkan, El Moussaoui, Van Wuytswinkel, Dehon, & Looze, 2003). After activation, the proteases become highly toxic to insects and fungi (Hou et al., 2018; Ramos et al., 2019).

Protease inhibitors are Pathogenesis-Related proteins (PR-6 group) widely distributed in plants, constitutively present in seeds and other organs or induced by pathogens or injury (van Loon, Rep, & Pieterse, 2006). In contrast, there are only a few studies related to purification and characterization of protease inhibitors in latex fluids (Azarkan et al., 2006). In general, these proteins interact with proteolytic enzymes of the aggressors, inhibiting proteolysis and reducing the uptake of amino acids necessary for the growth and development of aggressors. Some studies indicate that protease inhibitors act as defense proteins since they can be strongly induced by wounds, as in *Ficus carica* L. latex (Kim et al., 2003), or exhibit insecticidal activity, as in *C. procera* (Ramos et al., 2010).

Chitin is an insoluble polysaccharide component of the cell wall of fungi and the exoskeleton and peritrophic matrix of insects. Chitinases are enzymes responsible for hydrolyzing β-1,4-glycosidic bonds of chitin, acting as defensive proteins against pathogenic fungi and insects. After proteases, the chitinases are the proteins most studied in latex fluids. In general, they present several isoforms that exhibit insecticidal and antifungal activities (Freitas et al., 2016a).

Osmotin-like proteins are proteins that have primary structures very close to osmotin. These proteins belong to family 5 of Pathogenesis-Related proteins. Osmotin was first identified in cell cultures of tobacco adapted to saline stress (Singh, Handa, Hasegawa, & Bressan, 1985). Since then, osmotin has been characterized in different plant species in response to osmotic stress. Besides this role in plant defense against water or salt stresses, osmotin has been reported as a cryoprotective molecule by providing protection at low temperatures, or as a protein related to plant defense against infection (Hong et al., 2004). It was reported that tobacco osmotin causes lysis of spores and inhibits growth (Kumar, Kumari, Kumar, Mohanalatha, & Kishor, 2015). Osmotin-like proteins have been reported in latex fluids. In the literature, there are only a few works that describe the purification, characterization and biological activities of latex osmotins (Freitas et al., 2011; Looze et al., 2009). The antifungal activity of an osmotin contained in the latex of *C. procera* was demonstrated by atomic force microscopy and fluorescence assays, suggesting that the protein induces damage to spore membranes (Ramos et al., 2015). Supplementary studies suggested the presence of osmotin-like proteins in the latex of *Cryptostegia grandiflora* R.Br., *Plumeria rubra* L. and *Himatanthus drasticus* (Mart.) Plumel (Freitas et al., 2015). Therefore, constitutive osmotin-like proteins seem to be produced and stored by laticifers of different species. However, the involvement of these proteins in plant defense still remains to be better clarified.

New proteins have been surprisingly reported in latex. A lipid transfer protein (9.5 KDa) isolated from the latex of *Chelidonium majus* L. was reported to be active against Gram-positive and Gram-negative bacteria (Nawrot et al., 2017). Also, two germin-like proteins with oxalate oxidase activity were reported, first in the latex of *C. procera* (Freitas et al., 2017) and later exhibiting proteolytic activity in the latex of *Thevetia peruviana* (Pers.) K. Schum. (Cruz et al., 2019). Therefore, although the implication of latex in plant defense is globally recognized, the comprehensiveness of such function is still underestimated. Seventeen distinct classes of PR-proteins are currently recognized (Ali et al., 2018). At least twelve classes possess representative members associated with laticifer products. Further transcriptomic and proteomic studies of laticifers should reveal novel PR-proteins in latex fluids, associated with the defensive role orchestrated by laticifers.

4. Exudate trapping

Since many latex compounds are toxic (Agrawal & Konno, 2009; Hagel et al., 2008; Wittstock & Gershenzon, 2002), the isolation of these substances in the laticifers can be observed in relation to adjacent tissues, and mature laticifers do not contain plasmodesmata. It is therefore difficult to affirm that a defined type of secondary metabolite is fully produced into laticifers. Under this perspective it is also hard to determine whether it is exclusively produced by the laticifers or can be found in other cells or plant tissues.

Although several alkaloids are found in latex, they are not necessarily synthesized by laticifers. In *Catharanthus roseus* (L.) G. Don, alkaloid precursors are synthesized in specialized parenchyma cells and sent to the laticifers via the apoplast pathway, to complete their biosynthesis (Yoder & Mahlberg, 1976). So, small compounds can be stored and found in laticifers, while the steps of their synthesis may begin outside laticifers.

The diversity of proteins already reported in laticifers indicates that many of them are of general interest to the cellular activities. However, information about proteins exclusively produced in laticifers is not available. Similar, traffic of proteins among laticifers and adjacent cells are unknown. The identification of true laticifer proteins would greatly help understanding about laticifer functions.

Looking for the responsiveness of the latex osmotin gene, in salt stressed cultured callus of *C. procera*, it was observed increased content of mRNA corresponding to osmotin. However, the protein was not detected (Souza et al., 2017). It is thought that specific requirements for osmotin synthesis such as post-translational modifications and post-transcriptional events are not available on callus. Whether they are only found in laticifers still remains undefined.

Lastly, to ensure the laticifer origin of a defined molecule, it is imposed highly precise protocols of collection in order to ensure the content evaluated is exclusively from laticifers. This limitation is to be overcome.

5. Evolution of latex and ecological implications

Latex is present in around 40 families of vascular plants, often without phylogenetic relation, showing that it has been developed independently during evolution (Prado & Demarco, 2018). Latescent families are found mainly among angiosperms, but they are also found in ferns, such as *Regnellidium diphyllum* Lindm (Marsileaceae), and gymnosperms, like *Gnetum gnemon,* (Gnetophyta) (Metcalfe, 1967). The convergent evolution of laticifers is observed from the presence of many similar components of phylogenetically distinct groups. Although the chemical classes of secondary metabolites are the same in the latex of the different lineages of latescent plants, the type of compound may vary among the species of the same family, such as in Moraceae, where only some species of *Ficus* produce cysteine proteases, while other species have other types of enzymes (Konno, 2011). In *Asclepias* (Apocynaceae), there is variation in cardenolides, with presence in *Asclepias curassavica* L. and absence in *A. speciose* Torr and *A. californica* Greene.

Latex has a plastic nature, showing that its production varies according to the surrounding environment. The biotic and abiotic variations, such as drought, light, inter- and intra-specific competition, pathogens and herbivory attack, result in the final concentration of compounds in latex (Dicke & van Loon, 2000). In some groups, such as Apocynaceae, production of latex is an ancestral condition, since it is conserved in all species of the family (Metcalfe, 1967). However, latex is usually present in only a few taxa of each family (Prado & Demarco, 2018). Because of its limited distribution, latex aspects can be an important characteristic for phylogenetic analysis, such as in Malpighiaceae, where latex is a synapomorphy of the Galphimieae tribe, supporting its monophyly (Vega, Castro, & Anderson, 2002).

6. Ecological relations

Even though latex is composed of large amounts of toxic substances, some specialist herbivores have strategies to prevent the plant from defending itself, and some are even capable of sequestering toxic components from latex and using them in their own defense (Dussourd, 2017; Konno, 2011). The occurrence of latex predominantly in tropical environments, where herbivory is more prevalent, indicates that predation is a factor that directs the evolution of this defense system (Prado & Demarco, 2018). By means of behavior strategies such as vein-cutting, vein-trenching and girdling, precise cuts are made in the plant for latex extravasation, reducing exudate ingestion by up to 90% (Agrawal & Konno, 2009; Dussourd, 2017; Konno, 2011).

When broken, the laticifers release latex, causing the loss of cell turgor. Consequently, the water from adjacent cells will cross the laticifers' walls, thus contributing to a larger extravasated flow (Pickard, 2008). Studies reveal that branched laticifers tend to have herbivore behaviors in vain-trenching and vein-cutting, since the plant's architecture allows for a higher extravasation of latex (Dussourd & Denno, 1991). These specialized herbivore strategies were described for 11 latescent plant families (Dussourd, 2017) and both behaviors have been shown to have evolved independently in several lineages of insects, and in several environments (Dussourd & Denno, 1991). Some specialist herbivores of the Dilophonotini tribe may feed on several latescent species from phylogenetically distant groups, such as Apocynaceae, Caricaceae, Euphorbiaceae and Moraceae (Farrell, Dussourd, & Mitter, 1991), showing that the behavior to prevent the ingestion of latex is effective according to the structure of the laticifer system and does not depend on latex composition.

On the other hand, some species have developed physiological strategies to counteract latex's toxic action, as observed in the monarch butterfly, *D. plexippus* L. (Lepidoptera: Danaidae). The larvae ingest the cardenolides present in latex and use them in their own defense against predators (Malcolm & Zalucki, 1996) (Fig. 3). By ingesting these cardenolides, which inhibit the potassium sodium pump in cells, the caterpillars manage to overcome this toxicity through their insensitivity to it. This insensitivity was found to be related to the mutation of a single amino acid at the cardenolide binding site to the plasma membrane (Konno, 2011). Opitz and Müller (2009) identified herbivores with the same strategy to sequester toxic components for their defense. However, in this case the compounds were the

Fig. 3 Fifth instars and adults of *Danaus gilippus* (Queen Butterfly) – left, and *Danaus plexippus* (Monarch Butterfly) – right. Butterflies were born under laboratory conditions and the caterpillars feed on fresh leaves of *Calotropis procera*.

isoprenoids of *Ipomoea batatas* (L.) Lam. (Convolvulaceae) and phorbol esters of *Jatropha curus* (Euphorbiaceae). Notably, monarch caterpillars also overcome the constitutive proteases of *C. procera* latex. The digestive apparatus of the caterpillar is efficient to digest the latex enzymes, eliminating the toxicity of chitinases and proteases in the latex (Pereira et al., 2010; Ramos et al., 2010).

7. Biotechnology of latex-borne substances

Based on elucidation of the purification steps, structure and activities associated with latex-borne substances, biotechnological applications have been proposed for latex compounds. Many latescent species are used in traditional medicine to treat different ailments. The shrub *C. procera,* native of India and widespread in tropical regions, has received much attention. The protein fraction obtained from this plant's latex showed anti-inflammatory activity against important human conditions. Protection against infection (Nascimento et al., 2016), acute inflammation (Alencar et al., 2017) and arthritis (Chaudhary, Ramos, Vasconcelos, & Kumar, 2016) have been described, among others. The capacity of latex proteins to reduce blood glucose levels was recently documented (Oliveira et al., 2019) and validated the folk use associated with this effect. The osmotin present in the latex, exhibiting antifungal activity against important plant pathogens, is a suitable candidate for further transgenic projects. Some antimetabolic chemistries derived from latex compounds have been reported and are under development (Syeda et al., 2018). The proteolytic activity of latex was shown to be efficient for cheese manufacture (Freitas et al., 2016b) and leather processing (Lopéz et al., 2017). Therefore, latex-borne substances are appropriate for biotechnology.

8. Conclusion

The ultrastructural characteristics of laticifers, their restricted distribution among different taxonomic groups and also in some species in each clade, make them very attractive for analysis by biologists. The intrinsic activities of laticifers, their highly specialized chemical and biochemical composition and the evidence that they are key elements in the co-evolutionary history between plants and insects have contributed to the understanding of many aspects of plant defense. In this context, latex-borne substances are promising tools for agricultural biotechnology for crop protection. The use of latescent plants in traditional systems of medicine for varied purposes also highlights latex fluids as valuable sources for pharmaceutical compounds.

Acknowledgments

Studies of latescent plants and latex-borne proteins are supported by grants from the Brazilian public agencies "CNPq", "CAPES", "FUNCAP" and "FAPESP". The authors are particularly grateful to the RENORBIO Program. This study is part of the consortium Molecular Biotechnology of Plant Latex.

References

Agrawal, A. A., & Konno, K. (2009). Latex: A model for understanding mechanisms, ecology, and evolution of plant defense against herbivory. *Annual Review of Ecology, Evolution and Systematics, 40*, 311–331.

Agrawal, A. A., Petschenka, G., Bingham, R. A., Weber, M. G., & Rasmann, S. (2012). Toxic cardenolides: Chemical ecology and coevolution of specialized plant–herbivore interactions. *New Phytologist, 194*, 28–45.

Alencar, N., Bitencourt, F., Figueiredo, I., Luz, P., Lima-Júnior, R., Aragão, K., et al. (2017). Side-Effects of Irinotecan (CPT-11), the clinically used drug for colon cancer therapy, are eliminated in experimental animals treated with latex proteins from *Calotropis procera* (Apocynaceae). *Phytotherapy Research, 31*, 312–320.

Ali, S., Ganai, B. A., Kamili, A. N., Bhat, A. A., Mir, Z. A., Bhat, J. A., et al. (2018). Pathogenesis-related proteins and peptides as promising tools for engineering plants with multiple stress tolerance. *Microbiological Research, 212–213*, 29–37.

Annett, H. E. (1920). Factors influencing alkaloidal content and yield of latex in the opium poppy (*Papaver somniferum*). *Biochemical Journal, 14*, 618–636.

Aref, H., Gaaliche, B., Fekih, A., Mars, M., Aouni, M., Chaumon, J., et al. (2011). *In vitro* cytotoxic and antiviral activities of *Ficus carica* latex extracts. *Natural Product Research, 25*, 310–319.

Aref, H., Mars, M., Fekih, A., Aouni, M., & Said, K. (2012). Chemical composition and antibacterial activity of a hexane extract of Tunisian caprifig latex from the unripe fruit of *Ficus carica*. *Pharmaceutical Biology, 50*, 407–412.

Azarkan, M., El Moussaoui, A., Van Wuytswinkel, D., Dehon, G., & Looze, Y. (2003). Fractionation and purification of the enzymes stored in the latex of *Carica papaya*. *Journal of Chromatography B, 790*, 229–238.

Azarkan, M., Garcia-Pino, A., Dibiani, R., Wyns, L., Loris, R., & Baeyans-Volant, D. (2006). Crystallization and preliminary x-ray analysis of a protease inhibitor from the latex of *Carica papaya*. *Acta Crystallographica Section F: Structural Biology Communications, 62*, 1239–1242.

Bauer, G., Gorb, S. N., Klein, M. C., Nellesen, A., von Tapavicza, M., & Speck, T. (2014). Comparative study on plant latex particles and latex coagulation in *Ficus benjamina, Campanula glomerata* and three Euphorbia species. *PLoS One, 9*, e113336.

Biesboer, D. D., & Mahlberg, P. G. (1978). Accumulation of non-utilizable starch in laticifers of *Euphorbia heterophylla* and *E. myrsinites*. *Planta, 143*, 5–10.

Cavalcante, G., Morais, S., Andre, W., Ribeiro, W., Rodrigues, A., Lira, F., et al. (2016). Chemical composition and in vitro activity of *Calotropis procera* (Ait.) latex on *Haemonchus contortus*. *Veterinary Parasitology, 226*, 22–25.

Chaudhary, P., Ramos, M., Vasconcelos, M., & Kumar, V. (2016). Protective effect of high molecular weight protein sub-fraction of *Calotropis procera* latex in monoarthritic rats. *Pharmacognosy Magazine, 12*, 147–151.

Cruz, W. T., Bezerra, E. H., Grangeiro, T. B., Lopes, J. L., Silva, M. Z., Ramos, M. V., et al. (2019). Structural and enzymatic characterization of Peruvianin-I, the first germin-like protein with proteolytic activity. *International Journal of Biological Macromolecules, 126*, 1167–1176.

Demarco, D. (2015). Micromorfología y histoquímica de los laticíferos de órganos vegetativos de especies de Asclepiadoideae (Apocynaceae). *Acta Biológica Colombiana, 20*, 57–65.

Demarco, D., Kinoshita, L. S., & Castro, M. D. M. (2006). Laticíferos articulados anastomosados: Novos registros para Apocynaceae. *Revista Brasileira de Botanica, 29*, 133–144.

Dghim, F., Bouaziz, M., Mezghani, I., Boukhris, M., & Neffati, M. (2015). Laticifers identification and natural rubber characterization from the latex of *Periploca angustifolia* Labill. (Apocynaceae). *Flora, 217*, 90–98.

Dicke, M., & van Loon, J. J. (2000). Multitrophic effects of herbivore-induced plant volatiles in an evolutionary context. *Entomologia Experimentalis et Applicata, 97*, 237–249.

Dussourd, D. E. (2017). Behavioral sabotage of plant defenses by insect folivores. *Annual Review of Entomology, 62*, 15–34.

Dussourd, D. E., & Denno, R. F. (1991). Deactivation of plant defense: Correspondence between insect behavior and secretory canal architecture. *Ecology, 72*, 1383–1396.

Farrell, B. D., Dussourd, D. E., & Mitter, C. (1991). Escalation of plant defense: Do latex and resin canals spur plant diversification? *The American Naturalist, 138*, 881–900.

Freitas, C. D. T., Freitas, D. C., Cruz, W. T., Porfirio, C. T., Silva, M. Z., Oliveira, J. S., et al. (2017). Identification and characterization of two germin-like proteins with oxalate oxidase activity from *Calotropis procera* latex. *International Journal of Biological Macromolecules, 105*, 1051–1061.

Freitas, C. D. T., Leite, H. B., Oliveira, J. P. B., Amaral, J. L., Egito, A. S., Vairo-Cavalli, S., et al. (2016b). Insights into milk-clotting activity of latex peptidases from *Calotropis procera* and *Cryptostegia grandiflora*. *Food Research International, 87*, 50–59.
Freitas, C. D. T., Nogueira, F. C., Vasconcelos, I. M., Oliveira, J. T., Domont, G. B., & Ramos, M. V. (2011). Osmotin purified from the latex of *Calotropis procera*: Biochemical characterization, biological activity and role in plant defense. *Plant Physiology and Biochemistry, 49*, 738–743.
Freitas, C. D. T., Oliveira, J. S., Miranda, M. R. A., Macedo, N. M. R., Sales, M. P., Villas-Boas, L. A., et al. (2007). Enzymatic activities and protein profile of latex from *Calotropis procera*. *Plant Physiology and Biochemistry, 45*, 781–789.
Freitas, C. D. T., Silva, M. Z., Bruno-Moreno, F., Monteiro-Moreira, A. C., Moreira, R. A., & Ramos, M. V. (2015). New constitutive latex osmotin-like proteins lacking antifungal activity. *Plant Physiology and Biochemistry, 96*, 45–52.
Freitas, C. D. T., Viana, C. A., Vasconcelos, I. M., Moreno, F. B., Lima-Filho, J. V., Oliveira, H. D., et al. (2016a). First insights into the diversity and functional properties of chitinases of the latex of *Calotropis procera*. *Plant Physiology and Biochemistry, 108*, 361–371.
Groeneveld, H. W., Hageman, J., & Vellenga, A. T. N. (1982). The involvement of sucrose, glucose and other metabolites in the synthesis of triterpenes and dopa in the laticifers of *Euphorbia lathyris*. *Phytochemistry, 21*, 1589–1597.
Hagel, J. M., Yeung, E. C., & Facchini, P. J. (2008). Got milk? The secret life of laticifers. *Trends in Plant Science, 13*, 631–639.
Hernandez-Hernandez, A., Alarcon-Aguilar, F., Almanza-Perez, J., Nieto-Yañez, O., Olivares-Sanchez, J., Duran-Diaz, A., et al. (2017). Antimicrobial and anti-inflammatory activities, wound-healing effectiveness and chemical characterization of the latex of *Jatropha neopauciflora* Pax. *Journal of Ethnopharmacology, 204*, 1–7.
Hong, J. K., Jung, H. W., Lee, B. K., Lee, S. C., Lee, Y. K., & Hwang, B. K. (2004). An osmotin-like protein gene, CAOSM1, from pepper: Differential expression and in situ localization of its mRNA during pathogen infection and abiotic stress. *Physiological and Molecular Plant Pathology, 64*, 301–310.
Hou, S., Jamieson, P., & He, P. (2018). The cloak, dagger, and shield: Proteases in plant-pathogen interactions. *Biochemical Journal, 475*, 2491–2509.
Hua, J., Liu, Y., Xiao, C., Jing, S., Luo, S., & Li, S. (2017). Chemical profile and defensive function of the latex of *Euphorbia peplus*. *Phytochemistry, 136*, 56–64.
Huber, M., Triebwasser-Freese, D., Reichelt, M., Heiling, S., Paetz, C., Chandran, J., et al. (2015). Identification, quantification, spatiotemporal distribution and genetic variation of major latex secondary metabolites in the common dandelion (*Taraxacum officinale* agg.). *Phytochemistry, 115*, 89–98.
Khatoon, S. (2017). A novel histological approach for identification of alkaloid bearing plants. *International Journal of Botany, 13*, 28–36.
Kim, J. S., Kim, Y. O., Ryu, H. J., Kwak, Y. S., Lee, J. Y., & Kang, H. (2003). Isolation of stress-related genes of rubber particles and latex in fig tree (*Ficus carica*) and their expressions by abiotic stress or plant hormone treatments. *Plant and Cell Physiology, 44*, 412–414.
Kitajima, S., Aoki, W., Shibata, D., Nakajima, D., Sakurai, N., Yazaki, K., et al. (2018). Comparative multi-omics analysis reveals diverse latex-based defense strategies against pests among latex-producing organs of the fig tree (*Ficus carica*). *Planta, 247*, 1423–1438.
Kitajima, S., Taira, T., Oda, K., Yamato, K., Inukai, Y., & Hori, Y. (2012). Comparative study of gene expression and major proteins' function of laticifers in lignified and unlignified organs of mulberry. *Planta, 235*, 589–601.
Konno, K. (2011). Plant latex and other exudates as plant defense systems: Roles of various defense chemicals and proteins contained therein. *Phytochemistry, 72*, 1510–1530.

Konno, K., Ono, H., Nakamura, M., Tateishi, K., Hirayama, C., Tamura, Y., et al. (2006). Mulberry latex rich in antidiabetic sugar-mimic alkaloids forces dieting on caterpillars. *Proceedings of the National Academy of Sciences, 103*, 1337–1341.
Krstić, G., Anđelković, B., Choi, Y., Vajs, V., Stević, T., Tešević, V., et al. (2016). Metabolic changes in *Euphorbia palusrtis* latex after fungal infection. *Phytochemistry, 131*, 17–25.
Kumar, S. A., Kumari, P. H., Kumar, G. S., Mohanalatha, C., & Kishor, P. B. K. (2015). Osmotin: A plant sentinel and a possible agonist of mammalian adiponectin. *Frontiers of Plant Science, 6*, 163.
van Loon, L. C., Rep, M., & Pieterse, C. M. J. (2006). Significance of inducible defense-related proteins in infected plants. *Annual Review of Plant Physiology, 44*, 135–162.
Looze, Y., Boussard, P., Huet, J., Vandenbussche, G., Raussens, V., & Wintjens, R. (2009). Purification and characterization of a wound-inducible thaumatin-like protein from the latex of *Carica papaya*. *Phytochemistry, 70*, 970–978.
Lopéz, L., Viana, C., Errasti, M., Garro, M., Martegani, J., Mazzilli, G., et al. (2017). Latex peptidases of *Calotropis procera* for dehairing of leather as an alternative to environmentally toxic sodium sulfide treatment. *Bioprocess and Biosystems Engineering, 40*, 1391–1398.
Mahlberg, P. G. (1975). Evolution of the laticifer in Euphorbia as interpreted from starch grain morphology. *American Journal of Botany, 62*, 577–583.
Mahlberg, P. G. (1993). Laticifers: An historical perspective. *The Botanical Review, 59*, 1–23.
Malcolm, S. B., & Zalucki, M. P. (1996). Milkweed latex and cardenolide induction may resolve the lethal plant defence paradox. In *Proceedings of the 9th international symposium on insect-plant relationships* (pp. 193–196). Dordrecht: Springer.
Mali, P., & Panchal, S. (2017). *Euphorbia tirucalli* L.: Review on morphology, medicinal uses, phytochemistry and pharmacological activities. *Asian Pacific Journal of Tropical Biomedicine, 7*, 603–613.
Matile, P., Jans, B., & Rickenbacher, R. (1970). Vacuoles of Chelidonium latex: Lysosomal property and accumulation of alcaloids. *Biochemie und Physiologie der Pflanzen, 161*, 447–458.
Metcalfe, C. R. (1967). Distribution of latex in the plant kingdom. *Economic Botany, 21*, 115–127.
Mooibroek, H., & Cornish, K. (2000). Alternative sources of natural rubber. *Applied Microbiology and Biotechnology, 53*, 355–365.
Moussaoui, A., Nijs, M., Paul, C., Wintjens, R., Vincentelli, J., Azarkan, M., et al. (2001). Revisiting the enzymes stored in the laticifers of *Carica papaya* in the context of their possible participation in the plant defence mechanism. *Cellular and Molecular Life Sciences, 58*, 556–570.
Nascimento, D., Ralph, M., Batista, J., Silva, D., Gomes-Filho, M., Alencar, N., et al. (2016). Latex protein extracts from *Calotropis procera* with immunomodulatory properties protect against experimental infections with Listeria monocytogenes. *Phytomedicine, 23*, 745–753.
Nawrot, R., Józefiak, D., Sip, A., Kuźma, D., Musidlak, O., & Goździcka-Józefiak, A. (2017). Isolation and characterization of a non-specific lipid transfer protein from *Chelidonium majus* L. latex. *International Journal of Biological Macromolecules, 104*, 554–563.
Nenaah, G. (2013). Potential of using flavonoids, latex and extracts from *Calotropis procera* (Ait.) as grain protectants against two coleopteran pests of stored rice. *Industrial Crops and Products, 45*, 327–334.
Oh, H. H., Ha, I. J., Lee, M. Y., Kim, E. O., Park, D., Lee, J. H., et al. (2018). Identification and metabolite profiling of alkaloids in aerial parts of *Papaver rhoeas* by liquid chromatography coupled with quadrupole time-of-flight tandem mass spectrometry Journal of Separation Science. *Journal of Separation Science, 41*, 2499–2696.
Oliveira, K., Gomes, M., Vasconcelos, R., Abreu, E., Fortunato, R., Loureiro, A., et al. (2019). Phytomodulatory proteins promote inhibition of hepatic glucose production

and favor glycemic control via the AMPK pathway. *Biomedicine & Pharmacotherapy, 109*, 2342–2347.

Oliveira, T., Munhoz, A., Lemes, B., Minozzo, B., Nepel, A., Barison, A., et al. (2013). Antitumoural effect of *Synadenium grantii* Hook f. (Euphorbiaceae) latex. *Journal of Ethnopharmacology, 150*, 263–269.

Opitz, S. E., & Müller, C. (2009). Plant chemistry and insect sequestration. *Chemoecology, 19*, 117–154.

Oppel, C., Dussourd, D., & Garimella, U. (2009). Visualizing a plant defense and insect counterploy: Alkaloid distribution in *Lobelia* leaves trenched by a plusiine caterpillar. *Journal of Chemical Ecology, 35*, 625–634.

Pereira, D., Ramos, M., Souza, D., Portela, T., Guimarães, J., Madeira, S., et al. (2010). Digestibility of defense proteins in latex of milkweeds by digestive proteases of monarch butterflies, *Danaus plexippus* L.: A potential determinant of plant–herbivore interactions. *Plant Science, 179*, 348–355.

Pickard, W. F. (2008). Laticifers and secretory ducts: Two other tube systems in plants. *New Phytologist, 177*, 877–888.

Prado, E., & Demarco, D. (2018). Laticifers and secretory ducts: Similarities and differences. In *Ecosystem services and global ecology*. IntechOpen. https://doi.org/10.5772/intechopen.75705.

Ramos, M. V., Demarco, D., da Costa Souza, I. C., & Freitas, C. D. T. (2019). Laticifers, latex, and their role in plant defense. *Trends in Plant Science, 24*, 553–567.

Ramos, M. V., Grangeiro, T. B., Freire, E. A., Sales, M. P., Souza, D. P., Araújo, E. S., et al. (2010). The defensive role of latex in plants: Detrimental effects on insects. *Arthropod-Plant Interactions, 4*, 57–67.

Ramos, M. V., Oliveira, R. S., Pereira, H. M., Moreno, F. B., Lobo, M. D., Rebelo, L. M., et al. (2015). Crystal structure of an antifungal osmotin-like protein from *Calotropis procera* and its effects on *Fusarium solani* spores, as revealed by atomic force microscopy: Insights into the mechanism of action. *Phytochemistry, 119*, 5–18.

Rasmann, S., Johnson, M., & Agrawal, A. (2009). Induced responses to herbivory and jasmonate in three milkweed species. *Journal of Chemical Ecology, 35*, 1326–1334.

Samanani, N., Liscombe, D. K., & Facchini, P. J. (2004). Molecular cloning and characterization of norcoclaurine synthase, an enzyme catalyzing the first committed step in benzylisoquinoline alkaloid biosynthesis. *The Plant Journal, 40*, 302–313.

Sessa, R., Bennett, M., Lewis, M., Mansfield, J., & Beale, M. (2000). Metabolite profiling of sesquiterpene lactones from lactuca species: Major latex components are novel oxalate and sulfate conjugates of lactucin and its derivatives. *Journal of Biological Chemistry, 275*, 26877–26884.

Shivkar, Y., & Kumar, V. (2003). Anthelmintic activity of latex of *Calotropis procera*. *Pharmaceutical Biology, 41*, 263–265.

Singh, N. K., Handa, A. K., Hasegawa, P. M., & Bressan, R. A. (1985). Proteins associated with adaptation of cultured tobacco cells to NaCl. *Plant Physiology, 79*, 126–137.

Souza, D. P., Freitas, C. D. T., Pereira, D. A., Nogueira, F. C., Silva, F. D., Salas, C. E., et al. (2011). Laticifer proteins play a defensive role against hemibiotrophic and necrotrophic phytopathogens. *Planta, 234*, 183–193.

Souza, I. C. C., Ramos, M. V., Costa, J. H., Freitas, C. D. T., Oliveira, R. S. B., Moreno, F. B., et al. (2017). The osmotin of *Calotropis procera* latex is not expressed in laticifer-free cultivated callus and under salt stress. *Plant Physiology and Biochemistry, 119*, 312-218.

Syeda, S., Sánchez, G., Hong, K., Hawkinson, J., Georg, G., & Blanco, G. (2018). Design, synthesis, and *in vitro* and *in vivo* evaluation of ouabain analogues as potent and selective Na,K-ATPase α4 isoform inhibitors for male contraception. *Journal of Medicinal Chemistry, 61*, 1800–1820.

Vanparijs, J., Broekaert, W. F., Goldstein, I. J., & Peumans, W. J. (1991). Hevein - an antifungal protein from rubber-tree (*Hevea brasiliensis*) latex. *Planta, 183*, 258–264.
Vega, A. S., Castro, M. A., & Anderson, W. R. (2002). Occurrence and phylogenetic significance of latex in the Malpighiaceae. *American Journal of Botany, 89*, 1725–1729.
Wang, S., Liang, H., Zhao, Y., Wang, G., Yao, H., Kasimu, R., et al. (2016). New triterpenoids from the latex of *Euphorbia resinifera* Berg. *Fitoterapia, 108*, 33–40.
Wink, M. (2008). Plant secondary metabolism: Diversity, function and its evolution. *Natural Product Communications, 3*, 1205–1216.
Wittstock, U., & Gershenzon, J. (2002). Constitutive plant toxins and their role in defense against herbivores and pathogens. *Current Opinion in Plant Biology, 5*, 300–307.
Wood, C., Lee, K., Vaisberg, A., Kingston, D., Neto, C., & Hammond, G. (2001). A Bioactive spirolactone iridoid and triterpenoids from *Himatanthus sucuuba. Chemical and Pharmaceutical Bulletin, 49*, 1477–1478.
Yoder, L. R., & Mahlberg, P. G. (1976). Reactions of alkaloid and histochemical indicators in laticifers and specialized parenchyma cells of *Catharanthus roseus* (Apocynaceae). *American Journal of Botany, 63*, 1167–1173.

CHAPTER TWO

Structural diversity and distribution of laticifers

Simone Pádua Teixeira[a,*], Cristina Ribeiro Marinho[a] and Flávia Maria Leme[b]

[a]Departamento de Ciências Farmacêuticas, Faculdade de Ciências Farmacêuticas de Ribeirão Preto, Universidade de São Paulo (USP), São Paulo, Brazil
[b]Instituto de Biociências, Universidade Federal de Mato Grosso do Sul (UFMS), Cidade Universitária, C.P. 549, Campo Grande, Mato Grosso do Sul, Brazil
*Corresponding author: E-mail: spadua@fcfrp.usp.br

Contents

Abstract

The present chapter deals with the concept as well as the structure, ultrastructure and development of laticifer in plants. The morphological aspects of the laticifer development are used to better support the laticifer current classification. Some general functions attributed to the laticifer are discussed, especially those related to the plant protection. As a novelty we cited a role of laticifer presence and distribution in the pollination of the fig trees. An extensive bibliographic survey on taxonomic distribution of laticifers in plants is presented. Illustrations of anatomy, histochemistry and ultrastructure of laticifers are presented to exemplify the treated aspects. Unpublished or recently published aspects are highlighted in each topic. We showed that there is a great diversity of morphological types of laticifers and developmental processes involving the laticifer formation in plants.

Advances in Botanical Research, Volume 93
ISSN 0065-2296
https://doi.org/10.1016/bs.abr.2019.09.003

1. Morphology and classification of laticifers

Laticifers are latex-producing cells or a series of connected cells that form systems which pass through various tissues of the plant body (Evert, 2006). Several classifications have been done to better understand the morphological diversity of laticifers. The most common is the classification based on laticifer ontogeny in which laticifers can be nonarticulated or articulated. Laticifers are called articulated or laticiferous vessels if their terminal walls may remain intact, become porous or even disappear completely during the laticifer development (Fig. 1A and B). If, more rarely, they are formed by a single multinucleated cell that elongates with the plant growth, they are called nonarticulated laticifers or laticiferous cells (Fahn, 1979; Evert, 2006; Ascensão, 2007; Hagel, Yeung, & Facchini, 2008). The multinucleated condition (Fig. 2A and B), found in both types of laticifers mentioned above, can have different origins. In the case of articulated laticifer, this condition is caused by the fusion of protoplasts of their cells while in nonarticulated laticifer the nuclei multiplicate resulting in a multinucleated laticifer cell (Evert, 2006).

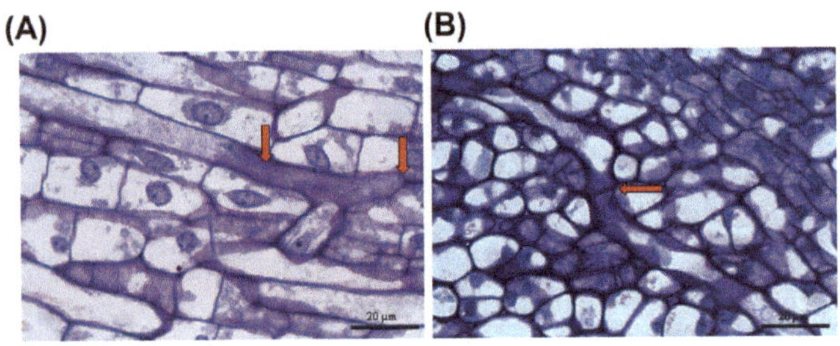

Fig. 1 Articulations in developing laticifers of two Moraceae species: *Dorstenia cayapia* Vell. (A) and *Ficus montana* Burm.f. (B). Note the thin transversal cell walls (arrows) that will be degraded later. Stain: toluidine blue.

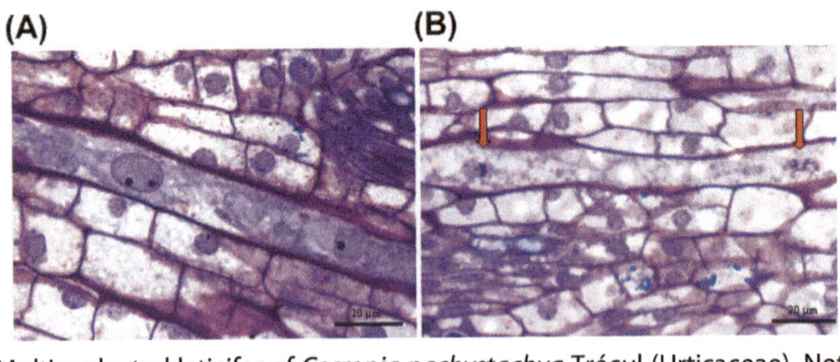

Fig. 2 Multinucleated laticifer of *Cecropia pachystachya* Trécul (Urticaceae). Note some mitotic figures (arrows) in (B) Stain: toluidine blue.

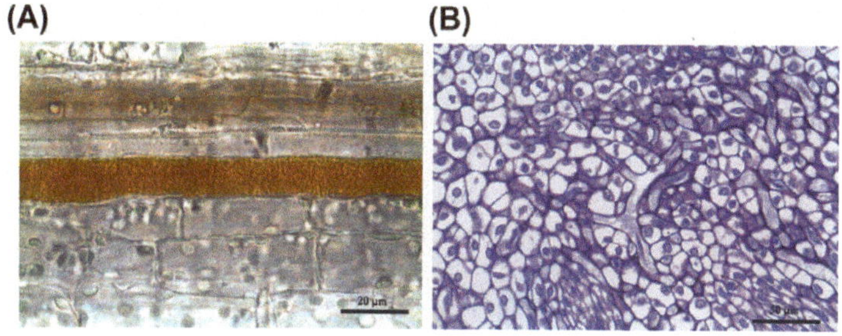

Fig. 3 Unbranched and branched articulated laticifers of *Cannabis sativa* L. (Cannabaceae) (A) and *Dorstenia cayapia* Vell. (Moraceae) (B). Note the brownish natural color of the latex of (*C sativa* (A). Stain: toluidine blue in (B)

Both nonarticulated and articulated types of laticifers may form a system of linear tubes, extending longitudinally along with the growth of the organs, or may form a network of interconnected tubes, extending also radially. Thus, these laticifers can be subdivided into four subtypes: nonarticulated unbranched, nonarticulated branched, articulated nonanastomosed (Fig. 3A) and articulated anastomosed (Fahn, 1979; Evert, 2006; Ascensão, 2007; Hagel et al., 2008). The branching of the nonarticulated laticifers is caused by the growth of lateral projections in the laticifer cell wall, whereas the formation of anastomoses is a process involving the degeneration of parts of the laticifer lateral wall and its adjacent cells (Fahn, 1979; Canaveze & Machado, 2016; Canaveze, Mastroberti, Mariath, & Machado, 2019). In the processes of formation of an anastomosing laticifer an action of laticifer inducing mechanism takes place, because the cells which are incorporated acquire laticiferous characteristics (Canaveze & Machado, 2016; Canaveze et al., 2019). Although the presence of branches is frequently associated with the nonarticulated laticifers, recent studies have shown that articulated laticifers can also branch out (Fig. 3B) (Canaveze et al., 2019; Canaveze & Machado, 2016; Marinho & Teixeira, 2019a).

The unequivocal classification of the laticifers requires ontogenetic study, which means the analysis of the number of precursor cells and the laticifer developmental phases in meristematic regions (Fig. 4A), observing the cell wall disintegration and the incorporations of new protoplasts and entire cells to the laticifer system (Canaveze & Machado, 2016; Fahn, 1979; Marinho & Teixeira, 2019a; Prado & Demarco, 2018). The scarcity of studies related to the origin and development of laticifers may lead to their misclassification in plant groups, since such studies are mainly necessary to distinguish the nonarticulated laticifers from the articulated laticifers that

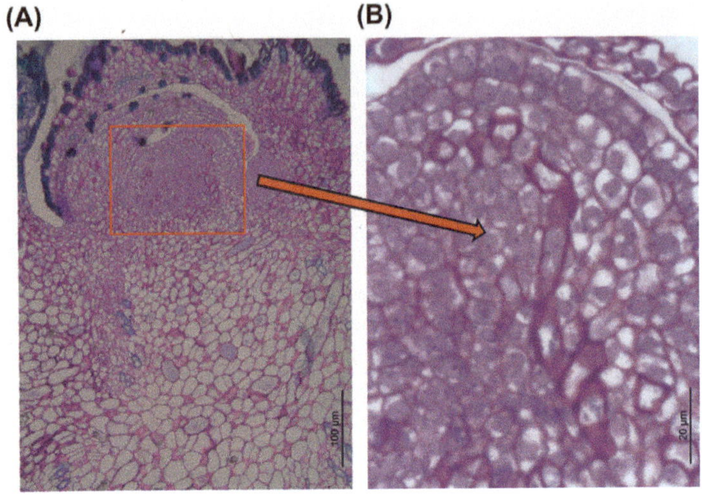

Fig. 4 Anatomical features of the laticifer precursor cells of *Ficus montana* Burm.f. (Moraceae). (A) Longitudinal view of an apical meristem. (B) Detail of laticifer precursor cells. Note the thick pecto-cellulosic cell walls. Stain: toluidine blue.

present complete degradation of their terminal cell walls (Ascensão, 2007; Fahn, 1979; Hagel et al., 2008). Example of a misinterpretation is found in some families of the urticalean clade, as Cannabaceae and Moraceae, in which laticifers were previously classified as nonarticulated (Metcalfe & Chalk, 1950; Fahn, 1979; Topper & Koek-Noorman, 1980; van Veenendaal & den Outer, 1990; Balaji, Subramanian, & Inamdar, 1993; Machado & Santos, 2004; Quintanar & Castrejón, 2004; Jacomassi, Moscheta, & Machado, 2010, 2007; Kitajima et al., 2012). Recent ontogenic studies have showed that, on the opposite, laticifers are articulated, because more than one precursor cell was detected in the laticifer formation (Leme, 2018; Marinho & Teixeira, 2019a) (Fig. 5).

The laticifer precursor cells are structurally distinguished from the neighboring cells by some general features, as their elongated shape, thicker lateral walls and thinner terminal walls (Fig. 4B) (Marinho & Teixeira, 2019a). Considering an articulated laticifer in a more advanced developmental phase the terminal walls disintegrate and the laticifer connected cells can exhibit several nuclei, forming a multinucleated structure that intrusively grows. This is evidenced by the sinuosity of the laticifer cell walls. If the laticifer is branched, lateral expansions of the cell are seen and the laticifer can assume several forms, as for example "Y" (Fig. 3B) or "H" (Canaveze & Machado, 2016; Marinho & Teixeira, 2019a).

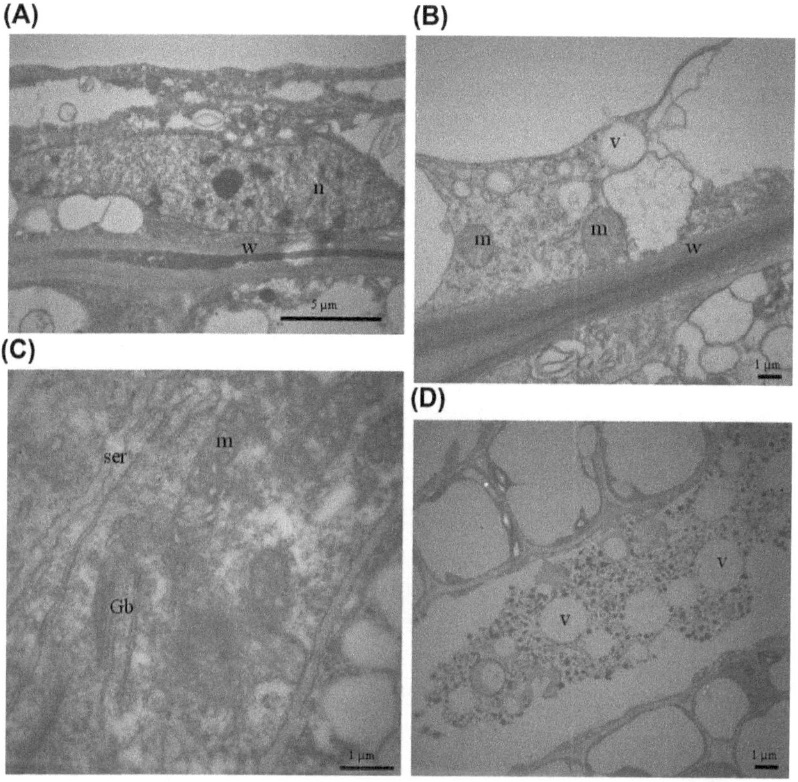

Fig. 5 Ultrastructural features of mature laticifers: thick walls (w), elongated peripheral nucleus (n), cytoplasm with mitochondria (m), smooth endoplasmic reticulum (ser), Golgi bodies (Gb), polyribosomes and several small vacuoles (v). (A–B) *Celtis pubescens* Spreng. (Cannabaceae) (C) *Trema micrantha* (L.). Blume (Cannabaceae) (D) *Maclura tinctoria* (L.). D. Don ex Steud. (Moraceae).

The type of mature laticifer has been associated with a certain process of laticifer formation. The intrusive growth, for example, is associated with the presence of nonarticulated laticifers (Mahlberg, 1993), while protoplast fusion and the addition of adjacent cells are associated with articulated laticifers (Lopes, Thadeo, Azevedo, Soares, & Meira, 2009; Milanez, 1959, 1977, pp. 343–379). However, novel ontogenic studies covering a larger number of species have shown that such processes can occur concomitantly in the formation of a type of laticifer in the same species, as in *Tabernaemontana catharinensis* (Apocynaceae, Canaveze & Machado, 2016) and in some Moraceae species (Marinho & Teixeira, 2019a).

The mature laticifer has walls that are in generally composed of pectin and cellulose (Figs. 1 and 2) (Marinho & Teixeira, 2019a). Such substances increase the laticifer elasticity and facilitate the laticifer continuing expansion throughout the plant tissues (Mahlberg, 1993). The intrusive apical growth of laticifers through the plant tissues can also be promoted by the actions of pectinases and cellulases. The pectinases can act on the dissolution of the middle lamella and cellulases operate in the degradation of the terminal cell walls in the case of an articulated laticifer (Allen & Nessler, 1984; Marinho & Teixeira, 2019b; Wilson, Nessler, & Mahlberg, 1976). Cytochemical reactions detected the synthesis of pectinases and cellulases in the endoplasmic reticulum and their transport to the laticifer wall by exocytosis (Marinho & Teixeira, 2019b).

Callose and suberin are other substances that can be present in the laticifer walls, playing a role as a sealant of the system, preventing the latex release to the adjacent tissues (Castro & Machado, 2012).

The subcellular composition of a laticifer is variable, depending on the species and especially on the chemical nature of the latex (Canaveze et al., 2019; Marinho & Teixeira, 2019b). The laticifer cytoplasm can be rich in mitochondria, plastids lacking grana, rough endoplasmic reticulum, Golgi bodies with dilated extremities, polyribosomes and several small vacuoles with content of different densities. The latex occupies the whole volume of the laticifer system (Leme, 2018; Canaveze et al., 2019; Marinho & Teixeira, 2019a, b). It is noteworthy the presence of plasmodesmata connecting the laticifer to adjacent cells in some species (Canaveze et al., 2019).

Latex is regarded as a suspension of various particles, such as polyisoprene, hydrocarbons, steroids, triterpenoids, fatty and aromatic acids, carotenes, phospholipids, proteins, alkaloids, inorganic compounds, among others, dispersed in a liquid with different refractive index. Some histochemical testes that can be applied to the latex are presented in Fig. 6. The color of the latex can be white and milky, yellow to brown (Fig. 3A), orange or colorless depending on the species analyzed (Fahn, 1979; Ascensão, 2007). The richer the latex is for a particular substance, the more an organelle or organelle clusters will predominate in the cytoplasm. An example is the presence of terpenes in latex and of plastids lacking grana associated with rough endoplasmic reticulum (Marinho & Teixeira, 2019a, 2019b). The presence of electron-dense plastids without thylakoids in plant secretory structures is often related to the production of terpenes (Marinho, Martucci, et al., 2018). Indeed, while sesquiterpenes are produced in cytosol with the action of the endoplasmic reticulum and peroxisomes, monoterpenes and diterpenes are known to be synthesized by plastids in plant cells (Muhlemann, Klempien, & Dudareva, 2014; Vranová, Coman, & Gruissem, 2013).

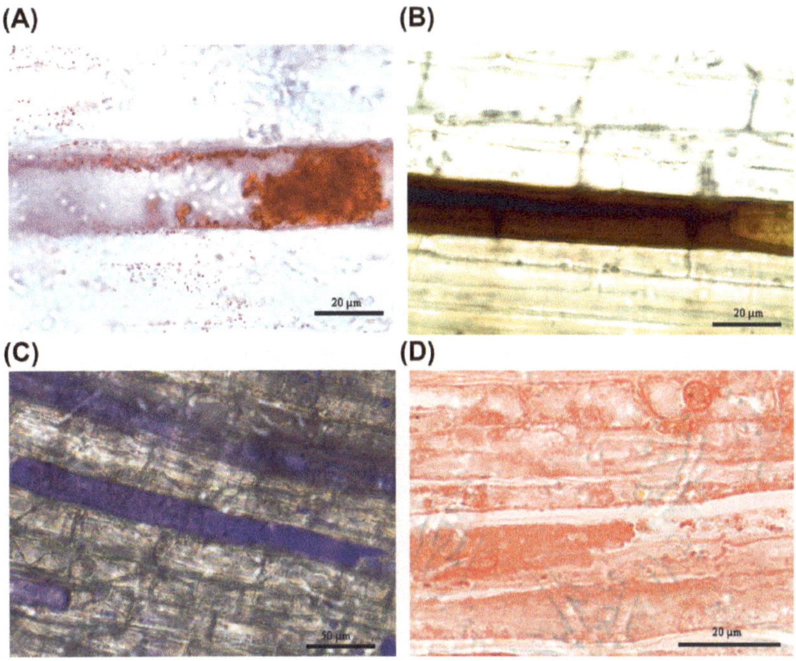

Fig. 6 General composition of the latex by applying histochemical tests to the laticifer. (A–B) Lipids detected with Sudan III (A) and phenolic compounds with ferric chloride (B) in *Cannabis sativa* L. (Cannabaceae). (C) Terpenes detected with Nadi reagent in *Sorocea bonplandii* (Baill.) W.C. Burger, Lanj. & de Boer (Moraceae). (D) Proteins detected with xylidine de Ponceau in *Pteroceltis tatarinowii* Maxim. (Cannabaceae).

2. Distribution of laticifers along the plant body

Laticifers have already been found in all the organs of the plant body, such as in the root, stem, leaf, sepal, petal, stamen, ovary and stigma. They can occur in all or in some particular organ of the same species. In *Nymphaea* L., for example, the occurrence of laticifer is reported for stamens and only for the outer tepals, although the similarity of both inner and outer tepals (Moseley, 1961). Other examples are found in Araceae (laticifers in the bracts, bracteoles and central axis of spadix), Leguminosae (laticifers in the pulvinulo) and Asteraceae (laticifers in the inflorescence) (Table 1).

Analyzing inside the plant organs, laticifers usually occur embedded in the cortex, vascular bundles and pith. Epidermal and subepidermal laticifers were only found in the fern *Regnellidium diphyllum* Lindman (Marsileaceae — Durairatnam, 1959) and in some species of Euphorbiaceae (*Croton* spp.,

Table 1 Taxonomic distribution of laticifer types and subtypes in vascular plants. Empty cells mean no available information.

Group	Family	Species	Laticifer morphtype	References
Fern	Marsileaceae	*Regnellidium diphyllum* Lindman	Articulated	Durairatnam (1959)
Gymnosperm	Gnetaceae	*Gnetum gnemon* L.	Articulated, anastomosed	Behnke and Herrmann (1978)
Ana-grade	Cabombaceae	*Brasenia* Schreb.		Les (1988)
Ana-grade	Cabombaceae	*Cabomba* Aubl.		Les (1988)
Ana-grade	Nymphaeaceae	*Barclaya* Wall.		Les (1988)
Ana-grade	Nymphaeaceae	*Nymphaea gardneriana* Planch.	Articulated	Zini, Galati, and Ferrucci (2017)
Ana-grade	Nymphaeaceae	*Nymphaea lotus* L.	Articulated	Zini et al. (2017)
Ana-grade	Nymphaeaceae	*Nymphaea* L.	Articulated, unbranched	Moseley (1961)
Ana-grade	Nymphaeaceae	*Victoria* Lindl.	Articulated, unbranched	Schneider and Williamson (1993)
Ana-grade	Nymphaeaceae	*Nuphar* Sm.		Les et al. (1999)
Ana-grade	Nymphaeaceae	*Euryale* Salisb.		Les et al. (1999)
Ana-grade	Nymphaeaceae	*Ondinea* Hartog	Articulated, unbranched	Williamson and Moseley (1989)
Ana-grade	Austrobaileyaceae	*Austrobaileya scandens* C.T. White		Carlquist (2001)
Magnoliid	Myristicaceae	*Myristica fragrans* Houtt.		Armstrong and Tucker (1986)
Monocot	Araceae	*Protarum* Engl.	Articulated, anastomosed	French (1988)
Monocot	Araceae	*Scaphispatha* Brongn. ex Schott	Articulated, anastomosed	French (1988)

Monocot	Araceae	*Zomicarpa* Schott	Articulated, anastomosed	French (1988)
Monocot	Araceae	*Ariopsis* Nimmo	Articulated, nonanastomosed	French (1988)
Monocot	Araceae	*Orontium* L.	Articulated, nonanastomosed	French (1988)
Monocot	Alismataceae	*Sagittaria* L.	Articulated	Govindarajalu (1967)
Monocot	Aponogetonaceae	*Aponogeton madagascariensis* (Mirb.) H. Bruggen	Articulated	Gunawardena, Greenwood, and Dengler (2007); Bruggen (1998)
Monocot	Cyclanthaceae	*Cyclanthus bipartitus* Poit ex A.Rich.	Nonarticulated, unbranched occasional branched	Wilder and Harris (1982)
Monocot	Amarylliaceae	*Allium* L.	Nonanastomosed	Huang and Sterling (1970)
Monocot	Musaceae		Articulated	Dahlgren, Clifford, and Yeo (1985)
Basal eudicot	Papaveraceae	*Papaver somniferum* L.	Articulated, anastomosed	Nessler and Mahlberg (1977, 1976); Fairbairn and Kapoor (1960)
Basal eudicot	Papaveraceae	*Chelidonium majus* L.	Articulated, nonanastomosed	Hagel et al. (2008)
Basal eudicot	Menispermaceae	*Tinomiscium* Miers ex Hook. f. & Thomson	Nonarticulated, anastomosed	Wilkinson (1986)
Basal eudicot	Menispermaceae	*Fibraurea* Lour.	Nonarticulated, anastomosed	Wilkinson (1986)
Basal eudicot	Lardizabalaceae	*Decaisnea insignis* (Griff.) Hook. f. & Thomson		Metcalfe and Chalk (1983)

(*Continued*)

Table 1 Taxonomic distribution of laticifer types and subtypes in vascular plants. Empty cells mean no available information.—cont'd

Group	Family	Species	Laticifer morphtype	References
Basal eudicot	Nelumbonaceae	*Nelumbo nucifera* Gaertn.	Articulated	Esau and Kosakai (1975); Williamson and Schneider (1993)
Rosid	Fabaceae	*Lonchocarpus lilloi* (Hassl.) Burk.	Articulated, unbranched	Lewinsohn (1991); Fernández-Dattoli et al. (2011)
Rosid	Ulmaceae	*Ampelocera glabra* Kuhlm.	Articulated, unbranched	Leme (2018)
Rosid	Ulmaceae	*Phyllostylon rhamnoides* (J. Poiss.) Taub.	Articulated, unbranched	personal observation
Rosid	Ulmaceae	*Ulmus parvifolia* Jacq.	Articulated, unbranched	personal observation
Rosid	Ulmaceae	*Zelkova serrata* (Thunb.) Makino.	Articulated, unbranched	Leme (2018)
Rosid	Cannabaceae	*Humulus lupulus* L.	Nonarticulated, unbranched	Hagel et al. (2008)
Rosid	Cannabaceae	*Cannabis sativa* L.	Articulated, unbranched	Leme (2018)
Rosid	Cannabaceae	*Celtis pubescens* (Kunth) Spreng.	Articulated, unbranched	Leme (2018)
Rosid	Cannabaceae	*Pteroceltis tatarinowii* Maxim.	Articulated, unbranched	Leme (2018)
Rosid	Cannabaceae	*Trema micrantha* (L.) Blume	Articulated, unbranched	Leme (2018)
Rosid	Moraceae	*Broussonetia* L'Hér. ex Vent.	Nonarticulated, branched	Evert (2006)
Rosid	Moraceae	*Morus* L.	Nonarticulated, branched	Evert (2006)
Rosid	Moraceae	*Sorocea bonplandii* (Baill.) W.C. Burger, Lanj. & de Boer	Articulated, branched	Marinho and Teixeira (2019a)

Rosid	Moraceae	*Artocarpus heterophyllus* Lam.	Articulated, branched	Marinho and Teixeira (2019a)
Rosid	Moraceae	*Maclura tinctoria* (L.) D.Don ex Steud.	Articulated, branched	Marinho and Teixeira (2019a)
Rosid	Moraceae	*Dorstenia cayapia* Vell.	Articulated, branched	Marinho and Teixeira (2019a)
Rosid	Moraceae	*Brosimum gaudichaudii* Trécul	Articulated, branched	Marinho and Teixeira (2019a)
Rosid	Moraceae	*Castilla elastica* Cerv.	Articulated, branched	Marinho and Teixeira (2019a)
Rosid	Moraceae	*Ficus citrifolia* Mill.	Articulated, branched	Marinho and Teixeira (2019a)
Rosid	Moraceae	*Ficus montana* Burm.f.	Articulated, branched	Marinho and Teixeira (2019a)
Rosid	Moraceae	*Ficus obtusiuscula* (Miq.) Miq.	Articulated, branched	Marinho and Teixeira (2019a)
Rosid	Moraceae	*Ficus caatingae* Castro	Nonarticulated	Araújo, Coelho, Ventrella, and Agra (2014)
Rosid	Moraceae	*Ficus cyclophylla* Miq.,	Nonarticulated	Araújo et al. (2014)
Rosid	Moraceae	*Ficus elliotiana* S. Moore	Nonarticulated	Araújo et al. (2014)
Rosid	Moraceae	*Ficus carica* L.	Nonarticulated, branched	Kajii, Morita, and Kuroda (2014)
Rosid	Urticaceae	*Cecropia pachystachya* Trécul	Articulated, branched	Marinho and Teixeira (2019a)
Rosid	Urticaceae	*Coussapoa microcarpa* (Schott) Rizzini	Articulated, branched	Marinho and Teixeira (2019a)
Rosid	Urticaceae	*Pourouma cecropiifolia* Mart.	Articulated, branched	Marinho and Teixeira (2019a)

(*Continued*)

Table 1 Taxonomic distribution of laticifer types and subtypes in vascular plants. Empty cells mean no available information.—cont'd

Group	Family	Species	Laticifer morphtype	References
Rosids	Urticaceae	*Laportea amplissima* = *Dendrocnide amplissima* (Blume) Chew	Nonarticulated, unbranched	Guérin (1923)
Rosids	Urticaceae	*Laportea longifolia* = *Dendrocnide longifolia* (Hemsl.) Chew	Nonarticulated, unbranched	Guérin (1923)
Rosids	Urticaceae	*Laportea moroides* = *Dendrocnide moroidea* (Wedd.) Chew	Nonarticulated, unbranched	Guérin (1923)
Rosids	Urticaceae	*Laportea platycarpa* (Wedd.) Wedd	Nonarticulated, unbranched	Guérin (1923)
Rosids	Urticaceae	*Laportea schomburgkii* W.Bull	Nonarticulated, unbranched	Guérin (1923)
Rosids	Urticaceae	*Laportea stimulans* = *Dendrocnide stimulans* (L.f.) Chew	Nonarticulated, unbranched	Guérin (1923)
Rosids	Urticaceae	*Urera baccifera* (L.) Gaudich. ex Wedd.	Nonarticulated, unbranched	Guérin (1923)
Rosids	Urticaceae	*Urera humblotii* = *Laportea humblotii* (Baill.) Friis	Nonarticulated, unbranched	Guérin (1923)
Rosid	Urticaceae	*Urtica* L.	Nonarticulated, unbranched	Metcalfe (1967); Evert (2006)
Rosid	Celastraceae	*Tontelea micrantha* (Mart. ex. Schult.) A. C. Sm.	Nonarticulated, nonanastomosed	Mercadante-Simões et al. (2014)

Rosid	Celastraceae	*Cheiloclinium anomalum* Miers		Gomes and Lombardi (2010)
Rosid	Celastraceae	*Cheiloclinium cognatum* (Miers) A.C. Sm.		Gomes and Lombardi (2010)
Rosid	Celastraceae	*Cheiloclinium hippocrateoides* (Peyr.) A.C. Sm.		Gomes and Lombardi (2010)
Rosid	Celastraceae	*Cheiloclinium serratum* (Cambess.) A.C. Sm.		Gomes and Lombardi (2010)
Rosid	Rhizophoraceae		Articulated	Dahlgren (1988)
Rosid	Malpighiaceae	*Galphimia brasiliensis* (L.) A. Juss.	Articulated, nonanastomosed, unbranched	Vega, Castro, and Anderson (2002)
Rosid	Malpighiaceae	*Verrucularia glaucophylla* A. Juss.	Articulated, nonanastomosed, unbranched	Vega et al. (2002)
Rosid	Salicaceae			Prado and Demarco (2018)
Rosid	Euphorbiaceae	*Omphalea* L.	Nonarticulated	Rudall (1994)
Rosid	Euphorbiaceae	*Adenocline acuta* (Thunb.) Baill.	Nonarticulated	Rudall (1994)
Rosid	Euphorbiaceae	*Aleurites moluccanus* (L.) Willd.	Nonarticulated	Rudall (1994)
Rosid	Euphorbiaceae	*Baloghia lucida* = *Baloghia inophylla* (G.Forst.) P.S.Green	Nonarticulated	Rudall (1994)
Rosid	Euphorbiaceae	*Bertya mitchellii* = *Bertya oleifolia* Planch.	Nonarticulated	Rudall (1994)
Rosid	Euphorbiaceae	*Beyeria* lechenaultii (DC.) Baill.	Nonarticulated	Rudall (1994)

(Continued)

Table 1 Taxonomic distribution of laticifer types and subtypes in vascular plants. Empty cells mean no available information.—cont'd

Group	Family	Species	Laticifer morphtype	References
Rosid	Euphorbiaceae	*Codiaeum variegatum* (L.) Rumph. ex A. Juss.	Nonarticulated	Rudall (1994)
Rosid	Euphorbiaceae	*Croton* L.	Nonarticulated	Rudall (1994)
Rosid	Euphorbiaceae	*Cyrtogonone argentea* (Pax) Prain	Nonarticulated	Rudall (1994)
Rosid	Euphorbiaceae	*Garcia nutans* Vahl ex Rohr	Nonarticulated	Rudall (1994)
Rosid	Euphorbiaceae	*Hevea confusa* = *Hevea pauciflora* var. *coriacea* Ducke	Articulated	Rudall (1994)
Rosid	Euphorbiaceae	*Jatropha curcas* L.	Nonarticulated	Rudall (1994)
Rosid	Euphorbiaceae	*Manihot* Mill.	Articulated	Rudall (1994)
Rosid	Euphorbiaceae	*Micrandra* Benth.	Nonarticulated	Rudall (1994)
Rosid	Euphorbiaceae	*Neoboutonia macrocalyx* Pax	Nonarticulated	Rudall (1994)
Rosid	Euphorbiaceae	*Pausandra macropetala* Ducke	Nonarticulated	Rudall (1994)
Rosid	Euphorbiaceae	*Ricinocarpus* Kuntze	Nonarticulated	Rudall (1994)
Rosid	Euphorbiaceae	*Ricinodendron heudelotii* (Baill.) Heckel	Nonarticulated	Rudall (1994)
Rosid	Euphorbiaceae	*Spirostachys venenifera* (Pax) Pax	Nonarticulated	Rudall (1994)
Rosid	Euphorbiaceae	*Vernicia cordata* (Thunb.) Airy Shaw	Nonarticulated	Rudall (1994)
Rosid	Euphorbiaceae	*Euphorbia lathyris* L.	Nonarticulated	Rudall (1994)
Rosid	Euphorbiaceae	*Synadenium grantii* = *Euphorbia umbellata* (Pax) Bruyns	Nonarticulated	Rudall (1994)

Rosid	Euphorbiaceae	*Sapium haematospermum* Müll.Arg.	Articulated, anastomosed	Demarco et al. (2013)
Rosid	Euphorbiaceae	*Hevea brasiliensis* (Willd. ex A. Juss.) Müll. Arg.	Articulated, anastomosed	Metcalfe (1967)
Rosid	Sapindaceae	*Paullinia* L.	Articulated, nonanastomosed	Cunha Neto, Martins, Somner, and Tamaio (2017)
Rosid	Sapindaceae	*Serjania* Mill.	Articulated, nonanastomosed	Cunha Neto et al. (2017)
Rosid	Bixaceae	*Bixa orellana* L.		Rivera and Flores (1988)
Rosid	Myrtaceae			Lewinsohn (1991)
Rosid	Caricaceae		Articulated, anastomosed	Mahlberg (1993)
Asterid	Loranthaceae			Lewinsohn (1991)
Asterid	Olacaceae	*Heisteria* Jacq.	Nonarticulated	Baas and Kool (1983); Baas, Oosterhoud, and Scholtes (1982)
Asterid	Olacaceae	*Chaunochiton* Benth.	Nonarticulated	Baas et al. (1982)
Asterid	Olacaceae	*Aptandra* Miers	Articulated	Baas et al. (1982)
Asterid	Olacaceae	*Harmandia* Baill.	Articulated	Baas et al. (1982)
Asterid	Cornaceae	*Alangium* Lam.	Articulated	Kubitzki (2004)
Asterid	Nyssaceae	*Camptotheca acuminata* Decne	Nonarticulated, unbranched	Monacelli, Valletta, Rascio, Moro, and Pasqua (2005)
Asterid	Sapotaceae	*Manilkara* Adans, *Palaquium* Blanco, *Chrysophyllum* L., *Pouteria* Aubl., *Diploon* Cronquist	Articulated	Almeida Jr. et al. (2013); Solereder (1908); Monteiro, Andreata, and Neves (2007)

(*Continued*)

Table 1 Taxonomic distribution of laticifer types and subtypes in vascular plants. Empty cells mean no available information.—cont'd

Group	Family	Species	Laticifer morphtype	References
Asterid	Aquifoliaceae			Lewinsohn (1991)
Asterid	Cardiopteridaceae	*Cardiopteris quinqueloba* (Hassk.) Hassk.	Articulated	Metcalfe and Chalk (1950)
Asterid	Asteraceae	*Cichorium intybus* L.	Articulated, branched, anastomosed	Melo-de-Pinna and Menezes (2003); Vertrees and Mahlberg (1978)
Asterid	Asteraceae	*Microliabum* Cabrera	Articulated, anastomosed	Gutiérrez and Luna (2013)
Asterid	Asteraceae	*Munnozia* Ruiz & Pav.	Articulated, anastomosed	Gutiérrez and Luna (2013)
Asterid	Asteraceae	*Paranephelius* Poepp.	Articulated, anastomosed	Gutiérrez and Luna (2013)
Asterid	Asteraceae	*Warionia saharae* Benthem ex Benth. & Cosson	Articulated, anastomosed	Katinas, Tellería, Susanna, and Ortiz (2008)
Asterid	Asteraceae	*Berkheya* Ehrh.	Articulated, anastomosed	Karis (2006)
Asterid	Asteraceae	*Cullumia* R. Br.	Articulated, anastomosed	Karis (2006)
Asterid	Asteraceae	*Cuspidia* Gaertn.	Articulated, anastomosed	Karis (2006)
Asterid	Asteraceae	*Didelta* L'Hér.	Articulated, anastomosed	Karis (2006)
Asterid	Asteraceae	*Gazania* Gaertn.	Articulated, anastomosed	Karis (2006)
Asterid	Asteraceae	*Gorteria* L.	Articulated, anastomosed	Karis (2006)

Asterid	Asteraceae	*Heterorhachis* Sch. Bip. ex Walp.	Articulated, anastomosed	Karis (2006)
Asterid	Asteraceae	*Hirpicium* Cass.	Articulated, anastomosed	Karis (2006)
Asterid	Campanulaceae	*Lobelia* L.	Articulated, anastomosed	Rosatti (1986); Metcalfe and Chalk (1950); Mahlberg (1993)
Asterid	Apocynaceae	*Allamanda blanchetii* A. DC.	Articulated, anastomosed	Gama et al. (2017)
Asterid	Apocynaceae	*Tabernaemontana catharinensis* A.DC.	Articulated, anastomosed	Canaveze and Machado (2016)
Asterid	Apocynaceae	*Asclepias curassavica* L.	Articulated, anastomosed	Demarco and Castro (2008)
Asterid	Apocynaceae	*Fischeria stellata* (Vell.) E.Fourn	Articulated, anastomosed	Demarco and Castro (2008)
Asterid	Apocynaceae	*Gonioanthela axillaris* (Vell.) Fontella & E.A. Schwarz	Articulated, anastomosed	Demarco and Castro (2008)
Asterid	Apocynaceae	*Matelea denticulata* (Vahl) Fontella & E.A. Schwarz	Articulated, anastomosed	Demarco and Castro (2008)
Asterid	Apocynaceae	*Oxypetalum banksii* R.Br. ex Schult.	Articulated, anastomosed	Demarco and Castro (2008)
Asterid	Apocynaceae	*Aspidosperma australe* Müll. Arg.	Articulated, anastomosed	Demarco et al. (2006)
Asterid	Apocynaceae	*Blepharodon bicuspidatum* E.Fourn.	Articulated, anastomosed	Demarco et al. (2006)
Asterid	Convolvulaceae	*Ipomoea pes-caprae* (L.) R. Br.	Articulated, nonanastomosed	Kuster, da Silva, Meira, and Azevedo (2016)

(Continued)

Table 1 Taxonomic distribution of laticifer types and subtypes in vascular plants. Empty cells mean no available information.—cont'd

Group	Family	Species	Laticifer morphtype	References
Asterid	Convolvulaceae	*Ipomoea imperati* (Vahl.) Griseb.	Articulated, nonanastomosed	Kuster et al. (2016)
Asterid	Convolvulaceae	*Cuscuta* L.	Articulated, nonanastomosed	Riviere, Clayson, Dockstader, Wright, and Costea (2013)
Asterid	Convolvulaceae	*Calystegia silvatica* (Kit.) Griesb.	Articulated, unbranched	Condon and Fineran (1989)
Asterid	Convolvulaceae	*Ipomoea purpurea* (L.) Roth	Articulated, nonanastomosed	Cortella (1989)
Asterid	Eucommiaceae	*Eucommia ulmoides* Oliv.	Branched, anastomosed	Magallon-Puebla and Cevallos-Ferriz (1994)
Asterid	Eucommiaceae	*Eucommia ulmoides* Oliv.	Articulated	Metcalfe (1967)
Asterid	Icacinaceae			Prado and Demarco (2018)

Beyeria leschenaultii and *Codiaeum variegatum*) in which the laticifers reach the trichome peduncle and epidermal cells (Rudall, 1994). In the majority of angiosperms laticifers are found externally associated to the phloem, but they may also occur in the cambium and xylem. In general laticifers accompany the vascular bundles along the plant organs (Evert, 2006; Fahn, 1979).

Leaf and stem are the organs with the highest number of records. On the contrary, floral organs are less recorded. The studies with laticifers especially occurring in the vegetative organs of the plant body showed the efficient roles exerted by them in sealing wounds by coagulating, blocking microorganism invasions and avoiding herbivory (Demarco & Castro, 2008; Demarco, Castro, & Ascensão, 2013). However, very little is proposed about the functions of laticifers located in the flower. A recently published study on *Ficus* species showed that laticifers can act helping the pollination mechanism of fig trees because their arrangement along the syconium seems to favor the mutualism between pollinating wasps and fig trees. Laticifers are absent in pistil and stamens that are organs with which pollinators interact, but they are present in organs that protect both seeds and pollinating larvae against natural enemies as non-pollinating wasps and other parasites of the mutualism (Marinho, Pereira, et al., 2018).

3. Taxonomic distribution of laticifers

Laticifers were reported in at least 43 families of vascular plants (Table 1). Angiosperm is the major group with 41 families, followed by ferns and gymnosperms with one family in each one. The laticifers are present in almost all the groups of angiosperm, in the basal clades (Ana-grade), magnoliids, monocots, basal eudicots, rosids and asterids (Table 1). The articulated laticifer is the morphotype most recorded among the species, occurring in 27 families. In some families both morphotypes articulated and nonarticulated are recorded, but this information should be treated with carefulness because such classification has not been accompanied of laticifer ontogeny in most studies.

A point that should be considered is that many species and even families are included in the list of latescent taxa by having an exudate similar to the latex. Such an exudate could be produced and stored by a laticifer or also by a secretory duct (Prado & Demarco, 2018). On the opposite, the latex in many species can be unnoticeable, which can attest to the absence of a laticifer system in a plant species. Thus, both the laticifer presence and the latex composition must be better investigated in order to avoid misinterpretations.

Although the occurrence and distribution of laticifers in plants is considered by many authors to be diagnostic for several taxonomic groups (see Table 1), there are few studies that establish relationships between the laticifer anatomy and ontogeny and the taxa in which they occur. It is most likely that different types of laticifers can be found in different species of the same family (Evert, 2006; Fahn, 1979; Hagel et al., 2008).

Among the few systematic comparative studies of laticifers, an example is observed in Araceae, whose exclusive presence of articulated anastomosed type of laticifer in the leaves and inflorescences in 48 species of the subfamily Colocasioideae revealed its systematic importance in the group (French, 1988). Other example is *Nymphaeaceae* in which only unbranched laticifers were reported (Schneider & Williamson, 1993). Other families, such as Euphorbiaceae and Apocynaceae, show both articulated and non-articulated laticifers (Canaveze & Machado, 2016; Demarco & Castro, 2008; Evert, 2006; Metcalfe, 1967; Rudall, 1994). Surprisingly, even in the same species there are reports of more than one type of laticifer which is the case of *Ficus goldmanii* (Moraceae), a species with articulated and nonarticulated laticifers in the secondary phloem (Quintanar & Castrejón, 2004). However, data on laticifer ontogeny was not found for this species, which may invalidate such information.

4. How to analyse the laticifer articulation

As previously stated, developmental studies are crucial for the determination of the presence or absence of articulations in the laticifers, allowing their correct classification in the plant groups. To this end, two approaches can be used, light microscopy and transmission electron microscopy. For both techniques, the plant material to be used must contain embryos or primary and/or secondary meristems, whether of vegetative or reproductive origin.

The study of the ontogeny of laticifers under light microscopy (LM) allows the use of various types of fixative reagents, colorants and inclusion media. Among the most common fixatives are FAA (formalin-acetic acid-alcohol), neutral buffered formalin and Karnovsky's solution. After dehydrated preferably in alcoholic series, the samples are embedded in synthetic resin or paraffin for sectioning in a rotating microtome. The thinner the sections obtained, the easier is the identification of possible transverse walls in the laticifers. The thickness of the cut obtained varies according to the chosen inclusion media, and while the synthetic resin allows the sectioning around 2–2.5 μm thick, the paraffin embedded material provides thicker sections (above 10 μm). After sectioning the samples and assembling them on histological slides, the material is stained with different dyes (Fig. 6), which also depends on the type of embedding medium used. Toluidine blue is the dye often used for materials embedded in resin, while paraffin allows the use of various types of reagents (astra blue and safranin O, alcoholic safranin, crystal violet and orange G, among others) (Canaveze & Machado, 2016; Demarco & Castro, 2008; Demarco et al., 2006, 2013; Gama, Rubiano, & Demarco, 2017; Leme, 2018; Lopes et al., 2009; Marinho & Teixeira, 2019a; Rudall, 1994). There is no consensus as to what would be the best dye for the visualization of the laticifer growth, since this is a factor that depends only on the species studied.

The protocols used to study the development of laticifers under transmission electron microscopy (MET) present fewer variables. It is important to consider that the plant samples to be analyzed in MET should already be collected and fixed in smaller sizes (around 1 mm^2). The fixative reagents include glutaraldehyde or Karnovsky's solution and the most commonly used post-fixative is the osmium tetroxide. The samples are dehydrated in an acetone series and included in a more rigid resin (Araldite or Spurr). This high-density resin allows the sectioning of the samples in very thin cuts (around 70 nm). After sectioning, the samples are arranged in small copper grids and contrasted with uranyl acetate and Reynolds lead citrate (Fineran, Condon, & Ingerfeld, 1988; Gama et al., 2017; Canaveze et al., 2019; Inamdar, Murugan, & Subramanian, 1988; Lee & Mahlberg, 1999; Leme, 2018; Marinho & Teixeira, 2019b; Wilson & Mahlberg, 1978).

The association of cytochemical analyzes with transmission electron microscopy is also an interesting strategy for the study of the laticifer development. It involves the localization of the activity of pectinases and cellulases that are two enzymes related to the degradation of cell walls. While these two enzymes are present in articulated laticifers that present degeneration of their transverse walls, only the activity of pectinases has been reported for the nonarticulated laticifers, acting in the degradation of the middle lamella of the cells. For that, the plant samples, after being fixed in Karnovsky, are part incubated in pectin and part incubated in carboxymethylcellulose and then transferred to Benedict's reagent heated to 80 °C.

Subsequently, the samples are washed in phosphate buffer and processed by usual TEM techniques (Allen & Nessler, 1984; Bal, 1974; Leme, 2018; Marinho & Teixeira, 2019b; Nessler & Mahlberg, 1981; Wilson et al., 1976).

To find out more about the incubation times and concentrations of each fixative, reagent and dye, please see the references cited in the previous paragraphs.

5. Outlook

The laticifer is a complex secretory structure, either from the structural and functional point of view. The recent discovery of articulated laticifers in plant families considered as having only nonarticulated laticifers evidences the importance of studies of development in plant secretory structures. In addition, it points out that the occurrence of the articulated type of laticifer in the plant groups is probably higher than what is currently recorded in the literature. As stated in the prior sections, the origin and development of laticifers can only be determined by anatomical studies on embryos and seedlings, as well as anatomical studies of regions of the apical or lateral meristems of adult individuals, when, in the latter case, laticifers are found in the secondary tissues (Hagel et al., 2008; Metcalfe, 1967).

Some researches on the ecology of laticifers indicate that the success of the latescent plants in the different environments is due to the functions commonly attributed to the latex of protection against herbivory and pathogens and sealing of injuries (Agrawal & Konno, 2009; Farrel, Dussourd, & Mitter, 1991; Konno, 2011; Prado & Demarco, 2018; Ramos, Demarco, Souza, & Freitas, 2019). Moreover, studies on the molecular biology of laticifers show that latex abundance may be induced by jasmonic acid, and therefore latex production may be enhanced under specific conditions. Since jasmonic acid is a well-known regulator involved in signaling in plant defense, this finding strengthens the value of laticifers and latex in facing enemies (Castelblanque, Balaguer, Martí, Orozco, & Vera, 2018).

However, much remains to be discovered about the ecological functions of the laticifers. Because of the great importance of latex to the human being, the advances in the uses of the substances extracted of the latex are immense compared to our understanding of its function in the plant body. In addition, despite the laticifer structure has been known for more than a century, disagreements about the origin, development and the lack of information about the chemical composition of latex in several groups still remains. An interesting example to consider is the occurrence of intrusive growth in laticifers, which has been questioned by some authors (Prado & Demarco, 2018; Ramos et al., 2019). Thus, the continuity in the studies of this intriguing and complex secretory structure is of extreme importance, which in the future could allow us to better understand the independent evolution of the laticifers in the plant kingdom.

Acknowledgments

This study was supported by Fapesp (grant numbers 2013/21794-5 and 2018/03691-8), Capes (finance code 001) and Cnpq (grant number 303493/2015-1).

References

Agrawal, A. A., & Konno, K. (2009). Latex: A model for understanding mechanisms, ecology, and evolution of plant defense against herbivory. *Annual Review of Ecology Evolution and Systematics, 40*, 311–331.

Allen, R. D., & Nessler, C. L. (1984). Cytochemical localization of pectinase activity in laticifers of *Nerium oleander* L. *Protoplasma, 119*, 74–78.

Almeida, E. B., Jr., Araújo, J. S., Santos-Filho, F. S., & Zickel, C. S. (2013). Leaf morphology and anatomy of *manilkara* adans. (Sapotaceae) from northeastern Brazil. *Plant Systematics and Evolution, 299*, 1–9.

Araújo, N. D., Coelho, V. P. M., Ventrella, M. C., & Agra, M. F. (2014). Leaf anatomy and histochemistry of three species of *Ficus* sect. Americanae supported by light and electron microscopy. *Microscopy and Microanalysis, 20*, 296–304.

Armstrong, J. E., & Tucker, S. C. (1986). Floral development in *myristica* (myristicaceae). *American Journal of Botany, 73*, 1131–1143.

Ascensão, L. (2007). Estruturas secretoras em plantas. Uma abordagem morfo-anatómica. In A. C. Figueiredo, J. G. Barroso, & L. G. Pedro (Eds.), *Potencialidades e aplicações das plantas aromáticas e medicinais. Curso teórico-prático (3ª ed.*, pp. 19–28). Lisboa: Faculdade da Universidade de Lisboa – Centro de Biotecnologia Vegetal.

Baas, P., & Kool, R. (1983). Comparative leaf anatomy of *heisteria* (olacaceae). *Blumea, 28*, 367–388.

Baas, P., Oosterhoud, E. van, & Scholtes, C. J. L. (1982). Leaf anatomy and classification of the *olacaceae, octoknema*, and *erythropalum*. *Allertonia, 3*, 155–210.

Bal, A. K. (1974). Cellulase. In M. A. Haya (Ed.), *Electron microscopy of enzymes* (Vol. 3, pp. 68–79). New York: Van Nostrand Reinhold.

Balaji, K., Subramanian, R. B., & Inamdar, J. A. (1993). Occurrence of non-articulated laticifers in *Streblus asper* lour. (Moraceae). *Phytomorphology, 43*, 235–238.

Behnke, H. D., & Herrmann, S. (1978). Fine structure and development of laticifers in *Gnetum gnemon* L. *Protoplasma, 95*, 371–384.

van Bruggen, H. W. E. (1998). Aponogetonaceae. In K. Kubitzki (Ed.), *The families and genera of vascular plants. IV flowering plants - monocotyledons alismatanae and commelinanae (except gramineae)*. Springer-Verlag Berlin Heidelberg GmbH.

Canaveze, Y., & Machado, S. R. (2016). The occurrence of intrusive growth associated with articulated laticifers in *Tabernaemontana catharinensis* A.DC., a new record for Apocynaceae. *International Journal of Plant Sciences, 177*, 458–467.

Canaveze, Y., Mastroberti, A. A., Mariath, J. E. A., & Machado, S. R. (2019). Cytological differentiation and cell wall involvement in the growth mechanisms of articulated laticifers in *Tabernaemontana catharinensis* A.DC. (Apocynaceae). *Protoplasma, 256*, 131–146.
Carlquist, S. (2001). Observations on the vegetative anatomy of *austrobaileya*: Habital, organographic and phylogenetic conclusions. *Botanical Journal of the Linnean Society, 135*, 1–11.
Castelblanque, L., Balaguer, B., Martí, C., Orozco, M., & Vera, P. (2018). LOL2 and LOL5 loci control latex production by laticifer cells in *Euphorbia lathyris*. *New Phytologist, 219*, 1467–1479.
Castro, M. M., & Machado, S. R. (2012). Células e tecidos secretores. In B. Appezzato-da-Glória, & S. M. Carmello-Guerreiro (Eds.), *Anatomia vegetal* (3^a ed., pp. 169–191). Viçosa: Editora UFV.
Condon, J. M., & Fineran, B. A. (1989). Distribution and organization of articulated laticifers in *Calystegia sylvatica* (convolvulaceae). *Botanical Gazette, 150*, 289–302.
Cortella, A. R. (1989). Secretory tissue in *Ipomoea purpurea* (convolvulaceae): Laticifers and glands. *Darwiniana, 29*, 17–23.
Cunha Neto, I. L., Martins, F. M., Somner, G. V., & Tamaio, N. (2017). Secretory structures in stems of five lianas of Paullinieae (Sapindaceae): Morphology and histochemistry. *Flora, 235*, 29–40.
Dahlgren, R. M. T. (1988). Rhizophoraceae and anisophylleaceae: Summary statement, relationships. *Annals of the Missouri Botanical Garden, 75*, 1259–1277.
Dahlgren, R. M. T., Clifford, H. T., & Yeo, P. F. (1985). *The families of the monocotyledons structure, evolution, and taxonomy*. New York: Springer-Verlag Berlin Heidelberg.
Demarco, D., & Castro, M. M. (2008). Laticíferos articulados anastomosados em espécies de Asclepiadeae (Asclepiadoideae, Apocynaceae) e suas implicações ecológicas. *Brazilian Journal of Botany, 31*, 701–713.
Demarco, D., Castro, M. M., & Ascensão, L. (2013). Two laticifer systems in *Sapium haematospermum* – new records for Euphorbiaceae. *Botany, 91*, 545–554.
Demarco, D., Kinoshita, L. S., & Castro, M. M. (2006). Laticíferos articulados anastomosados – novos registros para Apocynaceae. *Brazilian Journal of Botany, 29*, 133–144.
Durairatnam, M. (1959). The distribution of laticiferous tissue in different parts of *Regnellidium diphyllum* lindman. *Bot. Mag. Tokyo, 72*(851).
Esau, K., & Kosakai, H. (1975). Laticifers in *Nelumbo nucifera* gaertn.: Distribution and structure. *Annals of Botany, 39*, 713–719.
Evert, R. F. (2006). *Esau's plant anatomy – meristems, cells, and tissues of the plant body: Their structure, function, and development* (3rd ed.). Hoboken, New Jersey: John Wiley and Sons, Inc.
Fahn, A. (1979). *Secretory tissues in plants*. London/New York/San Francisco: Academic Press.
Fairbairn, J. W., & Kapoor, L. D. (1960). The laticiferous vessels of *Papaver somniferum* L. *Planta Medica, 8*, 49–61.
Farrel, B. D., Dussourd, D. E., & Mitter, C. (1991). Escalation of plant defense: Do latex/resin canals spur plant diversification? *The American Naturalist, 138*, 881–900.
Fernández-Dattoli, L. F., Loto, D. E., Flores, J. R., Luque, C. A., Debes, M. A., Albornoz, P. L., et al. (2011). Caracterización anatómica foliar de *Lonchocarpus lilloi* (Fabaceae, Papilionoideae). *Lilloa, 48*, 145–152.
Fineran, B. A., Condon, J. M., & Ingerfeld, M. (1988). An impregnated suberized wall layer in laticifers of the Convolvulaceae and its resemblance to that in walls of oil cells. *Protoplasma, 147*, 42–54.
French, J. C. (1988). Systematic occurrence of anastomosing laticifers in Araceae. *Botanical Gazette, 149*, 71–81.

Gama, T. S. S., Rubiano, V. S., & Demarco, D. (2017). Laticifer development and its growth mode in *Allamanda blanchetii* A. DC. (Apocynaceae). *Journal of the Torrey Botanical Society, 144*, 303–312.
Gomes, S. M. A., & Lombardi, J. A. (2010). Leaf anatomy as a contribution to the taxonomy of salacioideae N.Halle' ex thorne & reveal (celastraceae). *Plant Systematics and Evolution, 289*, 13–33.
Govindarajalu, E. (1967). Further contribution to the anatomy of the alismataceae: *Sagittaria guayanensis* H.B.K. ssp. *lappula* (D.Don) bogin. *Proceedings of the Indiana Academy of Science, 65*, 142.
Guérin, P. (1923). Les urticées: Cellules à mucilage, laticifères et canaux sécréteurs (suite et fin). *Bulletin de la Société Botanique de France, 70*, 255–263.
Gunawardena, A., Greenwood, J. S., & Dengler, N. G. (2007). Cell wall degradation and modification during programmed cell death in lace plant, *Aponogeton madagascariensis* (Aponogetonaceae). *American Journal of Botany, 94*, 1116–1128.
Gutiérrez, D. G., & Luna, M. L. (2013). A comparative study of latex-producing tissues in genera of Liabeae (Asteraceae). *Flora, 208*, 33–44.
Hagel, J. M., Yeung, E. C., & Facchini, P. J. (2008). Got milk ? The secret life of laticifers. *Trends in Plant Science, 13*, 1360–1385.
Huang, S., & Sterling, C. (1970). Laticifers in the bulb scales of *allium*. *American Journal of Botany, 57*, 1000–1003.
Inamdar, J. A., Murugan, V., & Subramanian, R. B. (1988). Ultrastructure of non-articulated laticifers in *Allamanda violacea*. *Annals of Botany, 62*, 583–588.
Jacomassi, E., Moscheta, I. S., & Machado, S. R. (2007). Morfoanatomia e histoquímica de *Brosimum gaudichaudii* Trécul (Moraceae). *Acta Botanica Brasilica, 21*, 575–597.
Jacomassi, E., Moscheta, I. S., & Machado, S. R. (2010). Morfoanatomia e histoquímica de órgãos reprodutivos de *Brosimum gaudichaudii* (Moraceae). *Brazilian Journal of Botany, 33*, 115–129.
Kajii, C., Morita, T., & Kuroda, K. (2014). Laticifers of *Ficus carica* and their potential role in plant defense. *IAWA Journal, 35*, 109–115.
Karis, P. O. (2006). Morphological data indicates two major clades of the subtribe Gorteriinae (Asteraceae-Arctotideae). *Cladistics, 22*, 199–221.
Katinas, L., Tellería, M. C., Susanna, A., & Ortiz, S. (2008). *Warionia* (Asteraceae): A relict genus of cichorieae? *Anales del Jardín Botánico de Madrid, 65*, 367–381.
Kitajima, S., Taira, T., Oda, K., Yamato, K. T., Inukai, Y., & Hori, Y. (2012). Comparative study of gene expression and major proteins function of laticifers in lignified and unlignified organs of mulberry. *Planta, 235*, 589–601.
Konno, K. (2011). Plant latex and other exudates as plant defense systems: Roles of various defense chemicals and proteins contained therein. *Phytochemistry, 72*, 1510–1530.
Kubitzki, K. (2004). Cornaceae. In K. Kubitzki (Ed.), *Flowering plants. Dicotyledons: Celastrales, oxalidales, rosales, cornales, ericales* (Vol. 6). Springer-Verlag Berlin Heidelberg.
Kuster, V. C., da Silva, L. C., Meira, R. M. S. A., & Azevedo, A. A. (2016). Glandular trichomes and laticifers in leaves of *Ipomoea pes-caprae* and *I. imperati* (convolvulaceae) from coastal restinga formation: Structure and histochemistry. *Brazilian Journal of Botany, 39*, 1117–1125.
Lee, K. B., & Mahlberg, P. G. (1999). Ultrastructure and development of nonarticulated laticifers in seedlings of *Euphorbia maculata* L. *Journal of Plant Biology, 42*, 57–62.
Leme, F. M. (2018). *Floral development and laticifers in species of Cannabaceae Martinov and ulmaceae mirb.* Campinas: Universidade Estadual de Campinas, Instituto de Biologia. Ph D Thesis.
Les, D. H. (1988). The origin and affinities of the ceratophyllaceae. *Taxon, 37*, 326–345.
Les, D. H., Schneider, E. L., Padgett, D. J., Soltis, P. S., Soltis, D. E., & Zanis, M. (1999). Phylogeny, classification and floral evolution of water lilies (Nymphaeaceae;

Nymphaeales): A synthesis of non-molecular, rbcL, matK, and 18S rDNA data. *Systematic Botany, 24*, 28–46.
Lewinsohn, T. M. (1991). The geographical distribution of plant latex. *Chemoecology, 2*, 64–68.
Lopes, K. L. B., Thadeo, M., Azevedo, A. A., Soares, A. A., & Meira, R. M. S. A. (2009). Articulated laticifers in the vegetative organs of *Mandevilla atroviolacea* (Apocynaceae, Apocynoideae). *Botany, 87*, 202–209.
Machado, A. V., & Santos, M. (2004). Morfo-anatomia foliar comparativa de espécies conhecidas como espinheira-santa: *Maytenus ilicifolia* (celastraceae), *Sorocea bonplandii* (Moraceae) e *Zollernia ilicifolia* (Leguminosae). *Insula, Florianópolis, 33*, 1–19.
Magallon-Puebla, S., & Cevallos-Ferriz, S. S. (1994). *Eucommia constans* n. sp. fruits from upper cenozoic strata of puebla, Mexico: Morphological and anatomical comparison with *Eucommia ulmoides* oliver. *International Journal of Plant Sciences, 155*, 80–95.
Mahlberg, P. G. (1993). Laticifers: An historical perspective. *The Botanical Review, 59*, 1–23.
Marinho, C. R., Martucci, M. E. P., Gobbo-Neto, L., & Teixeira, S. P. (2018). Chemical composition and secretion biology of the floral bouquet in legume trees. *Botanical Journal of the Linnean Society, 187*, 5–25.
Marinho, C. R., Pereira, R. A. S., Peng, Y.-Q., & Teixeira, S. P. (2018). Laticifer distribution in fig inflorescence and its potential role in the fig-fig wasp mutualism. *Acta Oecologica, 90*, 160–167.
Marinho, C. R., & Teixeira, S. P. (2019a). Novel reports of laticifers in Moraceae and urticaceae: Revisiting synapomorphies. *Plant Systematics and Evolution, 305*, 13–31.
Marinho, C. R., & Teixeira, S. P. (2019b). Cellulases and pectinases act together on the development of articulated laticifers in *Ficus montana* and *Maclura tinctoria* (Moraceae). *Protoplasma, 256*, 1093–1107.
Melo-de-Pinna, G. F. A., & Menezes, N. L. (2003). Meristematic endodermis and secretory structures in adventitious roots of Richterago Kuntze (Mutisieae-Asteraceae). *Brazilian Journal of Botany, 26*, 1–10.
Mercadante-Simões, M. O., Mazzottini-dos-Santos, H. C., Nery, L. A., Ferreira-Peracio, R. B., Ribeiro, L. M., Royo, V. A., et al. (2014). Structure, histochemistry and phytochemical profile of the bark of the sobol and aerial stem of *Tontelea micrantha* (Celastraceae - hippocrateoideae). *Anais da Academia Brasileira de Ciências, 86*, 1167–1179.
Metcalfe, C. R. (1967). Distribution of latex in the plant kingdom. *Economic Botany, 21*, 115–127.
Metcalfe, C. R., & Chalk, L. (1950). *Anatomy of the dicotyledons: Leaves, stem and wood in relation to taxonomy with notes on economic uses* (Vol. 2). Oxford: Clarendon Press.
Metcalfe, C. R., & Chalk, L. (1983). *Anatomy of the dicotyledons* (2 ed., Vol. 2). Oxford: Clarendon Press.
Milanez, F. R. (1959). Contribuição ao conhecimento anatômico de *Crytostegia grandiflora*. I Embrião. *Rodriguesia, 12–13*(33–34), 347.
Milanez, F. R. (1977). *Ontogênese dos laticíferos contínuos de* Neridium *(Nerium) oleander L* (pp. 343–379). Rio de Janeiro: *Trabalhos do XXVI* Congresso Nacional de Botânica.
Monacelli, B., Valletta, A., Rascio, N., Moro, I., & Pasqua, G. (2005). Laticifers in *Camptotheca acuminata* Decne: Distribution and structure. *Protoplasma, 226*, 155–161.
Monteiro, M. H. D. A., Andreata, R. H. P., & Neves, L. J. (2007). Estruturas secretoras em Sapotaceae. Pesquisas. *Botanica, 58*, 253–262.
Moseley, M. F. (1961). Morphological studies of the Nymphaeaceae II. The flowers of *nymphaea*. *Botanical Gazette, 122*, 233–259.
Muhlemann, J. K., Klempien, A., & Dudareva, N. (2014). Floral volatiles: From biosynthesis to function. *Plant, Cell and Environment, 37*, 1936–1949.
Nessler, C., & Mahlberg, P. (1977). Ontogeny and cytochemistry of alkaloidal vesicles in laticifers of *Papaver somniferum* L. (Papaveraceae). *American Journal of Botany, 64*, 541–551.

Nessler, C. L., & Mahlberg, P. G. (1981). Cytochemical localization of cellulase activity in articulated, anastomosing laticifers of *Papaver somniferum* L. (Papaveraceae). *American Journal of Botany, 68*, 730–732.

Prado, E., & Demarco, D. (2018). Chapter 6: Laticifers and secretory ducts: Similarities and Differences. In L. Hufnagel (Ed.), *Ecosystem services and global ecology*. London, United Kingdom: Szent István University. IntechOpen Limited.

Quintanar, A., & Castrejón, J. L. Z. (2004). Anatomía e histoquímica de la corteza de cinco espécies de Moraceae. *Polibotánica, 17*, 15–38.

Ramos, M. V., Demarco, D., Souza, I. C. C., & Freitas, C. D. T. (2019). Laticifers, latex, and their role in plant defense. *Trends in Plant Science, 24*, 553–567.

Rivera, D., & Flores, E. M. (1988). Morfología floral del achiote, *Bixa orellana* L. (Bixaceae). *Revista de Biologia Tropical, 36*, 499–509.

Riviere, S., Clayson, C., Dockstader, K., Wright, M. A. R., & Costea, M. (2013). To attract or to repel? Diversity, evolution and role of the "most peculiar organ" in the *cuscuta* flower (dodder, convolvulaceae)—the infrastaminal scales. *Plant Systematics and Evolution, 299*, 529–552.

Rosatti, T. J. (1986). The genera of Sphenocleaceae and Campanulaceae in the southeastern United States. *Journal of the Arnold Arboretum, 67*, 1–64.

Rudall, P. (1994). Laticifers in crotonoideae (Euphorbiaceae): Homology and evolution. *Annals of the Missouri Botanical Garden, 81*, 270–282.

Schneider, E. L., & Williamson, P. S. (1993). Nymphaeaceae. In K. Kubitzki (Ed.), The families and genera of vascular plants. *II Flowering Plants, Dicotyledons*. New York: Springer-Verlag.

Solereder, H. (1908). *Systematic anatomy of the dicotyledons, 2 vols*. Oxford: Clarendon Press.

Topper, S. M. C., & Koek-Noorman, J. (1980). The occurrence of axial latex tubes in the secondary xylem of some species of *Artocarpus* J.R. & G. Forster (Moraceae). *International Association of Wood Anatomists Bulletin, 1*, 113–119.

van Veenendaal, W. L. H., & den Outer, R. W. (1990). Distribution and development of the non-articulated branched laticifers of *Morus nigra* L. (Moraceae). *Acta Botanica Neerlandica, 39*, 285–296.

Vega, A. S., Castro, M. A., & Anderson, W. R. (2002). Occurrence and phylogenetic significance of latex in the malpighiaceae. *American Journal of Botany, 89*, 1725–1729.

Vertrees, G. L., & Mahlberg, P. G. (1978). Structure and ontogeny of laticifers in *Cichorium intybus* (Compositae). *American Journal of Botany, 65*, 764–771.

Vranová, E., Coman, D., & Gruissem, W. (2013). Network analysis of the MVA and MEP pathways for isoprenoid synthesis. *Annual Review of Plant Biology, 64*, 665–700.

Wilder, G. J., & Harris, D. H. (1982). Laticifers in *Cyclanthus bipartitus* poit. (Cyclanthaceae). *Botanical Gazette, 143*, 84–93.

Wilkinson, H. P. (1986). Leaf anatomy of *tinomiscium* and *fibraurea* (menispermaceae tribe fibraureae) with special reference to laticifers and astrosclereids. *Kew Bulletin, 41*, 153–169.

Williamson, P. S., & Moseley, M. F. (1989). Morphological studies of the Nymphaeaceae sensu lato XVII. Floral anatomy of *Ondinea purpurea* subspecies *purpurea* (Nymphaeaceae). *American Journal of Botany, 76*, 1779–1794.

Williamson, P. S., & Schneider, E. L. (1993). Nelumbonaceae. In K. Kubitzki (Ed.), *The families and genera of vascular plants. II Flowering Plants, Dicotyledons*. New York: Springer-Verlag.

Wilson, K., & Mahlberg, P. (1978). Ultrastructure of non-articulated laticifers in mature embryos and seedlings of *Asclepias syriaca* L. (Asclepiadaceae). *American Journal of Botany, 65*, 98–109.

Wilson, K. J., Nessler, C. L., & Mahlberg, P. G. (1976). Pectinase in *Asclepias* latex and its possible role in laticifer growth and development. *American Journal of Botany, 63*, 1140–1144.

Zini, L. M., Galati, B. G., & Ferrucci, M. S. (2017). Perianth organs in Nymphaeaceae: Comparative study on epidermal and structural characters. *Journal of Plant Research, 130*, 1047–1060.

CHAPTER THREE

Plant latex proteins and their functions

Oskar Musidlak, Sophia Bałdysz, Michalina Krakowiak and Robert Nawrot*
Department of Molecular Virology, Institute of Experimental Biology, Faculty of Biology, Adam Mickiewicz University in Poznań, Poznań, Poland
*Corresponding author: E-mail: rnawrot@amu.edu.pl

Contents

Advances in Botanical Research, Volume 93
ISSN 0065-2296
https://doi.org/10.1016/bs.abr.2019.11.001

Abstract

In this review we have collected information regarding latex proteins and their contribution to latex-borne defense among a variety of latex-bearing plants. We comprehensively describe the functions and properties of different pathogenesis- and defense-related proteins found in plant latex, including peroxidases, lipoxygenases, polyphenol oxidases, major latex proteins, β-1,3-glucanases, chitinases, osmotins, proteases, and others. The last section describes proteomic changes occurring during plant-virus interactions in latex-bearing papaya infected with *Papaya meleira* virus as an example.

1. Pathogenesis-related and defense-related proteins of laticiferous plants

Plants are constantly facing many dangers especially from different pathogens living in their environment. However, the infection is usually not lethal for them. The reason for that probably lies in many defense mechanisms which plants have developed during the course of evolution. Inside the cells they have evolved specialized molecules forming the plant defense system. It has been estimated that over 20,000 angiosperm plants from various families can produce latex (Kekwick, 2002; Lewinsohn, 1991). This convergent evolution of latex-bearing plants suggests that the latex offers great advantages when it comes to plant fitness and survival. The chemical and proteomic composition of latex may differ between plant species. The latex proteins found among various plant species belong to distinct families, therefore they do not have one common mechanism of action. In plant latex some proteins are produced constitutively (e.g. proteases, some PR proteins) and some are induced (produced *de novo*) upon stress or injury (e.g. PR proteins) (Ramos, Demarco, da Costa Souza, & de Freitas, 2019).

1.1 Latex-borne defense system

Plant latex comprises the cytoplasm of laticifers, which are extremely elongated cells distributed across the whole plant - in roots, stems, and leaves. Latex mostly comprises the content of the laticifers' vacuoles and forms an emulsion composed of different chemicals, such as various alkaloids, terpenoids, proteins, starches, oils, sugars, tannins, gums (Cho, Jo, Chu, Park, & Kim, 2014; Konno, 2011). It exudes from the plant immediately after mechanical damage of different origin, e.g. a bite of an insect and coagulates after air exposure (Cho et al., 2014; Konno, 2011; Zhou & Liu, 2010). The latex viscosity and clotting impedes herbivorous feeding because it limits the movement of their mouth apparatus and gums up their bodies

(Kitajima et al., 2010). Latex occurs in more than 20,000 plant species from over 40 families of angiosperms, which are referred to as the latex-bearing plants (Konno, 2011). Typical families include Euphorbiaceae, Asteraceae, Apocynaceae and Papaveraceae (Cho et al., 2014). The latex occurs in numerous phylogenetically unrelated plant families and orders, hence it represents a highly convergent trait, which has evolved independently many times (Hagel, Yeung, & Facchini, 2008; Konno, 2011). The main advantage of the evolutionary development of a laticiferous system is its direct action against herbivores. Such a conclusion originates from the fact of the frequent occurrence of latex-bearing plants in tropical regions, because of the high number of different herbivorous insects living in tropical areas, allowing for more intense interactions between them and plants than in temperate regions leading to evolutionary consequences (Konno, 2011).

The convergent evolution of laticifer systems in numerous phylogenetically distant plant species, suggests the existence of common advantages in latex-borne defenses (Konno, 2011). The main benefit is the speed of response of the system, which contains highly concentrated defense molecules across laticifers inside the whole plant. These defense substances are exuded immediately after the damage made by herbivorous insect, exactly at the point of damage, enabling the immediate transport of defense chemicals to the point of the attack. Therefore, latex-borne defense should be considered as a preformed defense, which is ready for action against pathogens even within a few seconds, because of the internal pressure inside laticifer cells. The process is much faster than in an inducible defense, in which it takes at least hours or even days for the concentrations of defense molecules to rise enough to act potently against pathogens (Karban & Kuc, 1999; Konno, 2011). Such mode of action is also economically beneficial in terms of production costs, as defense substances are directed exactly to their place of action (Farrell, Dussourd, & Mitter, 1991; Konno, 2011; Nawrot, 2017).

On the other hand latex is sticky and has the ability of clotting so it could trap the whole body of tiny insects or glue their mouthparts, which is less effective against larger insects (Dussourd, 1995; Konno, 2011; Zalucki et al., 2001). There are several mechanisms of latex coagulation in different latex-bearing plants. For example, highly elastic rubber particles and the protein hevein play important role in latex coagulation of *Hevea brasiliensis* (Gidrol, Chrestin, Tan, & Kush, 1994; d'Auzac, Prevot, & Jacob, 1995). In latex of *Carica papaya* the process resembles blood coagulation after wounding in mammals and is performed mainly thanks to cysteine endopeptidases (Moutim, Silva, Lopes, Fernandes, & Salas, 1999; Nawrot,

2017; Silva, Garcia, Lopes, & Salas, 1997). Other coagulation mechanisms could be related to polyphenol oxidase (PPO), like in *Taraxacum* spp. (Wahler et al., 2009) or could be evaporation-driven as in the latex of *Euphorbia* spp. (Bauer et al., 2014).

1.2 Defense-related proteins of different latex-bearing plants

Research on latex proteins of different plants and their functions revealed that they functionally resemble phloem proteins, however they are far more concentrated within laticifers than phloem proteins inside the vessels (Cho et al., 2014; Konno, 2011). Latex proteins comprise many functional types of proteins – from metabolic to energy and storage, however many of them have defense-related properties, as the function of latex is presumably connected with a range of defenses against different pathogens and stress conditions (Nawrot, 2017; Ramos et al., 2019). Plant latex of different plants contains different members of pathogenesis-related proteins, which are structurally unrelated, but they all exhibit activity against different aggressors and invaders, like herbivores, fungi, bacteria, viruses and parasites (Ramos et al., 2019).

Despite severe difficulties of plant latex proteomic analysis concerning latex collection and poor genome sequence data coverage, proteomic analyses of the latex of several plant species have been performed to date. Some examples are *H. brasiliensis*, *Lactuca sativa*, *Papaver somniferum*, and *Taraxacum brevicorniculatum* (Decker, Wanner, Zenk, & Lottspeich, 2000; Cho et al., 2009; D'Amato et al., 2010; Wahler et al., 2012).

The plant to be studied in a more comprehensive way thanks to the relative ease of collecting of a huge amount of latex is lettuce (*L. sativa*, Asteraceae), which is one of the most important leafy vegetables in the world. A proteomic analysis of latex in lettuce using MudPIT helped to identify 587 proteins (Cho et al., 2009). In the study by Cho and colleagues (Cho et al., 2014) all available sequence data of latex proteins was collected to establish an integrated dataset for the latex proteomes. These data comprised 714 lettuce proteins, 365 rubber tree proteins, 65 opium poppy proteins and 64 proteins of other latex-bearing plants. Functional annotations of the 1,208 latex proteins using Blast2GO revealed that many important biological processes occur in the latex.

Based on InterProScan annotation 546 non-redundant protein domains were detected. The most significant protein domains were the NAD(P)-binding, ATPase AAA type, small GTP-binding protein domains, the aldolase-type TIM barrel, and the armadillo-type fold. In addition, a proteasome, peptidase, Ras GTPase, ribosomal protein S5, Bet v 1 allergen, heat-shock protein 70, glutathione S-transferase, 14-3-3 protein, and ubiquitin were frequently identified in the latex proteomes (Cho et al., 2014).

Comparing the number of functionally annotated latex proteins (1,208) to the same number of phloem proteome (1,209) proteins we can observe close similarities in their number and also in their functions. The common proteins for both proteomes are involved in the translational machinery and proteasome complex, which suggests that such processes occur in both systems. It is worth to note that latex possesses many proteins involved in overall metabolism, such as glucose, carbohydrate, nitrogen, alcohol, and RNA metabolism. Moreover, latex possesses also transcription factors, which regulate expression of genes associated with development of the plant, like connected to flowering and development of the meristem. Some proteins could be found only in latex, like proteins targeted to mitochondria and plastids, which supports the conclusion that these organelles are present in laticifers of many species and are required during synthesis of various natural products, such as alkaloids, terpenoids, phenolic acids and others (Cho et al., 2014).

1.3 Pathogenesis-related and other defense-related proteins from the latex of laticiferous Papaveraceae

Papaveraceae family is the representative example of laticiferous plants from the North temperate region, however not all species produce latex (e.g. *Corydalis cava*). We focus on 2 most important species from the economic and pharmacological point of view — *Papaver somniferum* and *Chelidonium majus* — two close relatives producing benzylisoquinoline alkaloids.

The first proteomic study on the latex of *C. majus* using two-dimensional electrophoresis coupled with liquid chromatography-tandem mass spectrometry (LC-MS/MS) identified 21 proteins, mainly involved in plant defense, signal transduction, the tricarboxylic acid cycle, and nucleic acid binding (Nawrot et al., 2007a, 2007b). Subsequent quantitative proteomic studies with the use of *C. majus* transcriptome database expanded this number to 334 proteins and confirmed that *C. majus* latex contains among others predominant in terms of relative abundance major latex protein (MLP), PPO and enzymes responsible for alkaloid and phenylpropanoid biosynthesis (Nawrot, Barylski, Lippmann, Altschmied, & Mock, 2016). Moreover, comparative chemometric analysis of *C. majus* latex proteins showed that its protein content shifts during plant development from intense biosynthetic processes (mainly alkaloid biosynthesis) to plant defense against pathogens (Nawrot et al., 2017a).

The first study on the proteomic content of cytosolic and vesicle fractions of opium poppy latex was conducted using 2-DE and were identified by microsequencing and MALDI-TOF-MS. Both fractions were shown to contain 98 different proteins, from enzymes, chaperons, nucleases, putative transcription factors and signaling, storage to defense-related proteins, like a major-latex protein, which was more abundant in the cytosol (Decker et al., 2000). Only two alkaloid biosynthetic enzymes were identified in this study - codeinone reductase and reticuline 7-O-methyltransferase, using this approach (Decker et al., 2000).

A subsequent and more comprehensive study of elicitor-treated opium poppy cell cultures by 2-DE coupled with LC-MS/MS resulted in the annotation of up to eightfold more latex proteins (Zulak, Khan, Alcantara, Schriemer, & Facchini, 2009). It helped to identify 219 of 340 protein spots based on peptide fragment fingerprint searches of a combination of different databases. The largest category of proteins represented in the study were metabolic enzymes, including S-adenosylmethionine synthetase, several glycolytic, tricarboxylic acid cycle enzymes, an alkaloid, and several other secondary metabolic enzymes. Opium poppy cell cultures also contained many proteins involved in defense responses: the abundance of chaperones, protein degradation factors, heat shock proteins and pathogenesis-related proteins (Zulak et al., 2009).

To investigate more deeply the composition of defense-related proteins in members of Papaveraceae, 17 pathogenesis-related protein families together with several other groups of defense-related proteins were searched in the available literature data of *C. majus*, *P. somniferum* and a plant from family *Euphorbiaceae* - *H. brasiliensis* (Fister et al., 2016; Sels, Mathys, De Coninck, Cammue, & De Bolle, 2008; Van Loon, Rep, & Pieterse, 2006) (Table 1). Results have shown, that *C. majus* contains protein members from 12 pathogenesis-related protein families: PR-2, PR-3, PR-4, PR-5, PR-7, PR-9, PR-10, PR-11, PR-14, PR-15, PR-16 and PR-17, with 2 predominant families in terms of protein quantity: PR-9 (peroxidases) and PR-10 (ribonuclease-like/Bet v 1 protein family), with no members for PR-1, PR-6, PR-12 and PR-13 families (Table1). Moreover, several other defense-related proteins, which do not belong to PR proteins were recorded: overrepresented dirigent-like protein, PPO and lipoxygenase and others, like 14-3-3 proteins, glycine-rich RNA-binding proteins,

Table 1 Summary of PR gene families in the latex of *Chelidonium majus* L., *Papaver somniferum* and *Hevea brasiliensis*.

No.	Common name with conserved domain	Functions	*C. majus*	*P. somniferum*	*H. brasiliensis*
1	PR-2 β-1,3-glucanase/ glyco hydro 17 (pfam00332)	Aid in cell wall degradation.	Stem of the plant (Nawrot et al., 2016)	—	Spots in latex sample (D'Amato et al., 2010)
2	PR-3 Chitinase Class I, II, IV, VII/chitinase glyco hydro 19 (cd00325)	Aid in cell wall degradation.	Stem of the plant (Nawrot et al., 2016); *C.majus* whole plant extract (Nawrot, Zauber, & Schulze, 2014)	—	Spots in latex sample (D'Amato et al., 2010)
3	PR-4 Chitinase-Hevein-like/ barwin (pfam00967)	Aid in cell wall degradation. May have RNase and DNase activity.	Stem of the plant (Nawrot et al., 2016)	—	Spots in latex sample (D'Amato et al., 2010)
4	PR-5 Thaumatin-like/ thaumatin (pfam00314)	Degrade pathogen membranes.	Stem of the plant (Nawrot et al., 2016) *C.majus* whole plant extract (Nawrot et al., 2014)	—	Spots in latex sample (D'Amato et al., 2010)

(*Continued*)

Table 1 Summary of PR gene families in the latex of *Chelidonium majus* L., *Papaver somniferum* and *Hevea brasiliensis*.—cont'd

No.	Common name with conserved domain	Functions	*C. majus*	*P. somniferum*	*H. brasiliensis*
5	PR-6 Proteinase-inhibitor/ potato inhibitor family (pfam00280)	Inhibit proteolysis by herbivorous insects.	—	—	Spots in latex sample (D'Amato et al., 2010)
6	PR-7 Endoproteinase/PA subtilisin like (cd02120)	Aid in cell wall degradation.	Stem of the plant (Nawrot et al., 2016)	—	Cysteine proteinase - spots in latex sample (D'Amato et al., 2010)
7	PR-8 Chitinase Class III/ (cd02877)	GH18 hevamine XipI class III -Aid in cell wall degradation. May have lysozymal activity.	—	—	Spots in latex sample (D'Amato et al., 2010)
8	PR-9 Peroxidase/secretory peroxidase (cd00693)	Regulate reactive oxygen species concentration, contribute to cell wall lignification.	Overrepresented in latex (Nawrot, Lesniewicz, Pienkowska, & Gozdzicka-Jozefiak, 2007b); Whole plant (shoot) extract (Nawrot et al., 2014); Stem of the plant (Nawrot et al., 2016)	Visible spots in elicitor-treated opium poppy cell cultures (Zulak et al., 2009)	Spots in latex sample (D'Amato et al., 2010)

9	PR-10 Ribonuclease-like/ Bet v 1 (pfam00407)	Degrade RNA, may degrade viruses.	Major latex protein predominant in *C. majus* latex (Nawrot et al., 2016; 2017); Whole plant (shoot) extract (Nawrot et al., 2014)	Visible spots in latex serum – major latex protein (MLP) (Decker et al., 2000); Visible spots in elicitor-treated opium poppy cell cultures (Zulak et al., 2009)	–
10	PR-11 Chitinase class V/GH18 plant chitinase class v (cd02879)	Aid in cell wall degradation.	Stem of the plant (Nawrot et al., 2016)	–	–
11	PR-14 Lipid-transfer Protein/nsLTP1 (cd01960)	Degrade pathogen membranes, mechanism unclear.	Stem of the plant (Nawrot et al., 2016); *C.majus* whole plant extract (Nawrot et al., 2014)	–	Spots in latex sample (D'Amato et al., 2010)
12	PR-15 Germin/Oxalate Oxidase/Two cupin 1 (pfam00190) domains	Regulate reactive oxygen species production.	Stem of the plant (Nawrot et al., 2016)	–	–
13	PR-16 Germin-like/ Oxalate Oxidase-like/ Two cupin 1 (pfam00190) domains	Regulate reactive oxygen species production, catalyze monosaccharides.	Stem of the plant (Nawrot et al., 2016) Whole plant extract (Nawrot et al., 2014)	–	–

(Continued)

Table 1 Summary of PR gene families in the latex of *Chelidonium majus* L., *Papaver somniferum* and *Hevea brasiliensis*.—cont'd

No.	Common name with conserved domain	Functions	*C. majus*	*P. somniferum*	*H. brasiliensis*
14	PR-17 Unknown/BSP (pfam04450)	Proteinase function probable, mechanism unclear.	Stem of the plant (Nawrot et al., 2016)	—	—
15	Glycine-rich RNA-binding protein (CmGRP1)/- cd12449: RRM_CIRBP_RBM3 [RNA recognition motif in cold inducible RNA binding protein (CIRBP), RNA binding motif protein 3 (RBM3) and similar proteins]	Function in posttranscriptional regulation of gene expression by binding to different transcripts, thus allowing the cell to response rapidly to environmental signals.	Whole plant extract (Nawrot, Tomaszewski, Czerwoniec, & Goździcka-Józefiak, 2013); Visible spot in the latex sample (Nawrot, Kalinowski, & Gozdzicka-Jozefiak, 2007a); whole plant (shoot) extract (Nawrot et al., 2014)	Visible spots in elicitor-treated opium poppy cell cultures (Zulak et al., 2009)	Spots in latex sample (D'Amato et al., 2010)
16	PPO/-pfam00264, Common central domain of tyrosinase -pfam12142, Polyphenol oxidase middle domain -pfam12143, Protein of unknown function (DUF_B2219)	- oxidation of phenolic compounds, - involved in generation of ROS	Overrepresented in the latex (Nawrot et al., 2016; 2017)	Visible spot in latex serum (Decker et al., 2000)	—

17	Lipoxygenase/pfam00305: Lipoxygenase	oxygenation of polyunsaturated fatty acids	Overrepresented in the latex (Nawrot et al., 2016)	—	Spots in latex sample (D'Amato et al., 2010)
18	Dirigent-like protein/ pfam03018: Dirigent	induced during disease response in plants. Members of this family are involved in lignification	Overrepresented in the latex (Nawrot et al., 2016); Whole plant extract (Nawrot et al., 2014)	Visible spots in elicitor-treated opium poppy cell cultures (Zulak et al., 2009)	—
19	14-3-3 protein/ pfam00244: 14-3-3	regulatory, phosphoserine/ threonine-binding proteins; protein trafficking, metabolic regulation, apoptosis and others	Whole plant (shoot) extract (Nawrot et al., 2014)	Visible spots in latex serum (Decker et al., 2000)	Spots in latex sample (D'Amato et al., 2010)
20	Leucine-rich repeat transmembrane protein kinase e.g. plant NBS-LRRs/ COG4886: LRR Leucine-rich repeat (LRR) protein [Transcription]	- recognize pathogen Avr gene products, which induces effector-triggered immunity (ETI)	Whole plant (shoot) extract (Nawrot et al., 2017); Latex (Nawrot et al., 2014)	—	Spots in latex sample (D'Amato et al., 2010)

(*Continued*)

Table 1 Summary of PR gene families in the latex of *Chelidonium majus* L., *Papaver somniferum* and *Hevea brasiliensis*.—cont'd

No.	Common name with conserved domain	Functions	*C. majus*	*P. somniferum*	*H. brasiliensis*
21	superoxide dismutase/ cd00305: Cu-Zn_Superoxide_ Dismutase	- catalyze the conversion of superoxide radicals to molecular oxygen.	Cu/Zn SOD - Visible spot in the latex sample (Nawrot et al., 2007a); Whole plant (shoot) extract (Nawrot et al., 2014)	Cu/Zn SOD - Visible spot in latex serum (Decker et al., 2000)	Mn-SOD - spots in latex sample (D'Amato et al., 2010)
22	Glyoxalase I/cd07233: Glyoxalase_I	- part of the glyoxalase system for detoxifying methylglyoxal, a side product of glycolysis	Visible spot in the latex sample (Nawrot et al., 2007a); Whole plant (shoot) extract (Nawrot et al., 2014)	Visible spot in latex serum (Decker et al., 2000)	Spots in latex sample (D'Amato et al., 2010)
23	ATP-binding cassette transporter/cd00267: ABC_ATPase	transport of a wide variety of different compounds, like sugars, ions, peptides, and more complex organic molecules across membranes	Latex (Nawrot et al., 2017)	Visible spots in elicitor-treated opium poppy cell cultures (Zulak et al., 2009)	Spots in latex sample (D'Amato et al., 2010)

Classification and functions according to Sels et al. (2008) and Fister et al. (2016).

leucine-rich repeat protein kinases, glyoxalases, Cu/Zn superoxide dismutases, ABC transporters (Table 1).

In the case of *P. somniferum*, members from only 2 pathogenesis-related protein families were listed PR-9 (peroxidases) and PR-10/Bet v 1 (major latex proteins), however highly abundant, similarily as in *C. majus*. Moreover other defense-related proteins are present — glycine-rich protein, polyphenol oxidases, dirigent-like, 14-3-3 proteins, Cu/Zn-SOD, glyoxalase (lactoylglutathione lyase), ATP-binding cassette transporter, similarily as for *C. majus* (Table 1). To compare the protein content of Papaveraceae with the member of other family of latex-bearing plants, *H. brasiliensis* was chosen as a well studied and economically important species mainly due to rubber synthesis. *H. brasiliensis* latex is rich in PR proteins, which were also studied in the context of their allergenic properties. Its latex contains members of 9 PR protein families: PR-2, PR-3, PR-4, PR-5, PR-6, PR-7, PR-8, PR-9, PR-11, as well as other proteins, like GRP, lipoxygenase, 14-3-3, NBS-LRRs, Mn-SOD, glyoxalase, and ATP-binding cassette transporter. It is worth to note the lack of a member of the PR-10 family (major latex protein in the case of Papaveraceae), possibly because the highly abundant enzymes are connected with rubber synthesis (Table 1).

1.4 Proteins from *C. majus* latex connected with defense against herbivore attack and antiviral response

C. majus is a perennial plant from the family Papaveraceae and in traditional European and Chinese medicine it has been used to treat many diseases and health disorders (Benninger, Schneider, Schuppan, Kirchner, & Hahn, 1999; Colombo & Bosisio, 1996; Duke, 1985; Paris & Moyse, 1967; Zielińska et al., 2018). *C. majus* produces yellow latex which has been a remedy to treat visible symptoms of human papillomavirus infection such as warts, papillae, and condylomas (Culpeper, 1995; Mills, Bone, Corrigan, Duke, & Wright, 2008; Zevin, Altman, & Zevin, 1997; Zielińska et al., 2018). Studies on *C. majus* latex proteome revealed its defense-related proteins, mainly predominant and constantly present PR proteins (like MLPs - major latex proteins). These proteins, together with small-molecular compounds present within the latex, functionally comprise its constitutive defense-response system against microbial, fungal and viral pathogens, as well as herbivores. The system is mobilized within a few seconds after the mechanical damage or wounding (e.g. after the action of herbivorous insects), which is the prerequisite for the plant to be infected by the virus

(Konno, 2011; Mandadi & Scholthof, 2013; Nawrot et al., 2017, 2016; Souza et al., 2011). The latex exudes after the plant damage and due to its stickiness it can clot and either glue insect's mouthparts or trap its whole body, preventing the plant from further damages (Konno, 2011). Proteins present within the plant's milky sap could form three distinctive lines of defense (Nawrot, 2017). The first line of immediate defense response of *C. majus* latex is formed by highly abundant PPO and lipoxygenase (LOX), which act during clotting mechanism of latex and increase its stickiness (Nawrot, 2017). The second line of defense is connected with the oxidative burst, which often accompanies the cell wall damage (Wojtaszek, 1997). This line is assured by the action of oxidative proteins like peroxidase (POX) and lipoxygenase (LOX), which are both abundant in *C. majus* latex (Nawrot et al., 2007a, 2016). The third line of defense comprises direct antiviral defense with the action of constitutively expressed nucleic acid binding proteins, which are MLP and glycine-rich proteins (GRPs) (Nawrot et al., 2013, 2016, 2017). Both proteins potentially possess ribonucleic and deoxyribonucleic activities, and hence could be able to digest plant viral RNA or DNA (Huh & Paek, 2013) or act differently using unknown mechanisms (Nawrot, 2017).

Data known so far on the proteomic content of *C. majus* milky sap support the hypothesis that the plant possess strong defense system presenting the lack of disease symptoms even after infection, what could be also indirectly supported by the lack of studies on viruses or other pathogens infecting *C. majus* plants (Hrzenjak, Curkovic-Perica, Krajacic, & Mamula, 1999; Pospieszny, Borodynko, & Jonczyk, 2004). The only plant virus which was shown to infect *C. majus* in its natural habitats is the cucumber mosaic virus (CMV) (Brčák, 1979). Our recent results showed differences in 2D protein profiles between *C. majus* plants infected by a mixture of two potato virus Y (PVY) strains and healthy plants. Infected *C. majus* plants, comparing to the non-treated, could produce higher amounts of proteins connected with the plant defense response to the viral attack, identified using LC-ESI-MS/MS as MLPs, heat shock proteins or PR family members (unpublished data). These results contribute to a better understanding of the antiviral activity of *C. majus* latex. It could be therefore concluded that postulated antiviral plant latex mechanism of *C. majus* comprised of predominant and constantly expressed PR proteins, together with small-molecular compounds, like alkaloids, constitutes a novel type of immediate defense response against viruses (Nawrot, 2017).

2. Major latex proteins (MLPs) and other nucleic acid binding proteins

2.1 Major latex proteins (MLPs) and the history of their finding

The major latex proteins (MLPs) are frequently stored in laticifers, latex-filled tubular structures, which are spread throughout the plant (Hagel et al., 2008; Konno, 2011; Nawrot, 2017) and carry a viscous, milky sap called latex. These tubes give an ideal storage space for defense metabolites, including proteins, which are accumulated in these tubes in high concentrations and are used as part of the first line of defense. The other key function of laticifers is the compartmentation of the proteins and small molecular compounds to eliminate any toxic effect on other plant compartments and metabolites. Some MLPs begin to accumulate at the start of plant development and are present in the latex until plant maturation. Proteins highly expressed in the latex of opium poppy were found to have below 25% sequence identity to pathogenesis-related protein family 10 (PR-10) proteins. Despite this conclusion, the proteins were found to be homologous. The PR-10 protein class bears a resemblance to the MLP family based on similar function and structure, as indicated in Fig.1. A distinctive feature

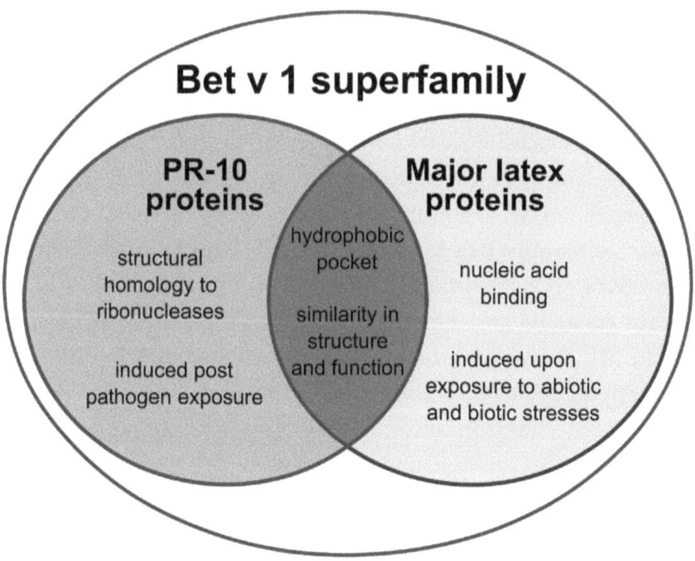

Fig. 1 Schematic representation of similarities of the PR-10 proteins and MLPs and their individual characteristics.

of the PR-10 protein group is their ability to bind ligands due to the presence of a hydrophobic pocket. The 3D structure of the pocket had been identified and was found to be Y-shaped (Lytle et al., 2009), which is correlated with its binding properties.

The PR-10 is a class of intracellular defense-related proteins, which can be found in different plant tissues and were also identified in laticifers. They show a structural homology to ribonuclease (Nawrot et al., 2007a), are induced post pathogen exposure and some are presumed to be expressed constitutively throughout plant development; however, this assumption still remains to be confirmed. They were first discovered in peas and parsley in the late 1980s (Fristensky, Horovitz, & Hadwiger, 1988; Somssich et al., 1988). They are acidic and their molecular weight ranges between 15 and 18 kDa (Radauer, Lackner, & Breiteneder, 2008). PR-10 proteins have the ability to cleave nucleic acids, which serves two functions: cleavage of pathogenic genomic material (e.g. virus DNA or RNA) and cleavage of its own genomic material, resulting in programmed cell death within the vicinity of the infection site. The ribonuclease activity has been proven for the Bet v 1 protein from birch (*Betula verrucosa*), a recombinant protein CaPR-10 from hot pepper (*Capsicum annum*) (Park et al., 2004), LaPR-10 protein from lupine (*Lupinus albus*) roots (Bantignies et al., 2000), PR-10c from birch (*Betula pendula*) (Koistinen et al., 2002). The structure of the Bet v 1 protein from birch and a PR-10 protein from yellow lupine enabled identification of 3 residues, which are likely to be involved in the ribonuclease activity: E96, E148, Y150. Their proximity likely translates to the presence of an active site for the catalytic reaction (Liu & Ekramoddoullah, 2006). Members of the PR-10 family, induced upon pathogen infection, have also been found in *C. majus* and *C. cava* (Nawrot et al., 2014). PR-10 proteins together with the major latex protein/ripening-related protein (MLP/RRP) subfamily belong to the Bet v 1 superfamily (Radauer et al., 2008). The MLP/RRP subfamily is the second in the number of members among plant proteins and includes major latex proteins (MLPs) (Nawrot et al., 2016). MLPs were first found in mature and developing laticifers of the opium poppy (*P. somniferum*) in 1985 (Nessler & Burett, 1992; Nessler et al., 1985, 1990). Their function is attributed to fruit and flower development and response to stress and plant defense.

The major latex protein of *C. majus* (CmMLP), which belongs to the MLP/RRP-like proteins, is associated with fruit ripening and its relative abundance is the highest in comparison to other latex proteins (Chruszcz et al., 2013; Nawrot, 2017). The latex of this plant has been used in traditional folk medicine (Colombo & Bosisio, 1996; Orvos et al., 2015) to treat the results of a human papilloma virus infection (HPV), such as warts and condylomas (Nawrot, 2017; Nawrot et al., 2014). It is assumed that the CmMLP has antiviral properties. The MLP-like protein 28 of *C. majus* was found in the milky sap, along with MLP-like protein 34, the latter being found in 1.73% of the protein content of the sap (Nawrot et al., 2016). The MLP-like protein 34 of *C. majus* shows both ribonucleic and deoxyribonucleic properties, thus suggesting the potential ability to digest the RNA and DNA of pathogens.

The typical molecular weight of MLP is 17 kDa. The MLP from *C. majus* has a molecular weight of 16.77 kDa. The protein is 147 amino acids (aa) in length and its isoelectric point is 5.88 (Nawrot, 2017). Within members of the MLP group the sequence similarity can vary. Between the VvMLP14 of *Vitis vinifera* and AT1G24020 of *Arabidopsis thaliana* the aa sequence similarity was approximately 56% (Zhang et al., 2018); however, the VvMLP10 and VvMLP13 showed 52% sequence similarity with AT1G70850. The length of MLPs can vary: the MLP28 of *C. majus* is of 147 aa in length, nine of the MLPs in *V. vinifera* were of 151 aa in length; however, MLP3 of *V. vinifera* is 224 aa in length. Several MLPs have been found in plants lacking latex, including *A. thaliana* (Wang et al., 2016), upland cotton (*Gossypium hirsutum*) (Yang et al., 2015) and ripening kiwifruit (*Actinidia deliciosa*). The latter was the first protein identified as a plant allergen (Chruszcz et al., 2013). Latex-bearing plants containing MLPs include melon, rubber tree (*H. brasiliensis*) (Konno, 2011; Wang et al., 2010), *C. majus* and opium poppy. Several MLPs have been overexpressed upon exposure to abiotic and biotic stress. For example the GhMLP28 was induced in upland cotton after exposure to *Verticillium dahliae* (Yang et al., 2015) and salt stress (Chen & Dai, 2010; Yang et al., 2015). Furthermore, VvMLP was induced after cold stress at 12 h but decreased at 24 h of treatment.

2.2 Other nucleic acid binding proteins - glycine-rich proteins (GRPs)

Another group of plant proteins often able to bind nucleic acids are glycine-rich proteins (GRPs). GRPs are a large superfamily of heterogenous proteins which contain conserved domains rich in glycine residues. Glycine can comprise 20%–70% of all amino acids of the GRP. Originally, they were discovered in the petunia over 30 years ago and until now they have been found in a wide range of species including cyanobacteria, plants, and animals (Cornels, Ichinose, & Barz, 2000; Park et al., 2000; Sachetto-Martins et al., 2000). Different functions for GRPs have been suggested including RNA binding, nucleolar targeting, interactions with other proteins, and participation in defense responses against biotic and abiotic stress stimuli (Nawrot et al., 2013). By controlling different stages of RNA post-transcriptional processing, they are implied to be important players in plant responses to various harmful situations (Yang et al., 2014). Furthermore, pathogen infections can modulate the levels of GRPs (Mousavi & Hotta, 2005). Additionally, plants employ RNA-binding proteins (RBPs) to protect themselves against viruses. For example, several host RBPs suppressed the replication, transport, and translation of viral RNA by specific interactions (Huh & Paek, 2013). They have been found in plant cell walls where along with proline-rich proteins and extensins they play a role of structural components (Keller & Baumgartner, 1991; Keller, Sauer, & Lamb, 1988; Ringli, Keller, & Ryser, 2001). GRPs can be categorized into different classes based on the presence of some structural features and the arrangement of glycine repeats. Five major classes of GPRs are distinguished in which the class IV can be further subdivided into four subclasses (IVa-IVd) (Czolpinska & Rurek, 2018; Mangeon, Junqueira, & Sachetto-Martins, 2010).

One of the plants in which GRP has been identified is *C. majus* L. (Nawrot et al., 2013). GRP from *C. majus* was found in its latex, named CmGRP1, categorized as a member of class IVa belonging to RBPs (Nawrot et al., 2007a, 2013). It means that apart from the glycine-rich domain, it also contains RNA-recognition motifs (RRM), making it a member of RNA-binding proteins (RBPs) (Fig. 2). RBPs control RNA processing and stabilize stress-induced mRNAs (Nawrot et al., 2013; Zhang et al., 2011). All these properties put CmGRP1 as a protein potentially involved in latex-borne defense.

Fig. 2 Schematic representation of CmGRP1 from *C. majus* primary protein structure. The aa sequence is representative for all class IVa GRPs.

3. Antimicrobial, antifungal, and insecticidal proteins in plant latex

As mentioned earlier, in plant latex some proteins are produced constitutively (e.g. proteases, some PR proteins) and some are induced (produced *de novo*) upon stress or injury (e.g. PR proteins) (Ramos et al., 2019). Proteins synthesized *de novo* play important roles in plant defense against bacterial and fungal pathogens. They can be found in different plant organs including flowers, seeds, leaves, stems, and roots from a broad spectrum of plant species (Nawrot et al., 2013). Apart from being potent anti-phytopathogenic agents, they are also detrimental to insects and work potently against bacteria infecting humans (Ramos et al., 2010). They usually act by degrading cell walls and many of these defense-related proteins have been isolated and their properties have been studied. They are usually categorized as pathogenesis-related (PR) proteins and belong to 5 of 17 families including osmotins or thaumatin-like proteins (PR-5), proteinases (PR-7), hevein-like proteins (etc.) (PR-2, -3, -4, -5, -14). Another group of proteins which sometimes intertwines with PR proteins are antimicrobial peptides (AMPs) including hevein-like proteins. Here we describe different groups of antifungal, antimicrobial, and insecticidal proteins which have been identified in the latex of different plants.

3.1 β-1,3-Glucanases (PR-2)

β-1,3-glucanases participate in plant defense against pathogenic fungi. They cleave β-1,3-glucosidic bonds in a major component of fungal cell walls, which is called β-1,3-glucan. Other functions of β-1,3-glucanases include cell division, cell elongation, seed germination, flower development, and fruit ripening (Nawrot, 2017). Furthermore, a synergistic effect was observed when β-1,3-glucanases and class I chitinases were combined, inhibiting the growth of particular pathogenic fungi. β-1,3-glucanases have been found in the latex of *C. majus* and *H. brasiliensis* (Chye & Cheung, 1995; Nawrot et al., 2016).

3.2 Chitinases (PR-3)

Chitinases are important players in plant defense against various pathogens including fungi, bacteria, viruses, and insects. They act by hydrolyzing the β-1,4-glycosidic bonds of chitin, which builds fungal cell walls. Chitinases have been found in the latex of *Morus* spp., *H. brasiliensis, Ficus microcarpa, Carica papaya, Euphorbia characias*, and *Calotropis procera* (Table 1). The latex protein fraction of *Plumeria rubra* exhibited strong chitinase activity (Freitas et al., 2010).

3.3 Chitinases hevein-like (PR-4)

Hevein is a chitin-binding antimicrobial peptide (AMP) with a molecular weight of 4.7 kDa, whose amino acid sequence is rich in cysteine residues that form disulfide bridges which stabilize the protein structure. The protein has been found in *H. brasiliensis* (rubber tree) latex and it inhibits the hyphal growth of fungi by binding to chitin (Van Parijs, Broekaert, Goldstein, & Peumans, 1991). Hevein also participates in latex coagulation (Gidrol et al., 1994). After the discovery of hevein many other similar AMPs containing chitin-binding domains have been identified among different mono- and dicotyledonous plant species, which now form a hevein-like protein family. The family includes approximately 20 members which all share common features. They are short (29−45 aa) basic peptides rich in cysteine and glycine residues (Slavokhotova, Shelenkov, Andreev, & Odintsova, 2017). The canonical hevein-like protein (e.g. hevein) contains eight cysteine residues forming four disulfide bridges however, the number of cysteines sometimes differs from that. Six of these cysteine residues are usually conserved and form a cysteine motif. Another feature which all hevein-like proteins have in common is a 20−40 aa conservative structural motif called the chitin-binding domain with the amino acid sequence SXFGY/SXYGY where "X" represents any aa (Slavokhotova et al., 2017). All hevein-like proteins have chitin-binding properties. Many plant pathogens including fungi, insects, and nematodes contain chitin, which is absent in plants. It has been suggested that chitin-binding proteins play a role in plant defense against different chitin-containing pathogens (Egorov and Odintsova, 2012).

Even though several hevein-like proteins displaying antimicrobial activity have been identified in different plant organs including fruits of *Sambucus nigra* (Van Damme et al., 1999), leaves of *Wasabia japonica* (Kiba, Saitoh, Nishihara, Omiya, & Yamamura, 2003), leaves of *Broussonetia papyrifera* (Zhao, Ma, Pan, Zhang, & Yuan, 2011), the bark of *Eucommia ulmoides* (Huang et al., 2002), and others, latex hevein-like proteins with similar properties are yet to be discovered. However, insecticidal activity has been demonstrated for two chitin-binding (hevein-like) proteins from *Morus* spp. latex (Kitajima et al., 2010).

Morus alba latex chitin-binding protein called MLX56 is toxic to caterpillars from the Lepidoptera order, including *Mamestra brassicae* and *Samia ricini*

(Wasano et al., 2009). Similarly, HMLX56, the protein from *Morus multicaulis*, containing two hevein-like chitin binding domains has insecticidal activity against *Plutella xylostella*. Furthermore, HMLX56 was also demonstrated to exert toxicity on fungal (*Botrytis cinerea*) and bacterial (*Pseudomonas syringae* pv. tomato DC3000) pathogens (Gai et al., 2017).

Out of all known chitin-binding proteins in plants class I and IV chitinases have the highest similarity to hevein-like AMPs. These two classes are comprised of 25–35 kDa monomeric enzymes which function as endochitinases and generate chitooligosaccharides.

3.4 Thaumatin-like proteins and osmotins (PR-5)

The name of thaumatin-like proteins (TLPs) comes from the homology (40%–70%) of their amino acid sequences to thaumatin isolated from the African berry bush (*Thaumatococcus daniellii*). They are also known as osmotins and the expression of several of them was increased during osmotic stress (Freitas et al., 2015). Unlike thaumatin, they are not sweet tasting and they have been found in the latex of different plant species, including *C. procera*, *Carica papaya* and in phloem exudates of hybrid poplar (*Populus trichocarpa* × *P. deltoides*) (Dafoe, Gowen, & Constabel, 2010). TLPs in plants play different roles, which are usually connected with stress response and defense against pathogens. Several TLPs were observed to exert antifungal activity in one of three different ways. Some act by permeabilizing membranes of fungi, others bind and degrade β-1,3-glucans or suppress fungal xylanases (Dafoe et al., 2010; Freitas et al., 2011; Looze et al., 2009). TLPs from plants without laticifers have insecticidal activity against phloem-feeding insects (Dafoe et al., 2010; Kempema, Cui, Holzer, & Walling, 2007; Zarate, Kempema, & Walling, 2007).

Several osmotins have been reported to play a role in antifungal plant defense (Husaini & Abdin, 2008; Rajam et al., 2007). Osmotins have been isolated from latex of *Carica papaya* (Looze et al., 2009). They have a molecular mass of 20–30 kDa and are rich in cysteine residues forming disulfide bonds. Purified osmotin from *C. procera* (CpOsm) latex exhibits antifungal activity (de Freitas et al., 2011). The protein inhibited germination of spores and mycelial growth of *Fusarium solani*, *Neurospora* spp. and *Colletotrichum gloeosporioides*. Its mechanism of action is based on fungi cell membrane permeabilization which is compatible with previously reported characteristics of osmotins and thaumatin-like proteins (Table 2).

A number of latex proteins from different plant species including *Carica papaya* and *H. brasiliensis* have been described to have lysozyme or chitinase

Table 2 Selected antibacterial, antifungal, and insecticidal proteins identified in latex of different plant species.

Latex protein/-s	Molecular mass	Protein family/ molecular function	Activity	Source plant	Reference
CpLP (protein fraction)	13 kDa	proteases/ostmotins	antifungal	*Calotropis procera*	de Freitas et al., 2011
P1G10 (protein fraction)	n/a	proteases	antifungal	*Carica candamarcensis*	Souza et al., 2011
Cg24-I	24 kDa	proteases	antifungal	*Cryptostegia grandiflora*	Ramos et al., 2014
AMP48	48 kDa	proteases	antifungal	*Artocarpus heterophyllus*	Siritapetawee et al., 2012
papain	23.4 kDa	proteases	insecticidal	*Carica papaya*	Konno et al., 2004; Konno, 2011
ficin	26 kDa	proteases	insecticidal	*Ficus carica*	Konno et al., 2004; Konno, 2011
CpOsm	22.3–22.5 kDa	osmotins/thaumatin-like proteins	antifungal	*Calotropis procera*	de Freitas et al., 2011
GLx Chi-A	33 kDa	chitinases	antifungal	*Ficus microcarpa*	Taira, Ohdomari, Nakama, Shimoji, & Ishihara, 2005
GLx Chi-B	32 kDa	chitinases	antifungal	*Ficus microcarpa*	Taira et al., 2005
GLx Chi-C	27 kDa	chitinases	antifungal	*Ficus microcarpa*	Taira et al., 2005
ELC	36,5 ± 2 kDa	chitinases	antifungal	*Euphorbia characias*	Spanò et al., 2015
hevein	4.7 kDa	chitin-binding proteins/chitinase	antimicrobial, antifungal	*Hevea brasiliensis*	Van Parijs et al., 1991
MLX56	56 kDa	chitin-binding	insecticidal	*Morus* spp.	Wasano et al., 2009
LA-a	50 kDa	chitinase-like proteins/hevein-like	insecticidal	*Morus* spp.	Kitajima et al., 2010
LA-b	46 kDa	chitinase-like proteins/hevein-like	insecticidal	*Morus* spp.	Kitajima et al., 2010
CmLTP 9.5	7.2 kDa	lipid transfer proteins	antimicrobial	*Chelidonium majus*	Nawrot et al., 2017

n/a – not available.

activities however, there is a lack of studies examining their antimicrobial potential (Azarkan et al., 1997, 2003; Huet et al., 2006; Patel, Singh, Yadav, Moir, & Jagannadham, 2010; Rozeboom, Budiani, Beintema, & Dijkstra, 1990; Sytwala, Günther, & Melzig, 2015). Furthermore, for some latex thaumatin-like/osmotin proteins researchers were not able to show antifungal activity (e.g. PTLP protein from papaya) (Looze et al., 2009) however, only two genera of yeast were analyzed. Similarly, no antifungal activity was observed for osmotin-like proteins from *Cryptostegia grandiflora*, *P. rubra*, and *Himatanthus drasticus* (Freitas et al., 2015).

3.5 Non-specific lipid transfer proteins (PR-14)

Plant non-specific lipid transfer proteins (nsLTPs) belong to the PR-14 family but due to their small size (c.a. 6.5–10.5 kDa) and participation in antimicrobial defense they are also often described as antimicrobial peptides (AMPs). By binding lipids they inhibit penetration of microbes through cell membranes (Tam, Wang, Wong, & Tan, 2015). nsLTPs were found in latex of *C. majus* and *H. brasiliensis* (Beezhold et al., 2003; Nawrot et al., 2017b). Recently, it has been shown that *C. majus* CmLTP 9.5 exhibits antimicrobial activity against Gram-negative (*Campylobacter jejuni*) and Gram-positive (*Listeria greyi*, *Clostridium perfringens*) bacteria (Nawrot et al., 2017b) (Table 2). Other known functions of nsLTPs include signaling, intracellular lipid transport, plant growth and development. They are also carriers of acyl monomers during wax and cutin accumulation in cell walls (Edstam, Viitanen, Salminen, & Edqvist, 2011; Yeats & Rose, 2008).

3.6 Proteases (PR-7)

Another group of antifungal and antimicrobial proteins in latex concerns proteases. These proteins are usually produced constitutively. Proteases are considered to be the most widespread and abundant of all the proteins in the latex of different plant species (Ramos et al., 2013; Torres et al., 2012). The most common in plant sap are cysteine and serine proteases (Baeyens-Volant, Matagne, El Mahyaoui, Wattiez, & Azarkan, 2015; Ekchaweng, Evangelisti, Schornack, Tian, & Churngchow, 2017; Konno, 2011; Ramos et al., 2019). Cysteine proteases were identified in *Carica papaya*, *Ficus carica*, *Morrenia brachystephana*, *C. procera*, *Asclepias barjonii*, and *Mangifera indica* latex, whereas serine proteases were found in the latex of *Ficus elastica*, *H. brasiliensis*, *Euphorbia sapina*, *Wrightia tinctoria*, *Ipomoea carnea*, and *M. indica*, and others (Konno, 2011). They inhibit proteolysis caused by herbivorous insects. They possess antifungal and antimicrobial activity. The presence of proteases was demonstrated in protein fractions from the latex of *C. grandiflora* (possibly proteases belonging to cysteine group) and *P. rubra* (Freitas et al., 2010). The jackfruit (*Artocarpus heterophyllus*) protease called AMP48 suppressed the growth of certain fungal (*Candida albicans*) and bacterial (*Pseudomonas aeruginosa*) pathogens at a minimum inhibitory concentration (MIC) of 2.2 mg/mL and minimum microbicidal concentration (MMC) of 8.8 mg/mL (Siritapetawee et al., 2012). Furthermore, antifungal activity was described for latex proteases identified in *C. grandiflora*, *C. procera*, and *Carica candamarcensis* (Ramos et al., 2014; Salas, Badillo-Corona, Ramírez-Sotelo, & Oliver-Salvador, 2015; Souza et al., 2011) (Table 2).

The first latex protein observed to exert detrimental effect in insects was papain from *Carica papaya* (Konno et al., 2004). It participates in plant defense against lepidopteran larvae including *S. ricini*, *M. brassicae*, and *Spodoptera litura*. Furthermore, other related cysteine proteases, namely ficin from *F. carica* latex and bromelain from pineapple stem were also toxic to *S. ricini* caterpillars (Konno et al., 2004). Papain and bromelain also showed antioxidant activity and scavenged free radicals (Manosroi et al., 2014). Other latex proteases have been also found to have insecticidal activities. Proteases from *Ficus virgatalatex* and *C. papaya* latex are poisonous to caterpillars of plant-feeding insects (Konno et al., 2004).

4. Plant oxidative/antioxidant proteins in response to herbivorous and pathogenic attack, and proteins involved in general stress response

4.1 Oxidative/antioxidant proteins against herbivores and pathogens

Reactive oxygen species (ROS) are produced continuously in healthy non-infected plants as a result of their metabolism. Their increase and burst inside the cell are usually the plant's way to tackle different harmful conditions such as biotic and abiotic stresses (Camejo, Guzmán-Cedeño, & Moreno, 2016). ROS in plants include hydrogen peroxide (H_2O_2), superoxide anion ($O_2^{\bullet -}$), singlet oxygen ($^1O_2$), or hydroxyl radical (•OH) and they are produced by peroxidases class III (POX), NADPH oxidase (NOX), oxalate oxidases (germins), amine oxidases, lipoxygenases, quinone reductases (Camejo et al., 2016). Hereafter, we will focus only on those ROS-generating plant enzymes which have been identified in plant latex.

Peroxidases (POX) (PR-9) belong to the most important antioxidant enzymes. Peroxidases occurring in plants usually belong to class III and they participate in ROS production during PAMP-triggered immunity in the apoplast and control H_2O_2 levels. Furthermore, they facilitate the polymerization of cell wall compounds. During the oxidative burst they are involved in the production of H_2O_2 and $O_2^{\bullet -}$(Bolwell et al., 2002). Ascorbate peroxidase is one of the most abundant proteins in *C. majus* latex (Nawrot et al., 2016) and its level in latex was increased during fruit-ripening phase almost twofold when compared to its level in the flowering phase (Nawrot et al., 2017). Ascorbate peroxidase activity was also determined for the latex protein fractions of *P. rubra*, *C. grandiflora* (Freitas et al., 2010), and *C. procera* (Freitas et al., 2007). A peroxidase was purified from the banyan tree (*Ficus benghalensis*) latex and the protein crystal structure has been solved (Palm et al., 2014; Sharma et al., 2012). Furthermore, a sycamore fig (*Ficus sycomorus*) latex is also a source of peroxidases (Mohamed, Abdel-Aty, Hamed, El-Badry, & Fahmy, 2011). Recently, a peroxidase from *Marsdenia megalantha* called *Mm*-POX has been purified. The protein exhibited antifungal activity via membrane permeabilization (Oliveira et al., 2017).

PPO is the enzymatic protein responsible for the generation of reactive oxygen species (ROS) and oxidation of *o*-diquinones and *o*-diphenols. These phenolic compounds participate in oxidative browning in wounded vegetables and fruits (Giribaldi et al., 2011; Wahler et al., 2009). Furthermore, PPO causes latex browning and coagulation in *Taraxacum* spp. and possibly in *C. majus* (Nawrot et al., 2016; Wahler et al., 2009). Latex coagulation is the first line of defense of latex-bearing plants against herbivorous insects and their larvae. As described earlier, depending on the insect size, the latex coagulation can completely immobilize the intruder or glue its mouthparts impeding movement (Nawrot, 2017; Konno, 2011). PPOs have been also found in the latex of *H. brasiliensis* and opium poppy (*P. somniferum*) (Bilka, Balazova, Bilkova, & Psenak, 2000; Wititsuwannakul, Chareonthiphakorn, Pace, & Wititsuwannakul, 2002).

Lipoxygenases (LOXs) are enzymes catalyzing the oxygenation of polyunsaturated fatty acids (PUFAs) including linoleic acid and arachidonic acid (Thaler, 1999). One of many functions of LOX metabolic pathway products is participation in defense against herbivores by recruiting elicitors and attracting natural enemies of particular herbivore agonists (Abdel-Mageed et al., 2016). Up-regulation of LOX genes upon pathogen attack was suggested to be linked with the hypersensitive response and increase in plant antimicrobial potential (antimicrobial activity) (Porta & Rocha-Sosa, 2002; Rustérucci et al., 1999; Weber, Chételat, Caldelari, & Farmer, 1999). LOX has been identified in *C. majus* latex (Nawrot et al., 2016).

Oxalate oxidases (OxO) (germins) (PR-15) are glycoproteins from plants which can degrade the oxalic acid (OA) to produce H_2O_2 in the course of fungal infection. It was observed that germins from barley (*Hordeum vulgare*) are one of the sources of H_2O_2 in response to powdery mildew fungus infection (Zhou et al., 1998). Increased activity of oxalate oxidase has been also observed in wheat (*Triticum aestivum*) during powdery mildew fungus infection (Dumas, Freyssinet, & Pallett, 1995; Hurkman & Tanaka, 1996; Zhang, Collinge, & Thordal-Christensen, 1995). Germin-like protein (GLP) from the latex of *Thevetia peruviana* was described to exhibit proteolytic activity (Freitas et al., 2016). Two germin-like proteins with oxalate oxidase activity have been identified in *C. procera* latex (Freitas et al., 2017).

Superoxide dismutases (SODs) which have been found in latex of such plant species like *C. majus* (Nawrot et al., 2016) and *C. procera* (Freitas et al., 2007) scavenge superoxide ($O_2^{\bullet -}$) radicals, thus protecting the cell from many types of oxidation-related cell damage (Karpinska et al., 2001). SOD from *E. characias* latex is thought to participate in anti-oxidative pathway initiated by oxidative stress resulting from environmental changes of plant surroundings (Atzori, Rescigno, & Padiglia, 2011).

4.2 Latex proteins involved in general stress response

Dirigent-like proteins comprise a family of proteins which are abundantly present in *C. majus* latex. Upon pathogen attack or abiotic stress, these proteins participate in cellular processes which lead to the deposition of lignan and lignin (Jin-Long et al., 2012). The lignification reinforces cell walls in the tissues attacked by the pathogen or after wounding (Burlat, Kwon, Davin, & Lewis, 2001).

ABC (ATP-binding cassette) transporters are mainly involved in transporting molecules through cellular membranes (Theodoulou, 2000). These proteins, are involved in many biological processes, one of which is the role in plant response to pathogen attack. They are also responsible for transport of small molecular weight compounds including alkaloids (Yazaki, 2006). Eight latex-specific ABC protein paralog families have been identified in *H. brasiliensis*, out of which ABCB, ABCG, and ABCI were the most predominant in terms of their abundance (Zhiyi, Guijuan, Yu, Longjun, & Rizhong, 2015).

4.3 Protein fractions with observed antioxidant and other activities

Another example of antioxidant properties of latex are protein fractions isolated from *C. grandiflora* and *P. rubra* latex. Potent antioxidant activity of superoxide dismutase (SOD) was detected in these two plant species, however, it is not certain which particular proteins present in these fractions are actively involved in the observed properties, because the individual proteins were not isolated (Freitas et al., 2010).

The latex of *C. procera* contains antioxidative proteins. The laticifer protein fraction (LP) from *C. procera* exhibited superoxide dismutase (SOD) and ascorbate peroxidase (APX) activities. It was suggested that they are the latex factors responsible for detoxification of reactive oxygen species such as superoxide and hydrogen peroxide with the major role of SOD as the main component responsible for ROS elimination (Freitas et al., 2007).

Plant latex proteins have also beneficial properties in animals. For example, protein fraction (called LP) from *C. procera* latex exhibits anti-inflammatory and analgesic activities. In rats with induced paw edema or monoarthritis LP fraction showed dose-dependent anti-inflammatory effect inhibiting paw edema and joint inflammation (Kumar, Chaudhary, Ramos, Mohan, & Matos, 2011). It was later demonstrated that LP fraction alleviates the functional limitations caused by arthritis in rats (Kumar, Chaudhary, Oliveira, & Ramos, 2014). These results show that the protein fraction from *C. procera* latex and perhaps latex can be potentially applied to relieve inflammation and pain often found in patients suffering from arthritis. It is possible that latex protein fractions from some other plants may present similar anti-inflammatory and soothing properties.

To conclude, the latex of different plant species contains enzymatic proteins which protect the plants from ROS-mediated degradation *in vivo*. On the other hand, the latex has also enzymes involved in generation of ROS to tackle various detrimental conditions such as biotic and abiotic stresses. It is thought that this system may be important in plant defense against infection. Maintaining the redox homeostasis holds the key to plant fitness and better adaptation to changes in its environment.

5. Proteomic insights into plant-virus interactions - changes in PMeV-infected latex-bearing *C. papaya* leaf proteins

Plant-virus interactions are crucial during viral infection. Research on the host-virus interplay will enable to gain information about and countermeasure the aftermath of virus-infected crop losses. The plant is vital for the replication, translation, and further spread of the virus. Gaining insights into all alterations within the plant caused by the virus and the diversity of plant viruses would enable prediction of viral infections in various ecosystems and possible prevention (Alexander, Mauck, Whitfield, Garrett, & Malmstrom, 2014).

Losses caused by plant viruses in agriculturally important crops can be estimated as high as US $ 30 billion per year (Nicaise, 2014). Viruses which carry (+)ssRNA genomes are responsible for most damages in crops. The viruses within this group can vary in organization of the genome, the structure of the capsid, specificity of the host, and mode of transmission (DeBlasio et al., 2015). Viral plant interactions can be identified as incompatible or compatible. The incompatible interaction occurs when the response results in plant resistance to viral infection (Xu et al., 2013). The compatible interaction is viewed as plant susceptibility to a particular virus.

The first borderline on which the virus-plant interaction occurs is the plant apoplast (Delaunois et al., 2014), which is an extracellular space outside the cellular membrane and is comprised of cell walls and spaces between cells. In order to successfully replicate, express their genes, and multiply, viruses utilize many host proteins (Martínez et al., 2016). The limited size of the virus genome often results in multiple functions of the encoded viral proteins. The size of the viral genome and encoded proteins is relatively small in comparison to the size of the plant genome which can be understood as many plant proteins taking part in the interaction and few viral proteins. However, the interaction is dynamic and several factors may be involved at different time points during infection.

Virus and plant proteins interact from beginning to end of the viral infection. These interactions take place in all organelles and subcellular compartments and are crucial for successful viral replication. The expression level and the total quantity of plant proteins can vary throughout viral infection. The symptoms of these changes are often alterations in physiology as an outcome of all plant and viral interactions in the forms of necrosis and chlorosis, which include lesion and patch formation, wilting, dwarfing (Di Carli, Benvenuto, & Donini, 2012). Viruses can influence the transcription of the plant proteins which in later stages of the virus replication cycle will become interacting partners with viral proteins (DeBlasio et al., 2015). The plant protein groups involved in plant-virus interactions are listed in Table 3.

The key factors influencing the outcome of the plant-virus interaction from a proteomic perspective include the virus and host plant species, the

Table 3 Plant protein groups involved in plant-virus interactions.

Functional protein group	Interacting viruses	References
Translational factors	Tomato bushy stunt virus (TBSV)	(Nagy, Pogany, & Lin, 2014; Sanfaçon, 2015)
RNA-binding proteins	Tomato bushy stunt virus (TBSV)	(Nagy et al., 2014)
proteins connected to the chloroplast	Potato virus X (PVX)	(Bhat et al., 2013; Cheng, ChengHuangChenHsu, & Tsai, 2013; Lin, Ding, Hsu, & Tsai, 2007; Heinlein, 2015; Zhao, Zhang, Hong, & Liu, 2016; Soares et al., 2017; Lin et al., 2015)
proteins involved in photosynthetic activity	Cucumber green mottle mosaic virus (CGMMV)	(Liu, Liang, Liu, Luo, & Li, 2015; Stare et al., 2015)
proteins involved in carbon metabolism	Odontoglossum ringspot virus (ORSV)	(Alexander & Cilia, 2016)
pathogenesis-related proteins (PR-proteins)	Potato virus Y (PVY)	(Alexander & Cilia, 2016; Otulak-Kozieł, Kozieł, & Lockhart, 2018)
kinases involved in the MAPK pathway	Tomato yellow leaf curl China virus (TYLCCNV)	(Hu et al., 2019)
cell wall associated kinases which interact with viral plant movement proteins	Tobacco mosaic virus (TMV)	(Citovsky, McLean, Zupan, & Zambryski, 1993)
SUMO-conjugating enzymes	Tomato golden mosaic virus (TGMV)	(Sánchez-Durán et al., 2011)
vesicular transport proteins	Cabbage leaf curl virus (CabLCV)	(Kumar, 2019)
protein biosynthesis and metabolism	Tobacco rattle virus (TRV)	(Brizard, Carapito, Delalande, Van Dorsselaer, & Brugidou, 2006; Fernández-Calvino et al., 2014)

(Continued)

Table 3 Plant protein groups involved in plant-virus interactions.—cont'd

Functional protein group	Interacting viruses	References
lipid metabolism proteins	Rice yellow mottle virus (RYMV)	(Brizard et al., 2006)
stress response proteins	Sugarcane mosaic virus (SCMV)	(Brizard et al., 2006; Wu et al., 2013)
proteins involved in RNA binding and transport dependent on vesicles	Tomato bushy stunt virus (TBSV)	(Nagy et al., 2014)
cell death and defense-related proteins	Tobacco etch virus (TEV)	(Carmo, Resende, Silva, Ribeiro, & Mehta, 2013; Mandadi & Scholthof, 2013; Musidlak, Nawrot, & Goździcka-Józefiak, 2017; Wang, Hajano, Ren, Lu, & Wang, 2015)
chaperones, including heat shock proteins (HSPs)	Red clover necrotic mosaic virus (RCNMV)	(Hyodo, Hashimoto, Kuchitsu, Suzuki, & Okuno, 2017)
proteins involved in production of reactive oxygen species	Red clover necrotic mosaic virus (RCNMV)	(Hyodo et al., 2017)

age of the plant and its health condition at the moment of infection. Plant proteins involved in the plant-virus interaction include key host proteins with a central position in the cellular network, called hub proteins (Nagy et al., 2014).

The *Papaya meleira virus* causes the Papaya Sticky Disease (PSD) in papaya plants (*Carica papaya* L.), which results with latex bursts from fruits and leaves. The virus particles can bind to the polymers present within the latex and show tropism to laticifers. The photosynthetic machinery is a crucial player in viral infection due to its role in energy and carbon molecules production. The proteins within the photosynthetic machinery, including the photosystem I subunit O, Mog1/PsbP/DUF1795-like photosystem II reaction center PsbP family protein, ATP synthase CF0 A subunit, photosystem II subunit R and photosystem II protein H, ATP synthase CF1 epsilon subunit, PsbQ-like 1 were found to be upregulated in the viral infection of Papaya meleira virus (PMeV) in papaya plants. A change in expression of one protein group can cause a chain reaction as shown in Fig. 3.

57 proteins of the papaya plant proteome were upregulated and 54 proteins were downregulated after PMeV infection in comparison to 1333 of the entire plant proteome (Soares et al., 2017). Examples of the up- and downregulated proteins after PMeV infection are presented in Table 4.

The main group which showed alteration in expression was linked to photosynthetic machinery, mainly the photosystem II, as shown in Fig. 3. This group, along with ROS-detoxifying proteins, cell-wall modifying proteins, proteins involved in protein biosynthesis and degradation, proteins involved in amino acid metabolism and chaperones were found to be upregulated, suggesting mobilization of the plant to obtain greater amounts of resources to strengthen cell walls and recruit components involved in

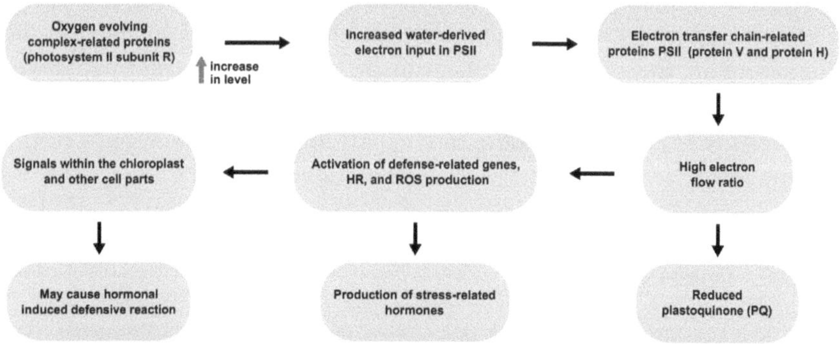

Fig. 3 Changes in protein levels during PMeV infection (Soares et al., 2017).

Table 4 Differentially regulated proteins during PMeV infection of papaya (*Carica papaya* L.) (Alexander & Cilia, 2016; Soares et al., 2017).

Upregulated proteins	Downregulated proteins
• Alpha-glucan phosphorylase 2 • nitrate reductase 2 • alanine-2-oxoglutarate aminotransferase 2 • chloroplast outer envelope protein 37 • NAD(P)-linked oxidoreductase superfamily protein • thioredoxin family protein • phosphoenolpyruvate carboxylase family protein • thioredoxin M-type 4 • photosystem I subunit I • glutamine synthetase 2 • pyrophosphorylase 6 • beta glucosidase 34 • peroxidase • peroxiredoxin • calreticulin • xyloglucan endo-transglycosylase	• importin alpha isoform 4 • beta-6 tubulin • xylem cysteine peptidase 1 • tubulin beta 8 • enolase • annexin 2 • pathogenesis-related 4 • phosphofructokinase family protein • serine carboxypeptidase-like 33 • annexin 8 • catalase 2 • calnexin 1 • ubiquitin-specific protease 21 • fibrillarin 2 • NAD(P)-binding Rossmann-fold superfamily protein • chymopapain • homolog of an HSP from the Hsp70 family • serine protease inhibitor

defense. The group of downregulated proteins included those involved in manipulation of translation, protein processing, and protein degradation, chaperones, proteins involved in the structure of the cytoskeleton (microtubules) and proteins involved in glycolysis.

Better understanding of the plant-virus interactions through a proteomic perspective will enable infection prevention and consequently improved crop yields. Other advantages of resistant versus susceptible plant cultivar studies include food safety optimization and adequate food quality controls. Alterations in the plant proteome can be used as diagnostic markers to identify virus-infected plants. The possible use of the latex and its protein components as a countermeasure to viral infection still remains to be further investigated.

6. Conclusions

It is recognized that plant latex plays an essential role in general plant defense against microbial and fungal pathogens as well as, it protects plants from herbivorous insects. Over the past few decades researchers have unraveled that the latex is composed of different defense compounds and proteins, many of which are stored therein at high concentrations, usually much higher than that found in plant organs including leaves. Latex composition including its protein profile can differ even between plants from the same genera. Several latex chitinase-binding proteins have been already proven to be potent insecticidal agents, however further studies are required to assess broader spectrum of other latex proteins with potential anti-insect, antimicrobial or antifungal activities. Also, other latex-bearing plants and latex proteomes await to be studied and analyzed in terms of potential biologically active proteins. The properties of latex proteins are promising in the light of development of novel antibiotic agents as well as for crop protection to prevent insect-feeding and pathogen-borne diseases. Furthermore, they can be used as natural food ingredients providing protection and for other applications (e.g. meat tenderization). There is a scarcity of virus-related studies on the latex-bearing plants because of the lack of viral disease symptoms, but hopefully in the future more studies regarding this matter will be conducted.

Acknowledgments

This work was financially supported by the National Science Center of the Republic of Poland (project no. 2016/21/N/NZ6/00997 to Oskar Musidlak).

References

Abdel-Mageed, W. M., Mohamed, N. H., Liu, M., El-Gamal, A. A., Basudan, O. A., Ismail, M. A., et al. (2016). Lipoxygenase inhibitors from the latex of *Calotropis procera*. *Archives of Pharmacal Research*. https://doi.org/10.1007/s12272-016-0725-9.

Alexander, M. M., & Cilia, M. (2016). A molecular tug-of-war: Global plant proteome changes during viral infection. *Current Plant Biology*. https://doi.org/10.1016/j.cpb.2015.10.003.

Alexander, H. M., Mauck, K. E., Whitfield, A. E., Garrett, K. A., & Malmstrom, C. M. (2014). Plant-virus interactions and the agro-ecological interface. *European Journal of Plant Pathology*. https://doi.org/10.1007/s10658-013-0317-1.

Atzori, E., Rescigno, A., & Padiglia, A. (2011). *Molecular analysis of a copper- and zinc-containing superoxide dismutase gene isolated from the latex of Euphorbia characias: Another piece in the molecular puzzle of Euphorbiaceae latex proteins*.

Azarkan, M., Amrani, A., Nijs, M., Vandermeers, A., Zerhouni, S., Smolders, N., et al. (1997). *Carica papaya* latex is a rich source of a class II chitinase. *Phytochemistry*.

Azarkan, M., Moussaoui, A. E. L., van Wuytswinkel, D., Dehon, G., & Looze, Y. (2003). Fractionation and purification of the enzymes stored in the latex of *Carica papaya*. *Journal of Chromatography B, 790*, 229–238.

Baeyens-Volant, D., Matagne, A., El Mahyaoui, R., Wattiez, R., & Azarkan, M. (2015). A novel form of ficin from *Ficus carica* latex: Purification and characterization. *Phytochemistry*.

Bantignies, B., Séguin, J., Muzac, I., Dédaldéchamp, F., Gulick, P., & Ibrahim, R. (2000). Direct evidence for ribonucleolytic activity of a PR-10-like protein from white lupin roots. *Plant Molecular Biology, 42*(6), 871–881.

Bauer, G., Gorb, S. N., Klein, M.-C., Nellesen, A., von Tapavicza, M., & Speck, T. (2014). Comparative study on plant latex particles and latex coagulation in *Ficus benjamina, Campanula glomerata* and three *Euphorbia* speciesR. G. Haverkamp (Ed.). *PLoS One, 9*(11), e113336. https://doi.org/10.1371/journal.pone.0113336.

Beezhold, D. H., Hickey, V. L., Kostyal, D. A., Puhl, H., Zuidmeer, L., van Ree, R., et al. (2003). Lipid transfer protein from *Hevea brasiliensis* (Hev b 12), a cross-reactive latex protein. *Annals of Allergy, Asthma, & Immunology, 90*(4), 439–445.

Benninger, J., Schneider, H. T., Schuppan, D., Kirchner, T., & Hahn, E. G. (1999). Acute hepatitis induced by greater celandine (*Chelidonium majus*). *Gastroenterology, 117*, 1234–1237.

Bhat, S., Folimonova, S. Y., Cole, A. B., Ballard, K. D., Lei, Z., Watson, B. S., et al. (2013). Influence of host chloroplast proteins on tobacco mosaic virus accumulation and intercellular movement. *Plant Physiology, 161*(1), 134–147.

Bilka, F., Balazova, A., Bilkova, A., & Psenak, M. (2000). Purification of polyphenol oxidase from opium poppy latex. *Pharmazie, 55*(2), 155–156.

Bolwell, G. P., Bindschedler, L. V., Blee, K. A., Butt, V. S., Davies, D. R., Gardner, S. L., et al. (2002). The apoplastic oxidative burst in response to biotic stress in plants: A three component system. *Journal of Experimental Botany, 53*, 1367–1376.

Brčák, J. (1979). Isolates of cucumber mosaic virus from spontaneously infected plants of *Chelidonium majus* and *Impatiens parviflora*. *Biologia Plantarum, 21*(3), 220–223.

Brizard, J.ean P., Carapito, C., Delalande, F., Van Dorsselaer, A., & Brugidou, C. (2006). Proteome analysis of plant-virus interactome: Comprehensive data for virus multiplication inside their hosts. *Molecular & Cellular Proteomics, 5*(12), 2279–2297.

Burlat, V., Kwon, M., Davin, L. B., & Lewis, N. G. (2001). Dirigent proteins and dirigent sites in lignifying tissues. *Phytochemistry, 57*(6), 883–897.

Camejo, D., Guzmán-Cedeño, Á., & Moreno, A. (2016). Reactive oxygen species, essential molecules, during plant-pathogen interactions. *Plant Physiology and Biochemistry, 103*, 10–23.

Carmo, L. S. T., Resende, R. O., Silva, L. P., Ribeiro, S. G., & Mehta, A. (2013). Identification of host proteins modulated by the virulence factor AC2 of tomato chlorotic mottle virus in *Nicotiana benthamiana*. *Proteomics, 13*(12–13), 1947–1960.

Chen, J.-Y., & Dai, X.-F. (2010). Cloning and characterization of the gossypium hirsutum major latex protein gene and functional analysis in *Arabidopsis thaliana*. *Planta, 231*(4), 861–873.

Cheng, S.-F., Cheng, S.-F., Huang, Y.-P., Chen, L.-H., Hsu, Y.-H., & Tsai, C.-H. (2013). Chloroplast phosphoglycerate kinase is involved in the targeting of bamboo mosaic virus to chloroplasts in *Nicotiana benthamiana* plants. *Plant Physiology*. https://doi.org/10.1104/pp.113.229666.

Cho, W. K., Chen, X. Y., Uddin, N. M., Rim, Y., Moon, J., Jung, J. H., et al. (2009). Comprehensive proteome analysis of lettuce latex using multidimensional protein-identification technology. *Phytochemistry, 70*, 570–578.

Cho, W. K., Jo, Y., Chu, H., Park, S. H., & Kim, K. H. (2014). Integration of latex protein sequence data provides comprehensive functional overview of latex proteins. *Molecular Biology Reports, 41*(3), 1469–1481.

Chruszcz, M., Ciardiello, M. A., Osinski, T., Majorek, K. A., Giangrieco, I., Font, J., et al. (2013). Structural and bioinformatic analysis of the kiwifruit allergen act D 11, a member of the family of ripening-related proteins. *Molecular Immunology, 56*(4), 794–803.

Chye, M. L., & Cheung, K. Y. (October 1995). beta-1,3-Glucanase is highly-expressed in laticifers of *Hevea brasiliensis*. *Plant Molecular Biology, 29*(2), 397–402.

Citovsky, V., McLean, B. G., Zupan, J. R., & Zambryski, P. (1993). Phosphorylation of tobacco mosaic virus cell-to-cell movement protein by a developmentally regulated plant cell wall-associated protein kinase. *Genes & Development, 7*(5), 904–910.

Colombo, M. L., & Bosisio, E. (1996). Pharmacological activities of *Chelidonium majus* L. (Papaveraceae). *Pharmacological Research, 33*(2), 127–134.

Cornels, H., Ichinose, Y., & Barz, W. (2000). Characterization of cDNAs encoding two glycine-rich proteins in chickpea (*Cicer arietinum* L.): Accumulation in response to fungal infection and other stress factors. *Plant Science, 154*, 83–88.

Culpeper, N. (1995). *Culpeper's complete herbal. A book of natural remedies for ancient ills*. Wordsworth: Hertfordshire.

Czolpinska, M., & Rurek, M. (2018). Plant glycine-rich proteins in stress response: An emerging, still prospective story. *Frontiers of Plant Science, 9*, 302.

D'Amato, A., Bachi, A., Fasoli, E., Boschetti, E., Peltre, G., Sénéchal, H., et al. (2010). In-depth exploration of *Hevea brasiliensis* latex proteome and "hidden allergens" via combinatorial peptide ligand libraries. *Journal of Proteomics, 73*, 1368–1380.

d'Auzac, J., Prevot, J. C., & Jacob, J. L. (1995). What's new about lutoids? A vacuolar system model from Hevea latex. *Plant Physiology and Biochemistry, 33*(6), 765–777.

Dafoe, N. J., Gowen, B. E., & Constabel, C. P. (2010). Thaumatin-like proteins are differentially expressed and localized in phloem tissues of hybrid poplar. *BMC Plant Biology, 10*, 191.

DeBlasio, S. L., Johnson, R., Mahoney, J., Karasev, A., Gray, S. M., MacCoss, M. J., et al. (2015). Insights into the polerovirus-plant interactome revealed by coimmunoprecipitation and mass spectrometry. *Molecular Plant-Microbe Interactions, 28*(4), 467–481.

Decker, G., Wanner, G., Zenk, M. H., & Lottspeich, F. (2000). Characterization of proteins in latex of the opium poppy (*Papaver somniferum*) using two-dimensional gel electrophoresis and microsequencing. *Electrophoresis, 21*, 3500–3516.

Delaunois, B., Jeandet, P., Clément, C., Baillieul, F., Dorey, S., & Cordelier, S. (2014). Uncovering plant-pathogen crosstalk through apoplastic proteomic studies. *Frontiers in Plant Science, 5*(June), 249.

Di Carli, M., Benvenuto, E., & Donini, M. (2012). Recent insights into plant-virus interactions through proteomic analysis. *Journal of Proteome Research, 11*(10), 4765–4780.

Duke, J. (1985). *Handbook of medicinal herbs*. CRC Press.

Dumas, B., Freyssinet, G., & Pallett, K. E. (1995). Tissue-specific expression of germin-like oxalate oxidase during development and fungal infection of barley seedlings. *Plant Physiology, 107*(4), 1091–1096.

Dussourd, D. E. (1995). Entrapment of aphids and whiteflies in lettuce latex. *Ecol. Popul. Biol., 88*, 163–172.

Edstam, M. M., Viitanen, L., Salminen, T. A., & Edqvist, J. (2011). Evolutionary history of the non-specific lipid transfer proteins. *Molecular Plant, 4*, 947–964.

Egorov, T. A., & Odintsova, T. I. (2012 Jan-Feb). Defense peptides of plant immune system. *Bioorganicheskaia Khimiia, 38*(1), 7–17.

Ekchaweng, K., Evangelisti, E., Schornack, S., Tian, M., & Churngchow, N. (2017). The plant defense and pathogen counter defense mediated by *Hevea brasiliensis* serine protease *HbSPA* and *Phytophthora palmivora* extracellular protease inhibitor *PpEPI10*. *PLoS One, 12*(5).

Farrell, B. D., Dussourd, D. E., & Mitter, C. (1991). Escalation of plant defense: Do latex and resin canals spur plant diversification? *The American Naturalist, 138*, 881–900.

Fernández-Calvino, L., Osorio, S., Hernández, M. L., Hamada, I. B., del Toro, F. J., Donaire, L., et al. (2014). Virus-induced alterations in primary metabolism modulate susceptibility to tobacco rattle virus in Arabidopsis. *Plant Physiology, 166*(4), 1821–1838.

Fister, A. S., Mejia, L. C., Zhang, Y., Herre, E. A., Maximova, S. N., & Guiltinan, M. J. (2016). *Theobroma cacao* L. pathogenesis-related gene tandem array members show diverse expression dynamics in response to pathogen colonization. *BMC Genomics, 17*, 363. https://doi.org/10.1186/s12864-016-2693-3. Erratum in: BMC Genomics. 2016;17: 715.

Freitas, C. D., da Cruz, W. T., Silva, M. Z., Vasconcelos, I. M., Moreno, F. B., Moreira, R. A., et al. (2016). Proteomic analysis and purification of an unusual germin-like protein with proteolytic activity in the latex of *Thevetia peruviana*. *Planta, 243*(5), 1115–1128.

Freitas, C. D. T., Freitas, D. C., Cruz, W. T., Porfírio, C. T. M. N., Silva, M. Z. R., Oliveira, J. S., et al. (2017). Identification and characterization of two germin-like proteins with oxalate oxidase activity from *Calotropis procera* latex. *International Journal of Biological Macromolecules*.

Freitas, C. D., Nogueira, F. C., Vasconcelos, I. M., Oliveira, J. T., Domont, G. B., & Ramos, M. V. (July 2011). Osmotin purified from the latex of *Calotropis procera*: Biochemical characterization, biological activity and role in plant defense. *Plant Physiology and Biochemistry, 49*(7), 738–743.

Freitas, C. D. T., Oliveira, J. S., Miranda, M. R. A., Macedo, N. M. R., Sales, M. P., Villas-Boas, L. A., et al. (2007). Enzymatic activities and protein profile of latex from *Calotropis procera*. *Plant Physiology and Biochemistry, 45*, 781–789.

Freitas, C. D., Silva, M. Z., Bruno-Moreno, F., Monteiro-Moreira, A. C., Moreira, R. A., & Ramos, M. V. (2015). New constitutive latex osmotin-like proteins lacking antifungal activity. *Plant Physiology and Biochemistry*.

Freitas, C. D. T., de Souza, D. P., Araújo, E. S., Cavalheiro, M. G., Oliveira, L. S., & Ramos, M. V. (2010). Anti-oxidative and proteolytic activities and protein profile of laticifer cells of *Cryptostegia grandiflora, Plumeria rubra* and *Euphorbia tirucalli*. *Brazilian Journal of Plant Physiology, 22*(1), 11–22.

Fristensky, B., Horovitz, D., & Hadwiger, L. A. (1988). cDNA sequences for pea disease resistance response genes. *Plant Molecular Biology, 11*(5), 713–715.

Gai, Y. P., Zhao, Y. N., Zhao, H. N., Yuan, C. Z., Yuan, S. S., Li, S., et al. (2017). The latex protein MLX56 from mulberry (*Morus multicaulis*) protects plants against insect pests and pathogens. *Frontiers of Plant Science, 8*, 1475.

Gidrol, X., Chrestin, H., Tan, H. L., & Kush, A. (1994). Hevein, a lectin-like protein from *Hevea brasiliensis* (rubber tree) is involved in the coagulation of latex. *The Journal of Biological Chemistry, 269*(12), 9278–9283.

Giribaldi, M., Purrotti, M., Pacifico, D., Santini, D., Mannini, F., Caciagli, P., et al. (2011). A multidisciplinary study on the effects of phloem-limited viruses on the agronomical performance and berry quality of *Vitis vinifera* cv. Nebbiolo. *Journal of Proteomics, 75*, 306e315.

Hagel, J. M., Yeung, E. C., & Facchini, P. J. (December 2008). Got milk? The secret life of laticifers. *Trends in Plant Science, 13*(12), 631–639. https://doi.org/10.1016/j.tplants.2008.09.005.

Heinlein, M. (2015). Plant virus replication and movement. *Virology, 479–480*(May), 657–671.

Hrzenjak, A., Curkovic-Perica, M., Krajacic, M., & Mamula, D. (1999). Spontaneous infection of *Chelidonium majus* and *Ligustrum* sp. with cucumber mosaic Cucumovirus (CMV) encapsidating satellite RNA. *Periodicum Biologorum, 101*, 229–232.

Huang, R. H., Xiang, Y., Liu, X. Z., Zhang, Y., Hu, Z., & Wang, D. C. (June 19, 2002). Two novel antifungal peptides distinct with a five-disulfide motif from the bark of *Eucommia ulmoides* Oliv. *FEBS Letters, 521*(1–3), 87–90.

Huet, J., Wyckmans, J., Wintjens, R., Boussard, P., Raussens, V., Vandenbussche, G., et al. (2006). Structural characterization of two papaya chitinases, a family GH19 of glycosyl hydrolases. *Cellular and Molecular Life Sciences*.

Huh, S. U., & Paek, K.-H. (2013). Plant RNA binding proteins for control of RNA virus infection. *Frontiers in Physiology, 4*, 397.

Hu, T., Huang, C., He, Y., Castillo-González, C., Gui, X., Wang, Y., et al. (2019). βC1 protein encoded in geminivirus satellite concertedly targets MKK2 and MPK4 to counter host defense. *PLoS Pathogens, 15*(4), e1007728.

Hurkman, W. J., & Tanaka, C. K. (July 1996). Germin gene expression is induced in wheat leaves by powdery mildew infection. *Plant Physiology, 111*(3), 735–739.

Husaini, A., & Abdin, M. Z. (2008). Overexpression of tobacco osmotin gene leads to salt stress tolerance in strawberry (Fragaria x ananassa Duch.) plants. *Indian Journal of Biotechnology*, 7, 465–471.

Hyodo, K., Hashimoto, K., Kuchitsu, K., Suzuki, N., & Okuno, T. (2017). Harnessing host ROS-generating machinery for the robust genome replication of a plant RNA virus. *Proceedings of the National Academy of Sciences*. https://doi.org/10.1073/pnas.1610212114.

Jin-Long, G., Li-Ping, X., Jing-Ping, F., Ya-Chun, S., Hua-Ying, F., You-Xiong, Q., et al. (2012). A novel dirigent protein gene with highly stem-specific expression from sugarcane, response to drought, salt and oxidative stresses. *Plant Cell Reports, 31*(10), 1801–1812.

Karban, R., & Kuc, J. (1999). Induced resistance against pathogens and herbivores: An overview. In A. A. Agrawal, S. Tuzun, & E. Bent (Eds.), *Inducible plant defenses against pathogens and herbivores: Biochemistry, ecology, and agriculture*. St. Paul, MN (USA): American Phytopathological Society Press.

Karpinska, B., Karlsson, M., Schinkel, H., Streller, S., Suss, K. H., Melzer, M., et al. (2001). A novel superoxide dismutase with a high isoelectric point in higher plants. Expression, regulation, and protein localization. *Plant Physiology, 126*, 1668–1677.

Kekwick, R. G. O. (2002). Latex and laticifers. *eLS*. https://doi.org/10.1038/npg.els.0000913.

Keller, B., & Baumgartner, C. (1991). Vascular-specific expression of the bean GRP 1.8 gene is negatively regulated. *Plant Cell*.

Keller, B., Sauer, N., & Lamb, C. J. (1988). Glycine-rich cell wall proteins in bean: Gene structure and association of the protein with the vascular system. *The EMBO Journal*.

Kempema, L. A., Cui, X., Holzer, F. M., & Walling, L. L. (2007). Arabidopsis transcriptome changes in response to phloem-feeding silverleaf whitefly nymphs. Similarities and distinctions in responses to aphids. *Plant Physiology, 143*(2), 849–865.

Kiba, A., Saitoh, H., Nishihara, M., Omiya, K., & Yamamura, S. (March 2003). C-terminal domain of a hevein-like protein from *Wasabia japonica* has potent antimicrobial activity. *Plant and Cell Physiology, 44*(3), 296–303.

Kitajima, S., Kamei, K., Taketani, S., Yamaguchi, M., Kawai, F., Komatsu, A., et al. (2010). Two chitinase-like proteins abundantly accumulated in latex of mulberry show insecticidal activity. *BMC Biochemistry*.

Koistinen, K. M., Kokko, H. I., Hassinen, V. H., Tervahauta, A. I., Auriola, S., & Karenlampi, S. O. (2002). Stress-related RNase PR-10c is post-translationally modified by glutathione in birch. *Plant, Cell and Environment*. https://doi.org/10.1046/j.1365-3040.2002.00858.x.

Konno, K. (2011). Plant latex and other exudates as plant defense systems: Roles of various defense chemicals and proteins contained therein. *Phytochemistry, 72*(13), 1510–1530.

Konno, K., Hirayama, C., Nakamura, M., Tateishi, K., Tamura, Y., Hattori, M., et al. (2004). Papain protects papaya trees from herbivorous insects: Role of cysteine proteases in latex. *The Plant Journal, 37*, 370–378.

Kumar, R. V. (2019). Plant antiviral immunity against geminiviruses and viral counter-defense for survival. *Frontiers in Microbiology, 10*(June), 1460.

Kumar, V. L., Chaudhary, P., Oliveira, R. M., & Ramos, M. V. (2014). *Calotropis procera* latex proteins ameliorate functional limitations associated with adjuvant induced inflammation in rat. *Musculoskeletal Biology*.

Kumar, V. L., Chaudhary, P., Ramos, M. V., Mohan, M., & Matos, M. P. V. (2011). Protective effect of proteins derived from the latex of *Calotropis procera* against inflammatory hyperalgesia in monoarthritic rats. *Phytotherapy Research*.

Lewinsohn, T. M. (1991). The geographical distribution of plant latex. *Chemoecology*, 64–68.

Lin, J.-W., Ding, M.-P., Hsu, Y.-H., & Tsai, C.-H. (2007). Chloroplast phosphoglycerate kinase, a gluconeogenetic enzyme, is required for efficient accumulation of bamboo mosaic virus. *Nucleic Acids Research, 35*(2), 424–432.

Lin, P.-C., Hu, W.-C., Lee, S.-C., Chen, Y.-L., Lee, C.-Y., Chen, Y.-R., et al. (2015). Application of an integrated omics approach for identifying host proteins that interact with odontoglossum ringspot virus capsid protein. *Molecular Plant-Microbe Interactions, 28*(6), 711–726.

Liu, J.-J., & Ekramoddoullah, A. K. M. (2006). The family 10 of plant pathogenesis-related proteins: Their structure, regulation, and function in response to biotic and abiotic stresses. *Physiological and Molecular Plant Pathology*. https://doi.org/10.1016/j.pmpp.2006.06.004.

Liu, H.-W., Liang, C.-Q., Liu, P.-F., Luo, L.-X., & Li, J.-Q. (2015). Quantitative proteomics identifies 38 proteins that are differentially expressed in cucumber in response to cucumber green mottle mosaic virus infection. *Virology Journal, 12*(December), 216.

Looze, Y., Boussard, P., Huet, J., Vandenbusche, G., Azarkan, M., Raussens, V., et al. (2009). Purification and characterization of a wound-inducible thaumatin-like protein from the latex of *Carica papaya*. *Phytochemistry, 70*(8), 970–978.

Lytle, B. L., Song, J., de la Cruz, N. B., Peterson, F. C., Johnson, K. A., Bingman, C. A., et al. (2009). Structures of two *Arabidopsis thaliana* major latex proteins represent novel helix-grip folds. *Proteins, 76*(1), 237–243.

Mandadi, K. K., & Scholthof, K. B. (May 2013). Plant immune responses against viruses: How does a virus cause disease? *Plant Cell, 25*(5), 1489–1505. https://doi.org/10.1105/tpc.113.111658. Review).

Mangeon, A., Junqueira, R. M., & Sachetto-Martins, G. (2010). Functional diversity of the plant glycine-rich proteins superfamily. *Plant Signaling & Behavior, 5*, 99–104.

Manosroi, A., Chankhampan, C., Pattamapun, K., Manosroi, W., Manosroi, J., & Manosroi, A. (2014). Antioxidant and gelatinolytic activities of papain from papaya latex and bromelain from pineapple fruits. *Chiang Mai Journal of Science*.

Martínez, F., Rodrigo, G., Aragonés, V., Ruiz, M., Lodewijk, I., Fernández, U., et al. (2016). Interaction network of tobacco etch potyvirus NIa protein with the host proteome during infection. *BMC Genomics, 17*(February), 87.

Mills, S., Bone, K., Corrigan, D., Duke, J. A., & Wright, J. V. (2008). *Principles and practice of phytotherapy: Modern herbal medicine*. Edinburgh: Churchill Livingstone.

Mohamed, S. A., Abdel-Aty, A. M., Hamed, M. B., El-Badry, M. O., & Fahmy, A. S. (2011). *Ficus sycomorus latex: A thermostable peroxidase*.

Mousavi, A., & Hotta, Y. (2005). Glycine-rich proteins: A class of novel proteins. *Applied Biochemistry and Biotechnology, 120*, 169–174.

Moutim, V., Silva, L. G., Lopes, M. T. P., Fernandes, G. W., & Salas, C. E. (March 29 , 1999). Spontaneous processing of peptides during coagulation of latex from *Carica papaya*. *Plant Science, 142*(Issue 2), 115–121. ISSN 0168-9452.

Musidlak, O., Nawrot, R., & Goździcka-Józefiak, A. (2017). Which plant proteins are involved in antiviral defense? Review on in vivo and in vitro activities of selected plant proteins against viruses. *International Journal of Molecular Sciences, 18*(11).

Nagy, P. D., Pogany, J., & Lin, J.-Y. (2014). How yeast can Be used as a genetic platform to explore virus—host interactions: From 'omics' to functional studies. *Trends in Microbiology*. https://doi.org/10.1016/j.tim.2014.02.003.

Nawrot, R. (2017). Defense-related proteins from *Chelidonium majus* L. as important components of its latex. *Current Protein & Peptide Science, 18*, 864—880.

Nawrot, R., Barylski, J., Lippmann, R., Altschmied, L., & Mock, H. P. (2016). Combination of transcriptomic and proteomic approaches helps to unravel the protein composition of *Chelidonium majus* L. milky sap. *Planta, 244*, 1055—1064.

Nawrot, R., Józefiak, D., Sip, A., Kuźma, D., Musidlak, O., & Goździcka-Józefiak, A. (2017b). Isolation and characterization of a non-specific lipid transfer protein from *Chelidonium majus* L. latex. *International Journal of Biological Macromolecules, 104*(PtA), 554—563.

Nawrot, R., Kalinowski, A., & Gozdzicka-Jozefiak, A. (2007a). Proteomic analysis of *Chelidonium majus* milky sap using two-dimensional gel electrophoresis and tandem mass spectrometry. *Phytochemistry, 68*, 1612—1622.

Nawrot, R., Lesniewicz, K., Pienkowska, J., & Gozdzicka-Jozefiak, A. (2007b). A novel extracellular peroxidase and nucleases from a milky sap of *Chelidonium majus*. *Fitoterapia, 78*, 496—501.

Nawrot, R., Lippmann, R., Matros, A., Musidlak, O., Nowicki, G., & Mock, H.-P. (2017). Proteomic comparison of *Chelidonium majus* L. latex in different phases of plant development. *Plant Physiology and Biochemistry, 112*, 312—325.

Nawrot, R., Tomaszewski, L., Czerwoniec, A., & Goździcka-Józefiak, A. (2013). Identification of a coding sequence and structure modeling of glycine-rich RNA-binding protein (CmGRP1) from *Chelidonium majus* L. *Plant Molecular Biology Reporter, 31*, 470—476.

Nawrot, R., Zauber, H., & Schulze, W. X. (2014). Global proteomic analysis of *Chelidonium majus* and *Corydalis cava* (Papaveraceae) extracts revealed similar defense-related protein compositions. *Fitoterapia, 94*, 77—87.

Nessler, C. L., Allen, R. D., & Galewsky, S. (1985). Identification and characterization of latex-specific proteins in opium poppy. *Plant Physiology, 79*(2), 499—504.

Nessler, C. L., & Burnett, R. J. (1992). Organization of the major latex protein gene family in opium poppy. *Plant Molecular Biology, 20*(4), 749—752.

Nessler, C. L., Kurz, W. G., & Pelcher, L. E. (1990). Isolation and analysis of the major latex protein genes of opium poppy. *Plant Molecular Biology, 15*(6), 951—953.

Nicaise, V. (2014). Crop immunity against viruses: Outcomes and future challenges. *Frontiers in Plant Science, 5*(November), 660.

Oliveira, H. P., Silva, R. G. G., Oliveira, J. T. A., Sousa, D. O. B., Pereira, M. L., Souza, P. F. N., et al. (2017). A novel peroxidase purified from *Marsdenia megalantha* latex inhibits phytopathogenic fungi mediated by cell membrane permeabilization. *International Journal of Biological Macromolecules*.

Orvos, P., Virág, L., Tálosi, L., Hajdú, Z., Csupor, D., Jedlinszki, N., et al. (2015). Effects of *Chelidonium majus* extracts and major alkaloids on hERG potassium channels and on dog cardiac action potential - a safety approach. *Fitoterapia, 100*(January), 156—165.

Otulak-Kozieł, K., Kozieł, E., & Lockhart, B. E. L. (2018). Plant cell wall dynamics in compatible and incompatible potato response to infection caused by potato virus Y (PVY). *International Journal of Molecular Sciences, 19*(3). https://doi.org/10.3390/ijms19030862.

Palm, G. J., Sharma, A., Kumari, M., Panjikar, S., Albrecht, D., Jagannadham, M. V., et al. (2014). Post-translational modification and extended glycosylation pattern of a plant latex peroxidase of native source characterized by X-ray crystallography. *FEBS Journal*.

Paris, R. R., & Moyse, H. (1967). *Précis de matière médicale* (Vol. II, pp. 207—208). Masson Ed., Paris.

Park, C.-J., Kim, K.-J., Shin, R., Park, J. M., Shin, Y.-C., & Paek, K.-H. (2004). Pathogenesis-related protein 10 isolated from hot pepper functions as a ribonuclease in an antiviral pathway. *The Plant Journal*. https://doi.org/10.1046/j.1365-313x.2003.01951.x.

Park, C. J., Park, C. B., Hong, S. S., Lee, H. S., Lee, S. Y., & Kim, S. C. (2000). Characterization and cDNA cloning of two glycine- and histidine-rich antimicrobial peptides from the roots of shepherd's purse, Capsella bursa-pastoris. *Plant Molecular Biology, 44*, 187–197.

Patel, A. K., Singh, V. K., Yadav, R. P., Moir, A. J. G., & Jagannadham, M. V. (2010). Purification and characterization of a new chitinase from latex of *Ipomoea carnea*. *Process Biochemistry, 45*(Issue 5).

Porta, H., & Rocha-Sosa, M. (2002). Plant lipoxygenases. Physiological and molecular features. *Plant Physiology, 130*(1), 15–21.

Pospieszny, H., Borodynko, N., & Jonczyk, M. (2004). Identification of cucumber mosaic virus (CMV) from *Chelidonium majus*. *Phytopathologia polonica, 34*, 93–96.

Radauer, C., Lackner, P., & Breiteneder, H. (2008). The Bet v 1 fold: An ancient, versatile scaffold for binding of large, hydrophobic ligands. *BMC Evolutionary Biology*. https://doi.org/10.1186/1471-2148-8-286.

Rajam, M. V., Chandola, N., Saiprasad Goud, P., Singh, D., Kashyap, V., Choudhary, M. L., et al. (2007). Thaumatin gene confers resistance to fungal pathogens as well as tolerance to abiotic stresses in transgenic tobacco plants. *Biologia Plantarum, 51*, 135.

Ramos, M. V., Araújo, E. S., Jucá, T. L., Monteiro-Moreira, A. C. O., Vasconcelos, I. M., Moreira, R. A., et al. (2013). New insights into the complex mixture of latex cysteine peptidases in *Calotropis procera*. *International Journal of Biological Macromolecules, 58*, 211–219.

Ramos, M. V., Demarco, D., da Costa Souza, I. C., & de Freitas, C. D. T. (June 2019). Laticifers, latex, and their role in plant defense. *Trends in Plant Science, 24*(6), 553–567.

Ramos, M. V., Grangeiro, T. B., Freire, E. A., Sales, M. P., Souza, D. P., Araujo, E. S., et al. (2010). The defensive role of latex in plants: Detrimental effects on insects. *Arthropod-Plant Interactions, 4*, 57–67.

Ramos, M. V., Souza, D. P., Gomes, M. T., Freitas, C. D., Carvalho, C. P., Júnior, P. A., et al. (2014). A phytopathogenic cysteine peptidase from latex of wild rubber vine *Cryptostegia grandiflora*. *The Protein Journal*.

Ringli, C., Keller, B., & Ryser, U. (2001). Glycine-rich proteins as structural components of plant cell walls. *Cellular and Molecular Life Sciences, 58*, 1430–1441.

Rozeboom, H. J., Budiani, A., Beintema, J. J., & Dijkstra, B. W. (1990). Crystallization of hevamine, an enzyme with lysozyme/chitinase activity from *Hevea brasiliensis* latex. *Journal of Molecular Biology*.

Rustérucci, C., Montillet, J.-L., Agnel, J.-P., Battesti, C., Alonso, B., Knoll, A., et al. (1999). Involvement of lipoxygenase-dependent production of fatty acid hydroperoxides in the development of the hypersensitive cell death induced by cryptogein of tobacco leaves. *The Journal of Biological Chemistry, 274*, 36446–36455.

Sachetto-Martins, G., Franco, L. O., & de Oliveira, D. E. (June 21, 2000). Plant glycine-rich proteins: A family or just proteins with a common motif? *Biochimica et Biophysica Acta, 1492*(1), 1–14.

Salas, C. E., Badillo-Corona, J. A., Ramírez-Sotelo, G., & Oliver-Salvador, C. (2015). Biologically active and antimicrobial peptides from plants. *BioMed Research International*.

Sánchez-Durán, M. A., Dallas, M. B., Ascencio-Ibañez, J. T., Reyes, M. I., Arroyo-Mateos, M., Ruiz-Albert, J., et al. (2011). Interaction between geminivirus replication protein and the SUMO-conjugating enzyme is required for viral infection. *Journal of Virology, 85*(19), 9789–9800.

Sanfaçon, H. (2015). Plant translation factors and virus resistance. *Viruses*. https://doi.org/10.3390/v7072778.

Sels, J., Mathys, J., De Coninck, B. M., Cammue, B. P., & De Bolle, M. F. (2008). Plant pathogenesis-related (PR) proteins: A focus on PR peptides. *Plant Physiology and Biochemistry, 46*, 941–950.
Sharma, A., Palm, G. J., Kumari, M., Panjikar, S., Jagannadham, M. V., & Hinrichs, W. (2012). Purification, crystallization and preliminary crystallographic analysis of banyan peroxidase. *Acta Crystallographica Section F.*
Silva, L. G., Garcia, O., Lopes, M. T., & Salas, C. E. (May 1997). Changes in protein profile during coagulation of latex from *Carica papaya*. *Brazilian Journal of Medical and Biological Research, 30*(5), 615–619.
Siritapetawee, J., Thammasirirak, S., & Samosornsuk, W. (January 2012). Antimicrobial activity of a 48-kDa protease (AMP48) from *Artocarpus heterophyllus* latex. *European Review for Medical and Pharmacological Sciences, 16*(1), 132–137.
Slavokhotova, A. A., Shelenkov, A. A., Andreev, Y. A., & Odintsova, T. I. (2017). Hevein-like antimicrobial peptides of plants. *Biochemistry Moscow, 82*, 1659.
Soares, E.de A., Werth, E. G., Madroñero, L. J., Ventura, J. A., Rodrigues, S. P., Hicks, L. M., et al. (2017). Label-free quantitative proteomic analysis of pre-flowering PMeV-infected *Carica papaya* L. *Journal of Proteomics, 151*(January), 275–283.
Somssich, I. E., Schmelzer, E., Kawalleck, P., & Hahlbrock, K. (1988). Gene structure and in situ transcript localization of pathogenesis-related protein 1 in parsley. *MGG Molecular & General Genetics*. https://doi.org/10.1007/bf00333403.
Souza, D. P., Freitas, C. D., Pereira, D. A., Nogueira, F. C., Silva, F. D., Salas, C. E., et al. (July 2011). Laticifer proteins play a defensive role against hemibiotrophic and necrotrophic phytopathogens. *Planta, 234*(1), 183–193.
Spanò, D., Pospiskova, K., Safarik, I., Pisano, M. B., Pintus, F., Floris, G., et al. (December 2015). Chitinase III in *Euphorbia characias* latex: Purification and characterization. *Protein Expression and Purification, 116*, 152–158.
Stare, T., Ramšak, Ž., Blejec, A., Stare, K., Turnšek, N., Weckwerth, W., et al. (2015). Bimodal dynamics of primary metabolism-related responses in tolerant potato-potato virus Y interaction. *BMC Genomics*. https://doi.org/10.1186/s12864-015-1925-2.
Sytwala, S., Günther, F., & Melzig, M. F. (2015). Lysozyme- and chitinase activity in latex bearing plants of genus Euphorbia–A contribution to plant defense mechanism. *Plant Physiology and Biochemistry.*
Taira, T., Ohdomari, A., Nakama, N., Shimoji, M., & Ishihara, M. (April 2005). Characterization and antifungal activity of gazyumaru (*Ficus microcarpa*) latex chitinases: Both the chitin-binding and the antifungal activities of class I chitinase are reinforced with increasing ionic strength. *Bioscience Biotechnology & Biochemistry, 69*(4), 811–818.
Tam, J. P., Wang, S., Wong, K. H., & Tan, W. L. (2015). Antimicrobial peptides from plants. *Pharmaceuticals, 8*(4), 711–757.
Thaler, J. (1999). Jasmonate-inducible plant defences cause increased parasitism of herbivores. *Nature, 399*, 686–688.
Theodoulou, F. L. (2000). Plant ABC transporters. *Biochimica et Biophysica Acta, 1465*, 79e103.
Torres, M. J., Trejo, S. A., Obregón, W. D., Avilés, F. X., López, L. M., & Natalucci, C. L. (2012). Characterization of the proteolytic system present in *Vasconcellea quercifolia* latex. *Planta.*
Van Damme, E. J., Charels, D., Roy, S., Tierens, K., Barre, A., Martins, J. C., et al. (1999). A gene encoding a hevein-like protein from elderberry fruits is homologous to PR-4 and class V chitinase genes. *Plant Physiology, 119*(4), 1547–1556.
Van Loon, L. C., Rep, M., & Pieterse, C. M. (2006). Significance of inducible defense-related proteins in infected plants. *Annual Review of Phytopathology, 44*, 135–162.
Van Parijs, J., Broekaert, W. F., Goldstein, I. J., & Peumans, W. J. (January 1991). Hevein: An antifungal protein from rubber-tree (*Hevea brasiliensis*) latex. *Planta, 183*(2), 258–264.

Wahler, D., Colby, T., Kowalski, N. A., Harzen, A., Wotzka, S. Y., Hillebrand, A., et al. (2012). Proteomic analysis of latex from the rubber-producing plant *Taraxacum brevicorniculatum*. *Proteomics, 12*, 901–905.

Wahler, D., Gronover, C. S., Richter, C., Foucu, F., Twyman, R. M., Moerschbacher, B. M., et al. (2009). Polyphenol oxidase silencing affects latex coagulation in Taraxacum species. *Plant Physiology, 151*, 334–346.

Wang, B., Hajano, J.-U.-D., Ren, Y., Lu, C., & Wang, X. (2015). iTRAQ-based quantitative proteomics analysis of rice leaves infected by rice stripe virus reveals several proteins involved in symptom formation. *Virology Journal, 12*(June), 99.

Wang, X., Shi, M., Lu, X., Ma, R., Wu, C., Guo, A., et al. (2010). A method for protein extraction from different subcellular fractions of laticifer latex in *Hevea brasiliensis* compatible with 2-DE and MS. *Proteome Science, 8*(June), 35.

Wang, Y., Yang, L., Chen, X., Ye, T., Zhong, B., Liu, R., et al. (2016). Major latex protein-like protein 43 (MLP43) functions as a positive regulator during abscisic acid responses and confers drought tolerance in *Arabidopsis thaliana*. *Journal of Experimental Botany, 67*(1), 421–434.

Wasano, N., Konno, K., Nakamura, M., Hirayama, C., Hattori, M., & Tateishi, K. (2009). A unique latex protein, MLX56, defends mulberry trees from insects. *Phytochemistry, 70*, 880–888.

Weber, H., Chételat, A., Caldelari, D., & Farmer, E. E. (1999). Divinyl ether fatty acid synthesis in late blight-diseased potato leaves. *Plant Cell, 11*, 485–493.

Wititsuwannakul, D., Chareonthiphakorn, N., Pace, M., & Wititsuwannakul, R. (2002). Polyphenol oxidases from latex of *Hevea brasiliensis*: Purification and characterization. *Phytochemistry, 61*(2), 115–121.

Wojtaszek, P. (1997). Oxidative burst: An early plant response to pathogen infection. *Biochemical Journal, 322*(Pt 3), 681–692.

Wu, L., Han, Z., Wang, S., Wang, X., Sun, A., Zu, X., et al. (2013). Comparative proteomic analysis of the plant-virus interaction in resistant and susceptible ecotypes of maize infected with sugarcane mosaic virus. *Journal of Proteomics, 89*(August), 124–140.

Xu, Q., Ni, H., Chen, Q., Sun, F., Tong, Z., Lan, Y., et al. (2013). Comparative proteomic analysis reveals the cross-talk between the responses induced by H_2O_2 and by long-term rice black-streaked dwarf virus infection in rice. *PLoS One, 8*(11), e81640.

Yang, D. H., Kwak, K. J., Kim, M. K., Park, S. J., Yang, K. Y., & Kang, H. (2014). Expression of Arabidopsis glycine-rich RNA-binding protein AtGRP2 or AtGRP7 improves grain yield of rice (*Oryza sativa*) under drought stress conditions. *Plant Science, 214*, 106–112.

Yang, C.-L., Liang, S., Wang, H.-Y., Han, L.-B., Wang, F.-X., Cheng, H.-Q., et al. (2015). Cotton major latex protein 28 functions as a positive regulator of the ethylene responsive factor 6 in defense against *Verticillium dahliae*. *Molecular Plant*. https://doi.org/10.1016/j.molp.2014.11.023.

Yazaki, K. (2006). ABC transporters involved in the transport of plant secondary metabolites. *FEBS Letters, 580*, 1183–1191.

Yeats, T. H., & Rose, J. K. (2008). The biochemistry and biology of extracellular plant lipid-transfer proteins (LTPs). *Protein Science, 17*, 191–198.

Zalucki, M. P., Malcolm, S. B., Paine, T. D., Hanlon, C. C., Brower, L. P., & Clarke, A. R. (2001). It's the first bites that count: Aurvival of first-instar monarchs on milkweeds. *Austral Ecology, 26*, 547–555.

Zarate, S. I., Kempema, L. A., & Walling, L. L. (2007). Silverleaf whitefly induces salicylic acid defenses and suppresses effectual jasmonic acid defenses. *Plant Physiology, 143*(2), 866–875.

Zevin, I. V., Altman, N., & Zevin, L. V. (1997). *A Russian herbal: Traditional remedies for health and healing*. Rochester: Healing Art Press.

Zhang, Z., Collinge, D. B., & Thordal-Christensen, H. (1995). Germin-like oxalate oxidase, a H_2O_2-producing enzyme, accumulates in barley attacked by the powdery mildew fungus. *The Plant Journal, 8*, 139–145.

Zhang, W., Zhao, F., Jiang, L., Chen, C., Wu, L., & Liu, Z. (2018). Different pathogen defense strategies in Arabidopsis: More than pathogen recognition. *Cells*. https://doi.org/10.3390/cells7120252.

Zhang, X., Zhen, J. B., Li, Z. H., Kang, D. M., Yang, Y. M., Kong, J., et al. (2011). Expression profile of early responsive genes under salt stress in upland cotton (*Gossypium hirsutum* L.). *Plant Molecular Biology Reporter, 29*, 626–637.

Zhao, M., Ma, Y., Pan, Y. H., Zhang, C. H., & Yuan, W. X. (August 2011). A hevein-like protein and a class I chitinase with antifungal activity from leaves of the paper mulberry. *Biomedical Chromatography, 25*(8), 908–912.

Zhao, J., Zhang, X., Hong, Y., & Liu, Y. (2016). Chloroplast in plant-virus interaction. *Frontiers in Microbiology,* 7(October), 1565.

Zhiyi, N., Guijuan, K., Yu, L., Longjun, D., & Rizhong, Z. (2015). Whole-transcriptome survey of the putative ATP-binding cassette (ABC) transporter family genes in the latex-producing laticifers of *Hevea brasiliensis*. *PLoS One*.

Zhou, Y.-F., & Liu, W.-Z. (2010). Laticiferous canal formation in fruit of decaisnea fargesii: A programmed cell death process? *Protoplasma*. https://doi.org/10.1007/s00709-010-0229-2.

Zhou, F., Zhang, Z., Gregersen, P. L., Mikkelsen, J. D., de Neergaard, E., Collinge, D. B., et al. (1998). Molecular characterization of the oxalate oxidase involved in the response of barley to the powdery mildew fungus. *Plant Physiology, 117*(1), 33–41.

Zielińska, S., Jezierska-Domaradzka, A., Wójciak-Kosior, M., Sowa, I., Junka, A., & Matkowski, A. M. (April 11, 2018). Greater celandine's ups and downs-21 centuries of medicinal uses of *Chelidonium majus* from the viewpoint of today's pharmacology. *Frontiers in Pharmacology, 9*, 299.

Zulak, K. G., Khan, M. F., Alcantara, J., Schriemer, D. C., & Facchini, P. J. (2009). Plant defense responses in opium poppy cell cultures revealed by liquid chromatography-tandem mass spectrometry proteomics. *Molecular & Cellular Proteomics, 8*(1), 86–98.

CHAPTER FOUR

Low-molecular compounds of latex-bearing plants and their biosynthesis

Peng Huang[a,b], Mengshan Sun[a,b] and Jianguo Zeng[a,b,c,*]
[a]Hunan Key Laboratory of Traditional Chinese Veterinary Medicine, Hunan Agricultural University, Changsha, Hunan, China
[b]College of Horticulture and Landscape, Hunan Agricultural University, Changsha, Hunan, China
[c]National and Local Union Engineering Research Center of Veterinary Herbal Medicine Resource and Initiative, Hunan Agricultural University, Changsha, Hunan, China
*Correspondence author: E-mail: zengjianguo@hunau.edu.cn

Contents

Advances in Botanical Research, Volume 93
ISSN 0065-2296
https://doi.org/10.1016/bs.abr.2019.09.004

Abstract

Laticifer cells are specialized plant cells that synthesize and accumulate latex. In fact, latex-bearing plants are only found in about 10% of flowering plant species, but many of them contain high-value secondary metabolites. In this review, we summarize the characteristics of latex-bearing plants. Additionally, the synthetic pathway and genes of secondary metabolites in *Macleaya cordata* and *Cannabis sativa* are described in detail.

Abbreviation

4OMT	4′−*O*-methyltransferase
6OMT	6′−*O*-methyltransferase
AGPs	antibiotic growth promoters
ALL	allocryptopine
BBE	berberine bridge enzyme
BIAs	benzylisoquinoline alkaloids
CBC	cannabichromene
CBCA	cannabichromenic acid
CBD	cannabidiol
CBDA	cannabidiolic acid
CBG	cannabigerol
CBGA	Cannabigerolic acid
CBN	cannabinol
CBND	cannabinodiol
CBNRA	cannabis phenolic acid
CFS	cheilanthifoline synthase
CHE	chelerythrine
CNMT	coclaurine-*N*-methyltransferase
COR	codeinone reductase
CRISPR	clustered regularly interspaced short palindromic repeats
CsTKS-CsOAC	tetraketide synthase-olivetolic acid cyclase
DBOX	dihydrobenzophenanthridine oxidase
DMAPP	dimethylallyl diphosphate
DOXP	deoxyxylulose-5-phosphate
E. coli	*Escherichia coli*
ECS	endocannabinoids system
EFSA	European Food Safety Authority
EFSA	European Food Safety Authority
FCR	feed conversion ratio
FDA	Food and Drug Administration
GPP	geranyl pyrophosphate
IPP	isopentenyl diphosphate
MEP	2-methylerythritol phosphate
MSH	*N*-methylstylopine 14′-hydroxylase
NCS	norcoclaurine synthase
NGPs	natural growth promoters
NMCH	*N*-methylcoclaurine 3′-hydroxylase

NPP	neryl pyrophosphate
OLA	olivetolic acid
P6H	6′-hydroxylase
PRO	protopine
SAN	sanguinarine
SPS	stylopine synthase
THC	tetrahydrocannabinol
THCA	Δ9 -tetrahydrocannabinolic acid
THCAS	Δ9-tetrahydrocannabinolic acid synthase
TNMT	tetrahydroprotoberberine-cis-*N*-methyltransferase

1. Introduction

Some plants emit latex-like substances when they are injured. Many latexes are easily recognizable, such as milky, and become sticky and quickly coagulate when exposed to air, while other latexes may also be yellow, orange, red or even without color. The diversity of latex appearance is closely related to diverse chemical composition, and often contains complex mixtures of terpenoids, phenolics, proteins, alkaloids, cardiac glycosides, lignans, cannabinoids and tannins (Evert, 2006). Laticifers are specialized plant cells that synthesize and accumulate latex. Based on their development and anatomy, 2 categories of laticifers were established: the articulated type and the nonarticulated type (De Bary, Bower, & Scott, 1884). The final tubing system is very similar in both types but the origin is completely different (Hagel, Yeung, & Facchini, 2008). Non-articulated laticifers are multinucleate, develop from single cells, which elongate extensively during plant growth, and form either unbranched vessels or intricately branched networks. By contrast, articulated laticifers have a compound origin, in that they develop from multiple cells. The nonarticulated laticifer was found in plants such as milkweed (*Asclepias* spp.) or hemp (*Cannabis* spp.) (Mahlberg, 1993), and articulated laticifer was widely present in plants such as dandelion (*Taraxacum* spp.) or moonflower (*Ipomoea alba*) or the rubber tree (*Hevea brasiliensis*) (Evert, 2006). Latex is produced in approximately 12,500 plant species, representing approximately 10% of all flowering plants (angiosperms), which belong to 900 genera of approximately 20 plant families (Kekwick, 2001). The chemical composition of latex is variable in different laticiferous plants, such as *Macleaya cordada* and *Papaver somniferum* (alkaloids), *Cannabis sativa* (phenolic glucosides), *H. brasiliensis* (terpenoids), *Garcinia mangostana* (terpenophenolic) (Dharmaratne, Piyasena, & Tennakoon, 2005). Some latexes have shown pharmaceutically active substances of high economic value, such as the narcotic drug morphine, the antimicrobial berberine (Zhang et al., 2012), the natural growth promoter sanguinarine (Huang et al., 2018c), the industrial raw material natural rubber (cis-1, 4-polyisoprene) (Wuyun et al., 2018). We have summarized some of the latex plants and their main chemical composition, see Table 1.

For a long time, these high economic value secondary metabolites can only be obtained from plants, and always affected by some factors, for example weather, environment, pests and diseases. With the advent of new technologies and dramatically decreased price, genome sequencing has become a basic but powerful tool for the studies of plant metabolisms in unexplored non-model organisms (Chae, Kim, Nilo-Poyanco, & Rhee, 2014; De Luca, Salim, Atsumi, & Yu, 2012; Zerikly & Challis, 2009). Sequenced plant genomes will serve as foundations for the discoveries of biosynthetic enzymes in plants (Boutanaev et al., 2015; Itkin et al., 2013; Nützmann & Osbourn, 2014; Shang et al., 2014). Some researchers have successfully clarified the synthetic pathways of these high-value compounds by combining high-quality genomic information with transcriptome and metabolomics multi-omics tools. In this survey, we firstly introduce two representative laticiferous plants (*Macleaya cordata* and *C. sativa*). Then, we describe the low-molecular compounds of latex-bearing plants and their functions. Thirdly, we analyse the biosynthetic pathways and anabolic patterns of these low molecular compounds. Finally, we compile successful metabolic engineering studies of these low molecular compounds and make future prospects.

2. Latex-bearing plants

2.1 *Macleaya cordata*

M. cordata (Willd.) R. Br. (Chinese name "Bo-luo-hui", Fig. 1 left) is a perennial herb that belongs to the Papaveraceae family and is typically prescribed as a traditional anti-bacterial medicine (Kosina et al., 2010; Sai et al., 2015), which effect and usage were well documented in Ben-Cao-Shi-Yi, a Chinese encyclopedia of botany and medicine from the early Tang dynasty. The color of *M. cordata* latex was yellowish-brown, which contains a large amount of benzylisoquinoline alkaloids, such as sanguinarine (SAN), chelerythrine (CHE), protopine (PRO) and allocryptopine (ALL). It flows out when the leaves, stems or capsules of *M. cordata* are broken. As an important

Table 1 Latex plants and their main chemical composition.

	Species	Family	Main chemical composition
1	*Macleaya cordata*	Papaveraceae	Alkaloids (sanguinarine, chelerythrine et al.)
2	*Cannabis sativa*	Cannabaceae	Phenolic glucosides
3	*Papaver somniferum*	Papaveraceae	Alkaloids (morphinan,codeine, thebaine et al.)
4	*Eschscholzia californica*	Papaveraceae	Alkaloids (morphinan,codeine, thebaine et al.)
5	*Chelidonium majus*	Papaveraceae	Alkaloids (sanguinarine, chelidonine, chelerythrine, berberine, coptisine et al.)
6	*Taraxacum koksaghyz*	Compositae	cis-1,4-Isoprene Triterpenes (lupeol, taraxerol, taraxasterol et al.)
7	*Cichorium intybus*	Compositae	Alkaloids, inulin, sesquiterpene lactones et al.
8	*Lactuca sativa*	Compositae	Sesquiterpene lactones, lactucin, deoxylactucin
9	*Taraxacum kok-saghyz*	Compositae	cis-1,4-Isoprene
10	*Calotropis gigantea*	Asclepiadaceae	Proteins (calotropain-FI, calotropain-FII et al.)
11	*Periploca sepium*	Asclepiadaceae	Pregnanes, polyisoprene et al.
12	*Cynanchum chinense*	Asclepiadaceae	Terpenes, proteins, carbohydrates, benzines et al.
13	*Asclepias curassavica*	Asclepiadaceae	Cardenolides (calactin, calotropin,voruscharin et al.)
14	*Euphorbia antiquorum*	Euphorbiaceae	Triterpene (euphol, cycloeucalenol, euphol 3-O-cinnamate et al.)
15	*Hevea brasiliensis*	Euphorbiaceae	cis-1,4-Isoprene
16	*Morus alba*	Moraceae	Alkaloids (1,4-dideoxy-1,4-imino-d-arabinitol,1-deoxynojirimycin, 1,4-dideoxy-1,4-imino-D-ribitol et al.) Protein (MLX56 et al.)
17	*Antiaris toxicaria*	Moraceae	Cardenolides, antiarosides et al.
18	*Ficus microcarpa*	Moraceae	Chitinases enzymes (microcarpain et al.)
19	*Broussonetia papyrifera*	Moraceae	Phenolic compounds, diterpenoids, alkaloids et al.
20	*Ficus carica*	Moraceae	Phenolic compounds (bergapten, psoralen et al.) Flavonoids (rutin et al.) Phenolic acids (feirulic acids et al.) Phytosterols (taraxasterol et al.)
21	*Thevetia peruviana*	Apocynaceae	Glycosides (thevetin et al.)

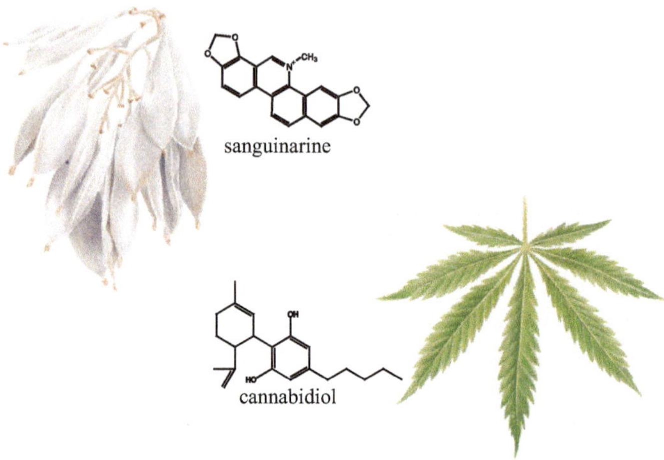

Fig. 1 Two laticiferous products and selected source plants. Sanguinarine (SAN) from *M. cordata* (left) and cannabidiol (CBD) from *C. sativa* (right).

phytogenic growth promoter, SAN have been used in feeding livestock for decades due to its beneficial effects on antimicrobial, anti-inflammatory and growth-promoting activities (Huang et al., 2018c). Nowadays, *M. cordata* has been approved by the European Food Safety Authority (EFSA) as a safe plant for the manufacture of feed additives (Franz et al., 2005). Currently, SAN-containing products has been marketed in over 70 countries (Ikezawa et al., 2003). With the completion of the genome project of *M. cordata* and discovery of the major SAN synthesis genes (Liu et al., 2017), there is hope to largely improve SAN production efficiency by genome-based crop breeding and microbial pathway reconstruction, which will decrease the costs and promote the application of SAN in global livestock farming.

2.2 *Cannabis sativa*

The *Cannabis* (Fig. 1 right) belongs to the family of Cannabinaceae and mostly dioecious. *Cannabis* is generally classified into three subspecies of *C. sativa*, *Cannabis indica*, and *Cannabis ruderalis*. However, according to the use, *C. sativa* is divided into two different groups: *marijuana* and *hemp* (Grotenhermen & Müller-Vahl, 2016; Hill, 2015; National Academies of Sciences and Medicine, 2017). *Marijuana*, primarily used entertainment for its psychoactive properties. In contrast to *Marijuana, hemp* is valued for its medicinal compounds, fiber, and seed (Johnson, 2018). However, due to the presence of tetrahydrocannabinol (THC), *cannabis* has been subject to strict control. Cultivars of hemp must have less than the threshold THC concentration to be grown under license in western countries permitting hemp cultivation. In fact, there is another important non-psychoactive chemical in *cannabis* (cannabidiol, CBD). In clinical studies, CBD is a potent antioxidative and anti-inflammatory agent (Lastres-Becker, Molina-Holgado, Ramos, Mechoulam, & Fernández-Ruiz, 2005; Johnson). Additionally, CBD also has therapeutic properties and is specifically being investigated as a treatment for both schizophrenia (Osborne, Solowij, Babic, Huang, & Weston-Green, 2017) and Alzheimer's disease (Watt & Karl, 2017). In particular, Epidiolex®, a product developed by GW pharmaceutics, has been used to treat epilepsy (Chen, Borgelt, & Blackmer, 2019).

Cannabinoids are toxic to plant cells, leading to mitochondrial permeability.transition and DNA degradation, ultimately induce apoptosis (Morimoto et al., 2007). Therefore, in order to avoid cell damage, cannabinoid synthesis and storage are concentrated on glandular hairs that are female flowers. Glandular hairs are composed of basal cells, stem cells, and secretory cells, surrounded by storage cavities. According to their color, the glandular hair of cannabis can be divided into different stages: the glandular hairs in the mature stage are translucent (the highest content of cannabinoids in this stage), while the glandular hairs in the old age are yellow, and the glandular hairs in the senescence stage are brown. Trichomes are the main organs for the synthesis and storage of cannabinoids in *C. sativa*, most located in the resin heads of capitate-stalked glandular trichomes (Mahlberg & Kim, 2004). In 2011, the whole genome sequencing of *C. sativa* has been completed (Van Bakel et al., 2011), and provided useful data for the identification and functional analysis of genes to biosynthesis of THC and CBD in *Cannabis*.

3. Low-molecular compounds of latex-bearing plants and their functions

3.1 The benzylisoquinoline alkaloids (BIAs) of *M. cordata*

Many BIAs have good pharmacological activity and clinical application value, such as morphine with analgesic effect, vinblastine with anti-cancer effect, ajmaline with antiarrhythmic effect, SAN with anti-inflammatory and animal growth effects (Hagel & Facchini, 2013). The main active ingredients in *M. cordata* are BIAs, and the highest content are the four BIAs alkaloids: SAN, chelerythrine (CHE), protopine (PRO), allocryptopine (ALL).

3.1.1 Application in animal health

Since the 1950s, low-dose antibiotics (Antibiotic growth promoters, AGPs) have been added in feed for the maintenance of animal health and application in the zootechny. Nowadays, the global consumption of antibiotics for livestock consumption is nearly twice that of humans (Aarestrup, 2012; Organization, 2012), problems such as bacterial resistance and super bacteria pose a serious threat to the environment and public health. Therefore, looking for "safe, effective, controllable, low-cost" feed antibiotic substitute is a hot topic in global research. In the past decade, probiotics, prebiotics and plant-derived natural products have been widely used as substitutes for AGPs, namely natural growth promoters (NGPs). SAN has been widely used in zootechny as important NGPs because of their antibacterial, anti-inflammatory and animal-promoting effects. Recent research indicates that chickens fed with SAN was found to grow faster, and improving either of food intake or FCR, alleviating host inflammation through down-regulating cytokines, and beneficial regulating of the intestinal microbiota mainly in the foregut. The taxonomic changes not only include the decreasing of transient and pathogenic bacteria, but also include the increasing of probiotic *Lactobacillus*. On the functional side, the pathways for amino acid metabolism and synthesis are significantly enriched.

3.1.2 Application in the human health

With modern pharmacological studies, SAN has antibacterial, antiseptic and antitumor activities (Kosina et al., 2010). Some studies have found SAN has potential anticancer activity for many cancer cells such as melanoma (Hammerová, Uldrijan, Táborská, & Slaninová, 2011; Serafim et al., 2008), leukemia (Han, Yoo, & Choi, 2008; Weerasinghe, Hallock, Tang, & Liepins, 2001) and lymphoma (Hussain et al., 2007). Additionally, SAN is also used for anti-schistosomiasis (Zhang & Coultas, 2013).

3.2 Cannabinoids in *C. sativa*

Currently, cannabis extracts worldwide are mainly used for medical and recreational purposes. *Cannabis* contains more than 400 chemical constituents, some of which are widely used in the medical field because of their good activity. In cannabis, the most prevalent compounds are Δ^9-THC acid,

CBD acid and CBN acid, followed by CBG acid, CBC acid and CBND acid. In addition, other secondary metabolites are also present in *Cannabis*, such as flavonoids, stilbenoids, terpenoids, lignans and alkaloids.

The endocannabinoid system (ECS) manages appetite, memory, emotional behavior, and perception of pain. The CBD relieves patients with neurological diseases by regulating signaling pathways and related receptors. In addition, the CBD can treat a variety of diseases such as anxiety disorders (such as post-traumatic stress disorder), obesity, epilepsy, dystonia, diabetes, cancer, neurodermatitis and Alzheimer's disease. A comprehensive review of the safety and side effects of the CBD showed that the CBD was safe and well tolerated, even at very high doses, without significant side effects. In June 2018, FDA approved Epidiolex® for the treatment of patients with lennox gastaut syndrome and dravet syndrome (Chen et al., 2019). In addition to medical use, *Cannabis* products are also used in the fields of entertainment, food.

4. The biosynthesis and metabolism of low-molecular compounds in latex-bearing plants

4.1 The biosynthesis of SAN in *M. cordata*

In the 1960s, researchers began to study the synthetic pathways of SAN in some Papaveraceae plants. SAN is produced by the condensation of tyrosine derivatives dopamine and 4′-hydroxyphenylacetaldehyde by norcoclaurine synthase (NCS) (Stadler et al., 1987; Stadler, Kutchan, & Zenk, 1989). Norcoclaurine converted to reticuline by 6′−*O*-methyltransferase (6OMT), coclaurine-*N*-methyltransferase (CNMT), *N*-methylcoclaurine 3′-hydroxylase (NMCH), 4′−*O*-methyltransferase (4OMT) and berberine bridge enzyme (BBE) converted to scoulerine (Frenzel & Zenk, 1990; Pauli & Kutchan, 2010; Sato et al., 1993). Scoulerine was transformed into stylopine under the action of two cytochrome P450 oxidases, cheilanthifoline synthase (CFS) and stylopine synthase (SPS) (Bauer & Zenk, 1989; Bauer & Zenk, 1991). The stylopine is N-methylated to tetramethylstylopine by tetrahydroprotoberberine-cis-*N*-methyltransferase (TNMT) (O''Keefe & Beecher, 1994; Takashi, Masanori, Tomoya, & Fumihiko, 2010). Followed *N*-methylstylopine is catalyzed to protopine by *N*-methylstylopine 14′-hydroxylase (MSH) (Beaudoin & Facchini, 2013; Rueffer & Zenk, 1985). Then protopine is catalyzed to 6′-hydroxylase-protopine by 6′-hydroxylase (P6H), and spontaneous intramolecular rearrangement forms

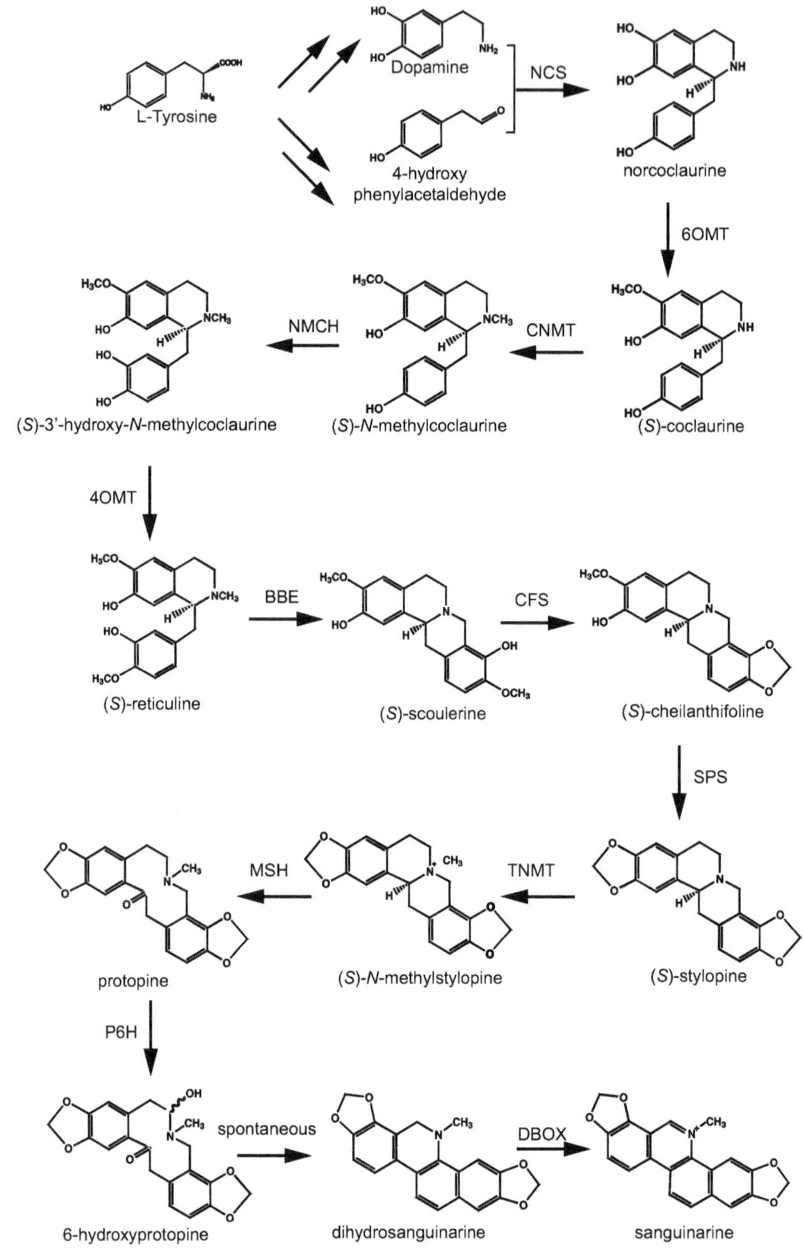

Fig. 2 The synthetic pathways of sanguinarine.

dihydrosanguinarine, (Takemura, Ikezawa, Iwasa, & Sato, 2013; Tanahashi & Zenk, 1990), and finally dihydrobenzophenanthridine is oxidized to SAN by dihydrobenzophenanthridine oxidase (DBOX). (Arakawa, Clark, Psenak, & Coscia, 1992; Schumacher & Zenk, 1988) (Fig. 2).

In 2017, the whole genome of *M. cordata* has been completed, which is the first species to complete genome-wide sequencing in the Papaveraceae. The 19-kmer analysis of these sequences suggested a genome size of 540.5 Mb and a heterozygous rate of 0.92%, a total of 22,328 protein-coding genes were predicted. Comparing the genome-wide data of *M. cordata* within biosynthetic genes of SAN and CHE in *Papaver bracteatum*, *Coptis japonica*, *Eschscholtzia californica* and *P. somniferum* for homologous gene alignment, 39 candidate gene sequences were obtained. Then, based on the transcriptome and metabolic profile data of different organs in *M. cordata*, it is concluded that the synthesis of SAN and CHE in *M. cordata* may have tissue-organ specificity, the protopine (PRO) and allocryptopine (ALL) which is the precursors of SAN and CHE synthesized in roots, while SAN and CHE are synthesized in capsules. According to this hypotheses, 16 genes were screened from 39 candidate genes. It includes 6 methyltransferase genes, 4 flavoprotein oxidase genes, 6 cytochrome P450 enzyme genes and 1 cytochrome P450 reductase gene. All candidate genes were transferred into *Saccharomyces cerevisiae* for heterologous expression verification, and finally, the genes which participated in 14 steps synthetic pathway from norcoclaurine to SAN and CHE were identified. Then, this study demonstrates the synthesis pathway of reticuline which is the important synthetic intermediate of SAN and CHE proposed by Battersby for the first time (Han et al., 2010). Mco2833 gene catalyzes the production of 6′−O-methylnorlaudanosoline by norlaudanosoline to catalyze the formation of norreticuline, and then both genes Mco769 and Mco8567 can catalyze the production of reticuline by norreticuline. This pathway has been confirmed to provide a new explanation for the biosynthesis of reticuline in higher plants, and provides a new route for the re-editing of SAN and CHE synthetic pathways.

4.2 The biosynthesis of CBD and THC in *Cannabis*

The biosynthetic pathways of CBD and THC have been elucidated and synthetic genes were identified. Zenk used ^{13}C isotope tracer technology to reveal that the alkylphenols in cannabinoids are derived from the polyketide pathway. The monoterpene fraction is derived from the deoxyxylulose-5-phosphate/2-methylerythritol phosphate (DOXP/MEP) pathway

(Fellermeier, Eisenreich, Bacher, & Zenk, 2001). In *Cannabis*, the polyketide synthase catalyzes the hexanoyl-CoA and binding the active site of enzyme. Then, the prolonged polyketide chain extension by a series of decarboxylation condensation of malonyl-CoA, leading to the ring closure and aromatization of the enzyme intermediate to form olivetolic acid (OLA), it is the substrate for cannabinoid synthesis. The DOXP/MEP pathway produces isopentenyl diphosphate (IPP) and its isomer dimethylallyl diphosphate (DMAPP), which form geranyl pyrophosphate (GPP) under the action of a synthetase. Under the action of prenyltransferase, OLA can accept GPP to form monoterpenoids-cannabigerol acid (CBGA), or accept a isomer of GPP-neryl pyrophosphate (NPP) to form another class of monoterpenoids - cannabis phenolic acid (CBNRA) (Fellermeier & Zenk, 1998). CBGA forms THCA, CBDA and CBCA after redox oxidation under the catalysis of THCA synthase, CBDA synthetase and CBCA synthetase, respectively (Fig. 3).

5. The metabolic engineering of low-molecular compounds in latex-bearing plants

5.1 Progress in synthetic biology of SAN

5.1.1 Plant metabolism engineering

Overexpression of some key genes in the synthetic pathway in plants often affects the content of downstream products, and using RNAi or CRISPR methods can also be targeted to increase the target metabolite content in the metabolic network (Allen et al., 2004; Cheng et al., 2014; Wijekoon & Facchini, 2012). In the poppy, RNAi is used to silence the codeinone reductase (COR) gene family, and finally, the content of the important intermediate reticuline is increased (Allen et al., 2004). Researchers have studied the functions of alkaloid biosynthesis genes in *P. somniferum* and *E. californica* through gene overexpression, virus-mediated gene silencing and gene editing technology, further demonstrating the potential of plant metabolic engineering technology to increase alkaloid content (Alagoz, Gurkok, Zhang, & Unver, 2016; Fujii, Inui, Iwasa, Morishige, & Sato, 2007).

Currently, *M. cordata* has successfully established a genetic transformation system (Huang et al., 2017), and the use of the system to overexpress key genes to increase alkaloid content (Huang et al., 2018a). In addition, in order to better develop and utilize the resources of *M. cordata*, a hairy root system

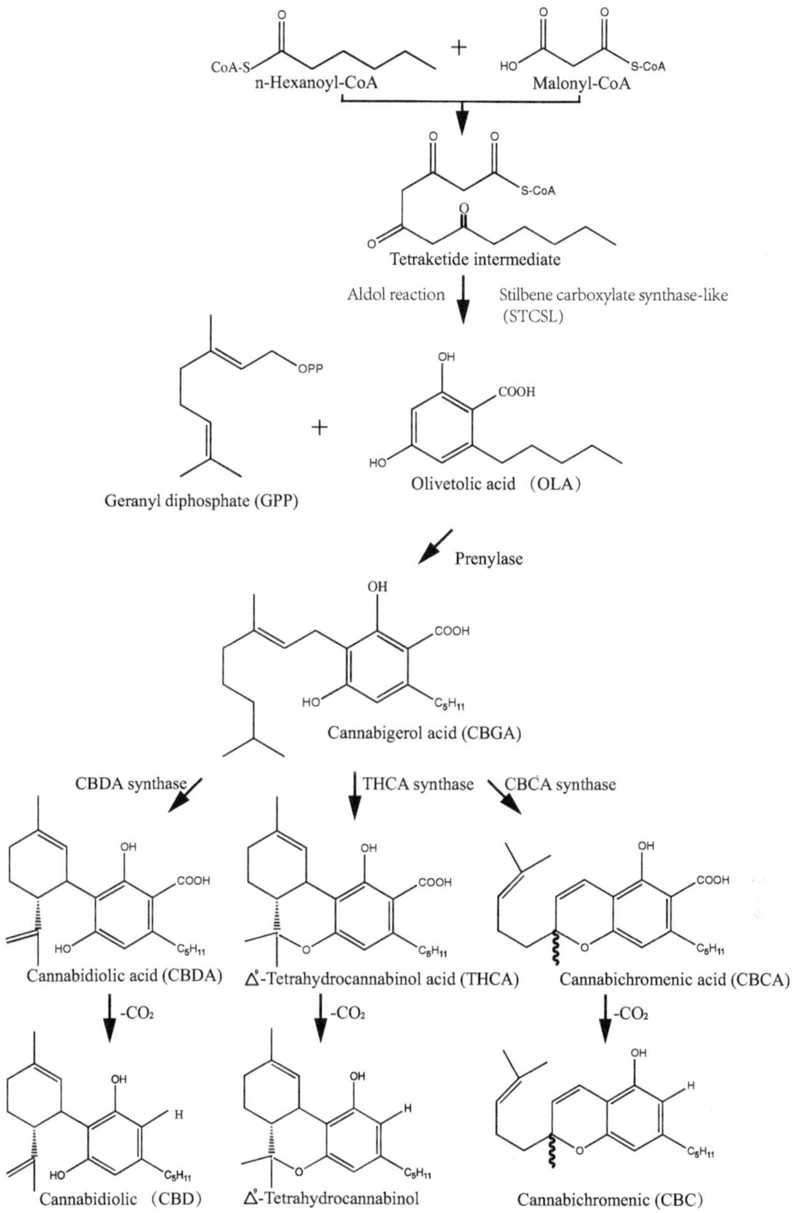

Fig. 3 The biosynthesis of cannabinoids.

has been established to lay the foundation for the future production of SAN and CHE using bioreactors (Huang et al., 2018b).

5.1.2 Microbial metabolism engineering

In order to produce SAN in industry-scales by microbes, Smolke team transferred some genes to yeast from *P. somniferum*, *E. californica* and *C. japonica* which can synthetize BIAs, and combining functional enzymes from different sources, finally made reticuline yield from 10 to 150 mg/L (Hawkins & Smolke, 2008). After it, Martin integrated 10 genes into yeast to produce SAN, which is the longest record of the SAN synthesis pathway in yeast (Fossati et al., 2014). In order to achieve the *de* novo production of SAN, the gene coming from betaine was integrated into the yeast, and finally realized the synthesis of reticuline from glucose (Deloache et al., 2015). In addition, *Escherichia coli* was also used to produce SAN precursor compound reticuline with a yield of 46 mg/L, and through further optimization to achieve a yield of 55 mg/L (Nakagawa et al., 2011). Although the industrial microorganisms can be used to produce SAN, the costs are still too high to industrial production. However, there are still many technical bottlenecks in the industrialized mass production of SAN through *de novo* synthesis.

5.2 Progress in synthetic biology of CBD

5.2.1 Plant metabolism engineering

It has been reported that hairy roots are induced in *Cannabis* explants by *Agrobacterium rhizogenes*-mediated exogenous gene transformation (Farag & Kayser, 2015; Wahby, Caba, & Ligero, 2013), However, no cannabinoids are currently detected in the hairy roots. In addition, researchers have not detected the expression of THC and CBD and synthetic genes after treating *Cannabis* cell suspension cultures with various elicitors (fungi, methyl jasmonate, salicylic acid, etc.) and ultraviolet irradiation stress. This result further explains the specificity of the synthesis sites of THC and CBD in glandular hairs (Flores-Sanchez et al., 2009). Although there are no successful *Cannabis* plants with high CDB content obtained by plant metabolism engineering, we can use gene editing technology to silence the THC synthesis branch to obtain high yield CBD breed, once the *Cannabis* transgenic plants are successful in the future.

5.2.2 Microbial metabolism engineering

At present, people can only extract active ingredients CBD and THC from *Cannabis* plants. In addition, since *Cannabis* is subject to the International

Drug Control Convention, most countries have a policy prohibiting the cultivation and use of *Cannabis* (Bifulco & Pisanti, 2015; Rehm & Fischer, 2015). In addition, due to the high cost of the synthesis precursor and the low conversion rate, it is difficult to obtain THC and CBD by chemical synthesis (Rehm & Fischer, 2015). In summary, the traditional method has been unable to provide a sufficient output of THC and CBD to meet market demand. Therefore, with the research on biosynthesis of CBD and THC functional genes in *Cannabis*, some researchers facilitate the use of microbial metabolic engineering techniques for heterologous production of CBD. For example, by heterologous production of THCA in the *Pichia* expression system, transgenic *Pichia pastoris* expressing THCAS can produce the THCA, and adding precursor CBGA to optimizing culture conditions, the final yield of THCA is 32.6 mg/L (Ahmad, Hirz, Pichler, & Schwab, 2014; Taura et al., 2007). In 2019, Jay D. Keasling et al. produced heterologous expression in different yeast engineering bacteria by optimizing and modifying related genes including the functional genes of cannabinoids in *Cannabis* plants (Luo et al., 2019). THCA and CBDA are produced by feeding different substrates, in which different strains with hexanoic acid as substrates can produce 1.1 mg/L THCA and 4.3 μg/L CBDA, respectively. Additionally, different strains with galactose as substrate can produce 2.3 mg/L THCA and 4.2 μg/L CBDA, respectively. After obtaining THCA and CBDA, it only needs to be heated to naturally decarboxylate to form THC and CBD. In addition, the overexpression of the CsTKS-CsOAC gene in the strain and the increase in the copy number of THCAS resulted in a THCA yield of 8.0 mg/L. This study is also the longest path and the highest yield of THCA and CBD synthesis in yeast. Although the CBD content produced by this research is still a long way from large-scale industrial production, the CBD can be produced quickly and exclusively by this method, which lays a research foundation for further industrial production in the future.

6. Conclusion

In this article, we selected *M. cordata* and *C. sativa* representing the articulated and nonarticulated laticifers, respectively. Since both plants are rich in pharmacologically active secondary metabolites, the most important secondary metabolite in *M. cordata* is SAN, while in *Cannabis* it is CBD. Interestingly, *M. cordata* has been proved to not contain addictive substances such as morphine and codeine, and was recorded on the European Food Safety Authority (EFSA) list as a component in feed requirements in animal production. In contrast, *Cannabis* is regulated by the government because it contains THC (even less than 0.3%). Therefore, only institutions that have obtained legal licenses can carry out planting. At present, the synthesis pathway of SAN in *M. cordata* and synthesis pathway of CBD in *Cannabis* were all clarified. Fortunately, due to the genetically transformation system has been established in *M. cordata*, it is possible to further study the synthesis and regulation mechanism of BIAs. Although *Cannabis* has been establishing a hairy root system, since cannabinoids are concentrated in glandular hairs, CBD is not available in hairy roots of *Cannabis*. With the further development of the research, if the *Cannabis* can break through the restrictions of transgenic seedlings in the future, it is expected that the varieties without THC and with high CBD will be obtained through CRISPR technology. Many latex plants contain high-value secondary metabolites and are widely used in industries, medicines and other fields, but research on latex plants is still lacking. With the development of genome sequencing technology and the reduction of costs, it is expected to provide a new opportunity for the research of milky plants.

Acknowledgments

We thank Yingying Zhang for assistance with Fig. 1 editing.

References

Aarestrup, F. (2012). Sustainable farming: Get pigs off antibiotics. *Nature, 486*, 465.

Ahmad, M., Hirz, M., Pichler, H., & Schwab, H. (2014). Protein expression in Pichia pastoris: Recent achievements and perspectives for heterologous protein production. *Applied Microbiology and Biotechnology, 98*, 5301–5317. https://doi.org/10.1007/s00253-014-5732-5.

Alagoz, Y., Gurkok, T., Zhang, B., & Unver, T. (2016). Manipulating the biosynthesis of bioactive compound alkaloids for next-generation metabolic engineering in opium poppy using CRISPR-cas 9 genome editing technology. *Scientific Reports, 6*, 30910.

Allen, R. S., Millgate, A. G., Chitty, J. A., Thisleton, J., Miller, J. A., et al. (2004). RNAi-mediated replacement of morphine with the nonnarcotic alkaloid reticuline in opium poppy. *Nature Biotechnology, 22*, 1559–1566. https://doi.org/10.1038/nbt1033.

Arakawa, H., Clark, W. G., Psenak, M., & Coscia, C. J. (1992). Purification and characterization of dihydrobenzophenanthridine oxidase from elicited *Sanguinaria canadensis* cell cultures. *Archives of Biochemistry and Biophysics, 299*, 1–7.

Bauer, W., & Zenk, M. H. (1989). Formation of both methylenedioxy groups in the alkaloid (*S*)-stylopine is catalyzed by cytochrome P-450 enzymes. *Tetrahedron Letters, 30*, 5257–5260.

Bauer, W., & Zenk, M. H. (1991). Two methylenedioxy bridge forming cytochrome P-450 dependent enzymes are involved in (*S*)-stylopine biosynthesis. *Phytochemistry, 30*, 2953–2961.

Beaudoin, G.a. W., & Facchini, P. J. (2013). Isolation and characterization of a cDNA encoding (*S*)- cis -*N*-methylstylopine 14-hydroxylase from opium poppy, a key enzyme in sanguinarine biosynthesis. *Biochemical and Biophysical Research Communications, 431*, 597–603.

Bifulco, M., & Pisanti, S. (2015). Medicinal use of cannabis in Europe: The fact that more countries legalize the medicinal use of cannabis should not become an argument for unfettered and uncontrolled use. *EMBO Reports, 16*, 130–132.

Boutanaev, A. M., Moses, T., Zi, J., Nelson, D. R., Mugford, S. T., et al. (2015). Investigation of terpene diversification across multiple sequenced plant genomes. *Proceedings of the National Academy of Sciences, 112*, E81–E88.

Chae, L., Kim, T., Nilo-Poyanco, R., & Rhee, S. Y. (2014). Genomic signatures of specialized metabolism in plants. *Science, 344*, 510–513.

Chen, J. W., Borgelt, L. M., & Blackmer, A. B. (2019). Epidiolex (cannabidiol): A new hope for patients with dravet or lennox-gastaut syndromes. *The Annals of Pharmacotherapy*, 1060028018822124.

Cheng, Q., Su, P., Hu, Y., He, Y., Gao, W., et al. (2014). RNA interference-mediated repression of SmCPS (copalyldiphosphate synthase) expression in hairy roots of *Salvia miltiorrhiza* causes a decrease of tanshinones and sheds light on the functional role of SmCPS. *Biotechnology Letters, 36*, 363–369. https://doi.org/10.1007/s10529-013-1358-4.

De Bary, A., Bower, F. O., & Scott, D. H. (1884). *Comparative anatomy of the vegetative organs of the phanerogams and ferns*.

De Luca, V., Salim, V., Atsumi, S. M., & Yu, F. (2012). Mining the biodiversity of plants: A revolution in the making. *Science, 336*, 1658–1661.

Deloache, W. C., Russ, Z. N., Narcross, L., Gonzales, A. M., Martin, V. J. J., et al. (2015). An enzyme-coupled biosensor enables (*S*)-reticuline production in yeast from glucose. *Nature Chemical Biology, 11*, 465.

Dharmaratne, H., Piyasena, K., & Tennakoon, S. (2005). A geranylated biphenyl derivative from *Garcinia mangostana*. *Natural Product Research, 19*, 239–243.

Evert, R. F. (2006). *Esau's plant anatomy: Meristems, cells, and tissues of the plant body: Their structure, function, and development*. John Wiley & Sons.

Farag, S., & Kayser, O. (2015). Cannabinoids production by hairy root cultures of *Cannabis sativa* L. *American Journal of Plant Sciences, 6*, 1874.

Fellermeier, M., Eisenreich, W., Bacher, A., & Zenk, M. H. (2001). Biosynthesis of cannabinoids. Incorporation experiments with ^{13}C-labeled glucoses. *European Journal of Biochemistry, 268*, 1596–1604.

Fellermeier, M., & Zenk, M. H. (1998). Prenylation of olivetolate by a hemp transferase yields cannabigerolic acid, the precursor of tetrahydrocannabinol. *FEBS Letters, 427*, 283–285.

Flores-Sanchez, I. J., Pec, J., Fei, J. N., Choi, Y. H., Dusek, J., et al. (2009). Elicitation studies in cell suspension cultures of *Cannabis sativa* L. *Journal of Biotechnology, 143*, 157–168. https://doi.org/10.1016/j.jbiotec.2009.05.006.

Fossati, E., Ekins, A., Narcross, L., Zhu, Y., Falgueyret, J. P., et al. (2014). Reconstitution of a 10-gene pathway for synthesis of the plant alkaloid dihydrosanguinarine in *Saccharomyces cerevisiae*. *Nature Communications, 5*, 3283.

Franz, C., Bauer, R., Carle, R., Tedesco, D., Tubaro, A., et al. (2005). Study on the assessment of plants/herb extracts and their naturally or synthetically produced components as "additives" for use in animal production. *EFSA J, 8*, 1694.

Frenzel, T., & Zenk, M. H. (1990). Purification and characterization of three isoforms of *S*-adenosyl-l-methionine: (*R,S*)-Tetrahydrobenzylisoquinoline-*N*-methyltransferase from *Berberis koetineana* cell cultures. *Phytochemistry, 29*, 3491–3497.

Fujii, N., Inui, T., Iwasa, K., Morishige, T., & Sato, F. (2007). Knockdown of berberine bridge enzyme by RNAi accumulates (*S*)-reticuline and activates a silent pathway in cultured California poppy cells. *Transgenic Research, 16*, 363–375.

Grotenhermen, F., & Müller-Vahl, K. (2016). Medicinal uses of *marijuana* and *cannabinoids*. *Critical Reviews in Plant Sciences, 35*, 378–405.

Hagel, J. M., & Facchini, P. J. (2013). Benzylisoquinoline alkaloid metabolism: A century of discovery and a brave new world. *Plant and Cell Physiology, 54*, 647–672.

Hagel, J. M., Yeung, E. C., & Facchini, P. J. (2008). Got milk? The secret life of laticifers. *Trends in Plant Science, 13*, 631–639.

Hammerová, J., Uldrijan, S., Táborská, E., & Slaninová, I. (2011). Benzo [c] phenanthridine alkaloids exhibit strong anti-proliferative activity in malignant melanoma cells regardless of their p53 status. *Journal of Dermatological Science, 62*, 22–35.

Han, X., Lamshöft, M., Grobe, N., Ren, X., Fist, A. J., et al. (2010). The biosynthesis of papaverine proceeds via (*S*)-reticuline. *Phytochemistry, 71*, 1305–1312.

Han, M. H., Yoo, Y. H., & Choi, Y. H. (2008). Sanguinarine-induced apoptosis in human leukemia U937 cells via Bcl-2 downregulation and caspase-3 activation. *Chemotherapy, 54*, 157–165. https://doi.org/10.1159/000140359.

Hawkins, K., & Smolke, C. (2008). Production of benzylisoquinoline alkaloids in *Saccharomyces cerevisiae*. *Nature Chemical Biology, 4*, 564–573.

Hill, K. P. (2015). Medical marijuana for treatment of chronic pain and other medical and psychiatric problems: A clinical review. *Jama, 313*, 2474–2483.

Huang, P., Liu, W., Xu, M., Jiang, R., Xia, L., et al. (2018a). Modulation of benzylisoquinoline alkaloid biosynthesis by overexpression berberine bridge enzyme in *Macleaya cordata*. *Scientific Reports, 8*, 17988.

Huang, P., Xia, L. Q., Liu, W., Jiang, R. L., Liu, X. B., et al. (2018b). Hairy root induction and benzylisoquinoline alkaloid production in *Macleaya cordata*. *Scientific Reports, 8*. doi: ARTN 1198610.1038/s41598-018-30560-0.

Huang, P., Xu, M., Xia, L. G., Qing, Z. X., Tang, Q., et al. (2017). Establishment of an efficient Agrobacterium-mediated genetic transformation method in *Macleaya cordata*. *Scientia Horticulturae, 226*, 302–306. https://doi.org/10.1016/j.scienta.2017.09.004.

Huang, P., Zhang, Y., Xiao, K., Jiang, F., Wang, H., et al. (2018c). The chicken gut metagenome and the modulatory effects of plant-derived benzylisoquinoline alkaloids. *Microbiome, 6*, 211.

Hussain, A. R., Al-Jomah, N. A., Siraj, A. K., Manogaran, P., Al-Hussein, K., et al. (2007). Sanguinarine-dependent induction of apoptosis in primary effusion lymphoma cells. *Cancer Research, 67*, 3888–3897. https://doi.org/10.1158/0008-5472.CAN-06-3764.

Ikezawa, N., Tanaka, M., Nagayoshi, M., Shinkyo, R., Sakaki, T., et al. (2003). Molecular cloning and characterization of CYP719, a methylenedioxy bridge-forming enzyme that belongs to a novel P450 family, from cultured Coptis japonica cells. *Journal of Biological Chemistry, 278*, 38557–38565.

Itkin, M., Heinig, U., Tzfadia, O., Bhide, A., Shinde, B., et al. (2013). Biosynthesis of antinutritional alkaloids in solanaceous crops is mediated by clustered genes. *Science, 341*, 175–179.

Johnson, R. (2018). "Hemp as an agricultural commodity": Library of Congress Washington DC Congressional Research Service).

Kekwick, R. G. (2001). *Latex and laticifers. Encyclopedia of life Sciences*. Wiley and Sons.

Kosina, P., Gregorova, J., Gruz, J., Vacek, J., Kolar, M., et al. (2010). Phytochemical and antimicrobial characterization of *Macleaya cordata* herb. *Fitoterapia, 81*, 1006–1012.

Lastres-Becker, I., Molina-Holgado, F., Ramos, J. A., Mechoulam, R., & Fernández-Ruiz, J. (2005). Cannabinoids provide neuroprotection against 6-hydroxydopamine toxicity in vivo and in vitro: Relevance to Parkinson's disease. *Neurobiology of Disease, 19*, 96–107.

Liu, X., Liu, Y., Huang, P., Ma, Y., Qing, Z., et al. (2017). The genome of medicinal plant *Macleaya cordata* provides new insights into benzylisoquinoline alkaloids metabolism. *Molecular Plant, 10*, 975–989.

Luo, X., Reiter, M. A., D'espaux, L., Wong, J., Denby, C. M., et al. (2019). Complete biosynthesis of cannabinoids and their unnatural analogues in yeast. *Nature, 567*, 123.

Mahlberg, P. G. (1993). Laticifers: An historical perspective. *The Botanical Review, 59*, 1–23.

Mahlberg, P. G., & Kim, E. S. (2004). Accumulation of cannabinoids in glandular trichomes of Cannabis (Cannabaceae). *Journal of Industrial Hemp, 9*, 15–36.

Morimoto, S., Tanaka, Y., Sasaki, K., Tanaka, H., Fukamizu, T., et al. (2007). Identification and characterization of cannabinoids that induce cell death through mitochondrial permeability transition in Cannabis leaf cells. *Journal of Biological Chemistry, 282*, 20739–20751.

Nakagawa, A., Minami, H., Kim, J. S., Koyanagi, T., Katayama, T., et al. (2011). Erratum: A bacterial platform for fermentative production of plant alkaloids. *Nature Communications, 2*, 326.

National Academies of Sciences, E., and Medicine. (2017). *The health effects of cannabis and cannabinoids: The current state of evidence and recommendations for research*. National Academies Press.

Nützmann, H.-W., & Osbourn, A. (2014). Gene clustering in plant specialized metabolism. *Current Opinion in Biotechnology, 26*, 91–99.

O"Keefe, B. R., & Beecher, C. (1994). Isolation and characterization of *S*-Adenosyl-L-Methionine:Tetrahydroberberine-cis-*N*-methyltransferase from suspension cultures of *Sanguinaria canadensis* L. *Plant Physiology, 105*, 395–403.

Organization, W.H. (2012). *The evolving threat of antimicrobial resistance: Options for action*. Geneva: World Health Organization.

Osborne, A. L., Solowij, N., Babic, I., Huang, X.-F., & Weston-Green, K. (2017). Improved social interaction, recognition and working memory with cannabidiol treatment in a prenatal infection (poly I: C) rat model. *Neuropsychopharmacology, 42*, 1447.

Pauli, H. H., & Kutchan, T. M. (2010). Molecular cloning and functional heterologous expression of two alleles encoding (*S*)-*N*-methylcoclaurine 3′-hydroxylase (CYP80B1), a new methyl jasmonate-inducible cytochrome P450-dependent monooxygenase of benzylisoquinoline alkaloid bios. *Plant Journal for Cell & Molecular Biology, 13*, 793–801.

Rehm, J., & Fischer, B. (2015). Cannabis legalization with strict regulation, the overall superior policy option for public health. *Clinical Pharmacology & Therapeutics, 97*, 541–544.

Rueffer, M., & Zenk, M. H. (1985). Berberine synthase, the methylenedioxy group forming enzyme inberberine synthesis. *Tetrahedron Letters, 26*, 201–202.

Sai, C.-M., Li, D.-H., Xue, C.-M., Wang, K.-B., Hu, P., et al. (2015). Two pairs of enantiomeric alkaloid dimers from Macleaya cordata. *Organic Letters, 17*, 4102–4105.

Sato, T., Fitchen, J., Fujiwara, H., et al. (1993). S-adenosyl-L-methionine : scoulerine-9-O-methyltransferase from cultured *Coptis japonica* cells. *Phytochemistry, 32*, 659–664.

Schumacher, H. M., & Zenk, M. H. (1988). Partial purification and characterization of dihydrobenzophenanthridine oxidase from *Eschscholtzia californica* cell suspension cultures. *Plant Cell Reports, 7*, 43–46.

Serafim, T. L., Matos, J. A., Sardao, V. A., Pereira, G. C., Branco, A. F., et al. (2008). Sanguinarine cytotoxicity on mouse melanoma K1735-M2 cells–nuclear vs. mitochondrial effects. *Biochemical Pharmacology, 76*, 1459–1475. https://doi.org/10.1016/j.bcp.2008.07.013.

Shang, Y., Ma, Y., Zhou, Y., Zhang, H., Duan, L., et al. (2014). Biosynthesis, regulation, and domestication of bitterness in cucumber. *Science, 346*, 1084–1088.

Stadler, R., Kutchan, T., Loeffler, S., Nagakura, N., Cassels, B., et al. (1987). Revision of the early steps of reticuline biosynthesis. *Tetrahedron Letters, 28*, 1251–1254.

Stadler, R., Kutchan, T. M., & Zenk, M. H. (1989). (*S*)-Norcoclaurine is the central intermediate in benzylisoquinoline alkaloid biosynthesis. *Phytochemistry, 28*, 1083–1086.

Takashi, M., Masanori, T., Tomoya, T., & Fumihiko, S. (2010). Molecular characterization of O-methyltransferases involved in isoquinoline alkaloid biosynthesis in *Coptis japonica*. *Proceedings of the Japan Academy, 86*, 757–768.

Takemura, T., Ikezawa, N., Iwasa, K., & Sato, F. (2013). Molecular cloning and characterization of a cytochrome P450 in sanguinarine biosynthesis from *Eschscholzia californica* cells. *Phytochemistry, 91*, 100–108.

Tanahashi, T., & Zenk, M. H. (1990). Elicitor induction and characterization of microsomal protopine-6-hydroxylase, the central enzyme in benzophenanthridine alkaloid biosynthesis. *Phytochemistry, 29*, 1113–1122.

Taura, F., Dono, E., Sirikantaramas, S., Yoshimura, K., Shoyama, Y., et al. (2007). Production of Delta(1)-tetrahydrocannabinolic acid by the biosynthetic enzyme secreted from transgenic Pichia pastoris. *Biochemical and Biophysical Research Communications, 361*, 675–680. https://doi.org/10.1016/j.bbrc.2007.07.079.

Van Bakel, H., Stout, J. M., Cote, A. G., Tallon, C. M., Sharpe, A. G., et al. (2011). The draft genome and transcriptome of *Cannabis sativa*. *Genome Biology, 12*, R102.

Wahby, I., Caba, J. M., & Ligero, F. (2013). Agrobacterium infection of hemp (*Cannabis sativa* L.): Establishment of hairy root cultures. *Journal of Plant Interactions, 8*, 312–320.

Watt, G., & Karl, T. (2017). In vivo evidence for therapeutic properties of cannabidiol (CBD) for Alzheimer's disease. *Frontiers in Pharmacology, 8*, 20.

Weerasinghe, P., Hallock, S., Tang, S.-C., & Liepins, A. (2001). Role of Bcl-2 family proteins and caspase-3 in sanguinarine-induced bimodal cell death. *Cell Biology and Toxicology, 17*, 371.

Wijekoon, C. P., & Facchini, P. J. (2012). Systematic knockdown of morphine pathway enzymes in opium poppy using virus-induced gene silencing. *The Plant Journal, 69*, 1052–1063. https://doi.org/10.1111/j.1365-313X.2011.04855.x.

Wuyun, T.-N., Wang, L., Liu, H., Wang, X., Zhang, L., et al. (2018). The hardy rubber tree genome provides insights into the evolution of polyisoprene biosynthesis. *Molecular Plant, 11*, 429–442.

Zerikly, M., & Challis, G. L. (2009). Strategies for the discovery of new natural products by genome mining. *ChemBioChem, 10*, 625–633.

Zhang, S. M., & Coultas, K. A. (2013). Identification of plumbagin and sanguinarine as effective chemotherapeutic agents for treatment of schistosomiasis. *International Journal for Parasitology-Drugs and Drug Resistance, 3*, 28–34. https://doi.org/10.1016/j.ijpddr.2012.12.001.

Zhang, X., Zhao, Y., Zhang, M., Pang, X., Xu, J., et al. (2012). Structural changes of gut microbiota during berberine-mediated prevention of obesity and insulin resistance in high-fat diet-fed rats. *PLoS One, 7*, e42529.

CHAPTER FIVE

Pharmacologically active compounds from latex-bearing plants

Alicja Warowicka[a,b,*], Robert Nawrot[c] and Anna Goździcka-Józefiak[c]
[a]Department of Animal Physiology and Developmental Biology, Institute of Experimental Biology, Faculty of Biology, Adam Mickiewicz University, Poznań, Poland
[b]NanoBioMedical Centre, Adam Mickiewicz University, Poznań, Poland
[c]Department of Molecular Virology, Institute of Experimental Biology, Faculty of Biology, Adam Mickiewicz University in Poznań, Poznań, Poland
*Corresponding author: E-mail: alicja@amu.edu.pl

Contents

Abstract

Plant latex is a milky-like fluid (sap) that is stored in plant specialized cells — laticifers. It contains a mixture of phytochemicals, proteins and enzymes, such as alkaloids, phenolics, terpenoids, defense proteins, proteases and chitinases. Most of these latex ingredients possess pharmacologic activity. Due to its curing effects, *Chelidonium majus* latex has been used in traditional folk medicine to treat skin warts and tubercles. Nowadays,

Advances in Botanical Research, Volume 93
ISSN 0065-2296
https://doi.org/10.1016/bs.abr.2019.11.002

natural plant latex derived molecules and their synthetic analogs are used in medicine as antimicrobial agents, analgesics, antiproliferative and anticancer drugs. Moreover, it has been reported that alkaloids and phenolics from the plant latex exhibit potential therapeutic effects against neurodegenerative diseases such as Alzheimer's and Parkinson's diseases. Most of natural plant latex derived products are present in low amounts, difficult to isolate, possess poor bioavailability and stability. Recent advancements in nanotechnology offer advanced nanomaterials as nanocarriers of pharmacologically active latex natural agents. Conjugation of small natural latex compounds with nanoparticles, liposomes, micelles, nanodiamonds and carbon nanotubes allows to improve bioavailability and stability of these natural pharmacologically active latex compounds. Furthermore, nanoscience approach provides time-controlled and site-specific delivery of potential therapeutics from latex, what gives new promising perspectives for their future use in medicine.

1. Introduction

Plant latex is a rich source of pharmaceuticals, pesticides and immune allergens. It contains phytosterols, flavonoids, acetogenins, alkaloids and phenols. In general, plant latex contains low-molecular weight compounds (previously referred to as secondary metabolites, such as alkaloids, flavonoids, terpenoids etc.) and macromolecular complexes, which are proteins. The plant latex compounds are responsible for numerous pharmacological properties of many plants like anti-tumors, anti-angiogenic, anti-diabetic, anti-proliferative, anti-arthritis, anti-inflammatory, antioxidant, antiasthmatic, anti-fertility, analgesic, immunomodulation, wound healing, cytotoxicity, vasodilatory activities. Most of them show diverse biological activities against bacteria, fungi, viruses, protozoans, nematodes, insects, and tumor cells. They are also used as disinfectant, anticoagulant, anti-inflammatory, antioxidant and antiproliferative agents that provide protection in wounds (Ramos, Demarco, & Cristina, 2019; Upadhyay, 2011; Zielińska, Jezierska-Domaradzka et al., 2018). Latex contains also wide variety of industrially important metabolic substances which can be harvested, modified, quenched, and polymerized easily for making goods and materials (Agrawal & Konno, 2016; Konno, 2011; Ramos et al., 2019). In plants various latex ingredients may serve as anti-feedants, against insect herbivory and play defensive role against pathogens.

Latex is a milky fluid secreted by elongated cells called laticifers and flows inside them throughout leaves, stems, fruits and roots of some flowering plants. Most of the biomolecules (secondary metabolites or low-molecular compounds) present in latex are produced in the plastids and endoplasmatic reticulum, and are later transported to the central vacuole. This is an organelle, where the lipophilic and hydrophilic compounds form an emulsion, typical for the latex. About 10% of flowering plants produce latex and they are found in over 40 families including, Papaveraceae, Euphorbiaceae, Apocynaceae, Caricaceae, Moraceae, Asclepidacae. Around 15–30% of plant species exude latex in tropical Africa, and 20%–35% in tropical America (Amazon), comparing to 8.9% worldwide (Chavan Bhagyashri, Hundiwale Jogendra, & Patil Avinash, 2015; Lewinsohn, 1991).

One of the latex-bearing medicinal plants with high pharmacological potential is Greater Celandine (*Chelidonium majus* L.). It is a perennial herbaceous plant from Papaveraceae family growing wild in Europe, a part of Asia and in North America, where it was introduced (Fig. 1).

C. majus was known in traditional herbal medicine for its wide activities, like antimicrobial, antiviral, anticancer, anti-inflammatory, antiparasitic, antifungal, analgesic, antioxidant, fungistatic, choleretic, hepatoprotective, immunomodulatory (Arora & Sharma, 2013; Biswas, 2013; Gilca, Gaman, Panait, Stoian, & Atanasiu, 2010; Maji & Banerji, 2015; Zielinska, Wójciak-Kosior et al., 2018). It was used to fight liver and biliary disorders, corns, eczema, tumors of the skin as internal tinctures, mixtures, fluids or

Fig. 1 *Chelidonium majus* L. in its natural habitat. (A) Adult plant in full of flowering phase; (B) Cut stem of the plant with exuding yellow milky sap (latex); (C) Mature *C. majus* fruits with seeds (Fot. R. Nawrot).

external preparations (Barnes, Anderson, & Phillipson, 2007; Etxenagusia, Anda, González-Mahave, Fernández, & Fernández de Corrès, 2000). Its yellow/orange latex, was used externally to treat warts, papillae and condylomas, caused by infections of human papilloma virus (HPV) (Etxenagusia et al., 2000). *C. majus* is rich in various biologically active substances, like benzylisoquinoline alkaloids (e.g. chelidonine, berberine, sanguinarine, coptisine, chelerythrine, protopine, stylopine, canadine), flavonoids (e.g. rutin, quercetin, luteolin), phenolic acids (gallic acid, chlorogenic acid), unsaturated fatty acids (linoleic acid, oleic acid), carotenoids, saponins, and some others (Arora & Sharma, 2013; Colombo & Bosisio, 1996; Nawrot et al., 2017; Tomè & Colombo, 1995). Recent research also showed that its latex is rich in proteins, which are connected with defense response and generation of precursor metabolites and energy (Nawrot, Barylski, Lippmann, Altschmied, & Mock, 2016; Nawrot, Lesniewicz, Pienkowska, & Gozdzicka-Jozefiak, 2007; Nawrot, Kalinowski, & Gozdzicka-Jozefiak, 2007; Nawrot, Zauber, & Schulze, 2014). Its latex was shown to contain 334 metabolic, structural and defense-related proteins (Nawrot et al., 2016). Proteins contained in *C. majus* latex in relatively large amounts are major latex protein (MLP, with high homology to Bet v1-like superfamily proteins), polyphenol oxidase (PPO), peroxidase (POX), lipoxygenase (LOX) and enzymes responsible for alkaloid and phenylpropanoid synthesis (Nawrot et al., 2016, 2017). Mostly, they belong to different pathogenesis-related proteins (PR), which are activated in stress- or defense-related conditions for the plant (Fister et al., 2016; Sels, Mathys, De Coninck, Cammue, & De Bolle, 2008; van Loon, Rep, & Pieterse, 2006).

2. Plant latex compounds

2.1 Low-molecular weight compounds of latex

2.1.1 Alkaloids

Many studies showed that various latex-bearing plants are a rich source of different secondary metabolites as alkaloids, which present strong biological activities. Alkaloids are a kind of the most abundant and diverse class of plant natural products. These low-molecular weight molecules were found in all laticiferous plants belonging to Papaveraceae, Campanulaceae, Apocynaceae and Moraceae families. Alkaloids are alkaline compounds containing nitrogen within a ring structure (Fig. 2). They are synthesized by plants from amino acids (e.g. tyrosine, tryptophan, ornithine, phenylalanine). Amino

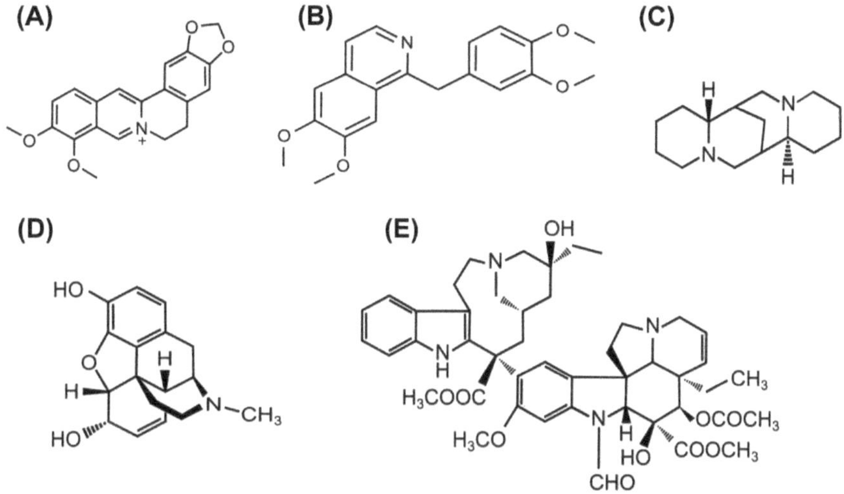

Fig. 2 The structures of selected pharmacologically active alkaloids isolated from the latex of different latex-bearing plants: berberine (A), papaverine (B), sparteine (C), morphine (D) and vinblastine (E).

acids are decarboxylated to amines and react with amine oxides to form aldehydes. Biosynthesis of alkaloids is complex and requires 20 or even more enzymatic steps (Takshak, 2018). They can be classified according to the biosynthetic pathway, chemical structure or biological activity. Depending on the precursor of synthesis, alkaloids can be classified into more than 20 different groups. Thus, there are isoquinoline alklaoids, indole alkaloids, pyrrolidine alkaloids, tropane alkaloids, piperidine alkaloids, vinca alkaloid etc. (Matsuura & Fett-neto, 2015). Due to their toxicity, alkaloids play an important role in plant defense mechanisms against different photogenic microorganisms and herbivores. They exert many pharmacological effects including antihypertensive (e.g. indole alkaloids), antimalarial (e.g. quinine), antiarrhythmic (e.g. quinidine, spareien), and anticancer (e.g. dimeric indoles, vincristine, vinblastine) (Matsuura & Fett-neto, 2015). Alkaloids exhibit numerous neuroprotective activities in neurological and neurodegenerative disorders, such as epilepsy, dementia, memory impairment, Alzheimer's and Parkinson's diseases. A large number of alkaloids extracted from various latex plants have been reported to possess strong pharmaceutical properties, including anticancer. For instance, vinca alkaloids, like vinblastine (VBL) isolated from *Catharanthus roseus* (Apocynaceae), have been recognized to bind with tubulin, a cytoskeleton protein, and inhibit mitosis. Thus, the alkaloid has been intensively studied as the anticancer drug. Nowadays, the semisynthetic congener of vinblastine — vinorelbine (VRL) is applied as chemotherapeutic agent for non-small cell lung cancer (NSCLC) and breast cancer (Cazzaniga et al., 2016). Vinca alkaloids are the oldest group of the plant alkaloids that have been used to treat cancer (Moudi, Go, Yien, & Nazre, 2013).

Papaver somniferum (Papaveraceae) (opium poppy) is one of the oldest plant with a long history of use in the folk and traditional medicine. *P. somniferum* latex has been used since ancient times for medicinal purposes and as psychoactive agent. Benzylisoquinoline alkaloids (BIAs) synthetized in *P. somniferum* are stored in specialized laticifers. Opium alkaloids, such as morphine, papaverine and codeine, are characteristic plant natural products isolated from the latex of the plant (Beaudoin & Facchini, 2014). These alkaloids show strong pharmacological activities and are used in medicine as analgesic (morphine), antispasmodic (papaverine), sedative, anti-cough (codeine), and cough-suppressing (noscapine) drugs. Morphine was the first medically useful plant-derived alkaloid (Gu et al., 2014). The white milky-like fluid from *P. somniferum* plant contains 25% (dry weight) and up to 5% (fresh mass) of morphine (Kotaro Konno, 2011).

The rich source of numerous bioactive plant alkaloids is also the latex of *C. majus* (Papaveraceae), which contains isoquinoline alkaloids such as chelidonine, sanguinarine, berberine, coptisine, making up to 20% of fresh weight of the latex (Tomè & Colombo, 1995; Zielińska, Jezierska-Domaradzka et al., 2018). Protoberberine alkaloids, a class of isoquinoline alkaloids, are widely known for their multiple pharmacological effects, like antioxidant, anti-inflammatory, anticancer and anti-microbial. Moreover, it has been reported that berberine (BBR), which is a representative for protoberberine alkaloid class, possess a wide range of pharmacological activities, including antiviral, anticancer and anti-neurodegenerative. BBR has been shown to suppress cell proliferation and induce apoptosis in human cancer cells, such as cervical (Jantova, Cipak, Cernakova, & Kostalova, 2003), breast (Ma et al., 2017) and lung cells (Xiao et al., 2018). In addition, BBR and coptisine (COP), another alkaloids from protoberberine class, showed a synergistic cytotoxic effect on cervical cancer cells (Warowicka et al., 2019). Previous studies reported that BBR has therapeutic potential in neurodegenerative diseases including Alzheimer's disease (Hussain et al., 2018). Moreover, BBR can act as a photosensitizer and might be used as a new photodynamic agent in antiglioblastoma therapy (Luiza Andreazza et al., 2016). BBR regulates EGFR/MEK/ERK, AMPK/mTOR and NF-κB cell signaling pathways (Wang et al., 2017). Scientific reports have also demonstrated that BBR isolated from latex-bearing plant *C. majus* interacts strongly with telomeric G-quadruplex and thus inhibits telomerase activity (Kazemi et al., 2020). Other alkaloids found in plant latex with pharmacological properties include lobeline from *Lobelia cardinalis* (Campanulaceae) acting on nicotinic acetylocholine receptors, sugar-mimic alkaloids (iminosugars) from Morus spp. (Moraceae) and phenanthroinolizidin alkaloids from *Ficus spp.* (Felpin & Lebreton, 2004; Konno, 2011).

Hence, naturally occurring plant compounds in latex have gained increasing attention for the treatment of various type of cancers (Sikander et al., 2016). Moreover, synthetic analogs of natural plant alkaloids are widely used in cancer chemotherapy as cytostatic drugs. Understanding the biosynthesis and mechanism of action of these low-molecular weight plant-derived compounds is required for the synthesis of new pharmacologically active biomolecules based on the structure of bioactive alkaloids and their derivatives.

2.1.2 Terpenoids

Terpenoids, also called isoprenoides, are modified terpenes. They were found in all latex-bearing plants, except *Euphorbia tirucalli, Ficus benghalensis* and *Ficus religiosa* (Manoorkar & Gachande, 2015). Terpenoids are the largest group of plant specialized secondary metabolites, and are highly diverse in chemical structure. They are composed of a combination of two or more isoprene units, which comprise a five carbon units, chemically known as 2-methyl-1,3-butadien (Fig. 3).

Terpenoids are subgrouped according to the number of five carbon units in their skeletons; i.e. hemiterpenes- (C5), monoterpenes- (C10), sesquiterpenes- (C15), diterpenes- (C20), sesterterpenes- (C25), triterpenes- (C30) and tetraterpenes- (C40:carotenoids). Their basic backbone structures are synthesized by the terpene synthase (TPS) enzyme. Subsequently, they are further modified by, for example, hydroxylation, dehydrogenation, acylation or glycosylation reactions, resulting in an array of chemically diverse terpenoid compounds (Singh & Sharma, 2015).

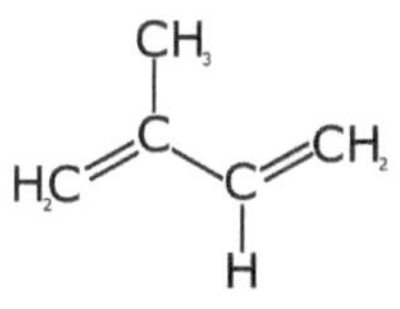

Fig. 3 Five-carbon isoprene unit that makes a terpenoid molecule.

Terpenoids from latex are used in industrial sector as flavors, fragrances, spices and are used in perfumery and cosmetics industry. Some of them have biological activities and have been shown to be available for pharmaceutical applications, like artemisinin (also called qinghaosu) and taxol, used as anti-malaria and anticancer medicines, respectively (Elfawal et al., 2012). Citral, menthol, camphor and cannabinoids belong to the best known terpenoides (Priya et al., 2018). Concerning their potential mechanism of action, terpenoids inhibit the isoprenylation of small G protein and gene expression (Singh and Sharma, 2014).

The latex of cultivated lettuce (*Lactuca sativa*) contains several sesquiterpene lactones which have antifungal activity. The diterpenoids, which are present in the latex of *Euphorbia species,* are toxic to insects and herbivores and also cause skin inflammation (Upadhyay, 2011). Biterpenoids and triterpenoids from the latex of *E. tirucalli* exhibit anticancer activities (Munro et al., 2015). Diterpene ingenol mebutate from latex sap of *Euphorbia peplus* L. has been reported to treat human non-melanoma skin cancer. Ingenol mebutate has been identified as the most active antitumor component obtained from the plant latex (Seca and Pinto, 2018).

It is worth to notice that terpenoids include also natural rubber, which is a cis-1,4 isoprene polymer, that is widely found in the latex of various plant species. Natural rubber latex (NRL) from *Hevea brasiliensis* has shown interesting biomedical properties, as improved wound healing, cell adherence, tissue formation, and angiogenesis (Men, Wang, Chen, Zhang, & Xian, 2019). The primary role of rubber in latex is to produce stickiness; sticky latex mires whole insects or their mouth parts. Rubber exists at a very high concentrations in plants belonging to: Euphorbiaceae (e.g. *H. brasilliensis*) - 44.3%, in *F.* spp. (Moraceae)- 15%—30%, in *Alstonia boonei* (Apocynaceae) - 15.5%, or in *Parthenium argentatum* (Asteraceae) - 8% (Men, Wang, Chen, Zhang, & Xian, 2019).

2.1.3 Phenolics: flavonoids and phenolic acids

Medicinal latex-bearing herbs are a rich source of natural biologically active plant compounds called phenolics. Phenolics have been divided into two groups: flavonoids and phenolic acids. Both of them are plant metabolites with broad range of medicinal activities. Flavonoids have a general polyphenolic structure and are classified, based on structural differences, into six major subgroups, such as flavonols, flavanones, flavanols (also known as catechins or dihydroflavonols), flavones, anthocyanidins and isoflavonoids. In nature, flavonoids can exist as glycosides (with attached sugar residues) or aglycones (with no attached sugar residues) (Abotaleb et al., 2019; Veeramuthu, Raja, Raja, & Al-dhabi, 2017). Flavonoids are among the largest classes of low molecular weight plant compounds with about 6000 members (Takshak, 2018). Flavonoids play an important role in plants. They act as signal molecules, protect plants from stresses and UV irradiation (as natural UV-B absorbers), protect some cell structures, play the role in antipathogenic plant defense and acclimatization (Takshak, 2018).

Flavonoids belong to plant secondary metabolites with strong, miscellaneous biological activities, including anticancer. These small natural molecules are involved in key cellular processes, such as cell division and proliferation. They interfere with cellular signal transduction pathways, limit angiogenesis, metastasis and induce apoptosis (Abotaleb et al., 2019). Moreover, flavonoids are considered as inhibitors for several cellular enzymes, such as lipoxygenase, cyclo-oxygenase (COX), xanthine oxidase (XO), acetylocholinoesterase (AChE) and butyrylcholinesterase (BChE) (Panche, Diwan, & Chandra, 2016). It has been reported that flavonoids show anticancer activities against breast, prostate and colon cancers. They are commonly present in fruits and vegetables, nuts, tea, wine or propolis (bee putty). Flavonoids are phytochemicals found in several plant parts, responsible for color and aroma of their flowers (Abotaleb et al., 2019). High concentration of flavonoids was also detected in the latex of medicinal plants, which have been used in folk medicine to treat various diseases for decades. For example latex from *Jatropha curcas* (Euphorbiaceae) and *Ficus carica* (Moraceae) are rich sources of phenolic constituents, which show anti-inflammatory and antioxidative effects (Upadhyay, 2011). In dietetics and food industries, flavonoids are used as phytonutrients and preservatives (Jakovljević, Stanković, & Topuzović, 2013). They belong to most active natural antioxidants, acting with phenolic acids as reducing agents and free radical scavengers (Ali Ghasemzadeh, 2011). Flavonoids exhibit anti-inflammatory, antimutagenic and anticancer effects (Panche et al., 2016). Consequently, in recent years, due to their pharmacological effects, flavonoids has been extensively studied as a source of medicinal natural drugs (Jucá et al., 2018).

Several studies support the antineurodegenerative potential of flavonoids (Dixon & Pasinetti, 2010). According to the literature data, certain flavonoids can reduce beta-amyloid (Aβ) accumulation and have been suggested to have positive effects against neurodegeneration process associated with Alzheimer's and Parkinson's diseases and dementia (Paris et al., 2012). The mechanism of such action is not fully understood, but it has been hypothesized that it is connected with antioxidant abilities of flavonoids and inhibition of Aβ oligomerization (Ono et al., 2003). Flavonoids isolated from the latex and aerial parts of *Euphorbia characias* L. (*Euphorbiaceae*), such as quercetin and its derivatives, show anticollagenase and antielastase activities and are responsible for the wound-healing ability of the plant (Özbilgin et al., 2018). Moreover, quercetin is a phytoestrogen, which induces apoptosis in breast cancer cells MCF-7 (Duo, Ying, Wang, & Zhang, 2012). Furthermore, several flavonoids such as quercetin, naringin, catechins, possess antiviral activities. These low-molecular weight plant compounds disturb replication and infectivity of DNA and RNA viruses (Panche et al., 2016).

Latex of *C. majus* contains several flavonoids with significant biological and pharmacological properties. In aerial parts of this plant, like stem and leaves, kaempferol, quercetin, isorhamnetin and their derivatives, as well as 5′-methoxy-flavonol and 6′- methoxy-flavonol, were detected (Grosso et al., 2014; Stancic-Rotaru, Mititelu, Crasmaru, & Balaban, 2003; Zielinska, Jezierska-Domaradzka et al., 2018; Zielinska, Wójciak-Kosior et al., 2018). Moreover, concentration of the various classes of plant secondary metabolites, including flavonoids, in extracts of *C. majus,* varies during different phenological stages of the plant (spring leaf rosette, the beginning of flowering, full of flowering, and fruit ripening stages). The highest concentration of flavonoids is observed during the month of May, just before full of flowering (Jakovljević et al., 2013).

Phenolic acids are known to play a role in plant growth and reproduction. These compounds are important molecules for plant-microbe interactions (e.g. mycorrhizae) and are involved in plant defense against stress. Phenolic acids are synthesized in plant via phenylpropanoid pathway, as products of the monolignol pathway and as breakdown products of lignin and cell wall polymers. These natural products are characterized by hydroxylated aromatic rings (Mandal, Chakraborty, & Dey, 2010).

2.1.4 Glycosides

Glycosides, particularly cardiac glycosides, are the most studied latex compounds. These substances are responsible for protecting plants against a vast number of insects. They are a group of cardiac-active compounds that inhibit Na^{+}/K^{+}-ATPase and are present in the latex of many Apocynaceae plants, including milkweeds (*Asclepias sp.*) and oleander. Cardiac glycosides are clinically used in the treatment of heart failure and as antiarrhythmic agents. They exhibit also potent antiproliferative and

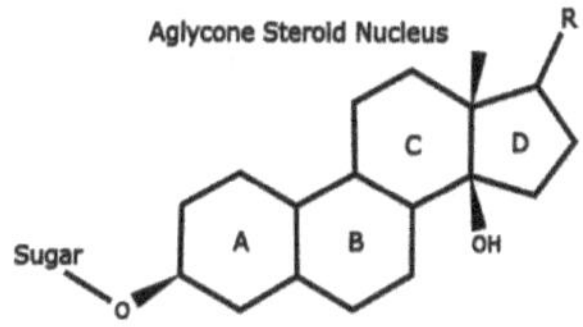

Fig. 4 Cardiac glycosides found in plant latex.

apoptotic effects on cancer cells through complex signal transduction mechanisms (El-Mallakh, Brar, & Yeruva, 2019; Groeneveld, 1982). Cardiac glycosides are composed of two structural features: the sugar (glycoside) and non-sugar (aglycone-steroid) moieties (Fig. 4).

The cardiac glycosides are a subgroup of cardioactive steroids (CAS), which are classified as cardenolides (with a 5-membered lactone ring) or a bufadienolides (with a 6-membered lactone ring). Most cardenolides were isolated from plants belonging to 11 families: Apocynaceae, Asclepiadaceae, Celastraceae, Cruciferae, Liliaceae, Leguminosae, Moraceae, Ranunculaceae, Scrophulariaceae, Sterculiaceae, and Tiliaceae. Recently, new cardiac glycosides were isolated from the latex of *Antiaris toxicaria* (Moraceae), which inhibited growth of various cancer cell lines at nanomolar concentrations. The studied glycosides exert their apoptotic effect through the Nur77-dependent apoptotic pathway (Liu et al., 2013).

2.1.5 Saponins

Saponins belong to active compounds often present in the latex of different plants. In most plants these bioactive constituents are a part of plant defense system against microorganisms, insects and molluscs (Hussain et al., 2019). Traditionally, saponins were used as natural detergents. They are broadly classified as triterpenoids, steroids or steroidal glycoalkaloids, based on the aglycone structure from which they are derived. Representative structure of saponine is presented on Fig. 5.

Two classes of saponins are recognized based on the structure of their aglycone or sapogenin: (1) steroidal saponins, which contain the characteristic four-ringed steroid nucleus, and (2) triterpenoid saponins, which belong to the group of saponin compounds. In plant latex saponins may serve as insecticides and to protect plants against pathogens (microbes and fungi) (Guclu-Ustundag & Mazza, 2007; Moses, Papadopoulou, & Osbourn, 2014; Shuntang, 2018). Saponins extracted from latex exert a wide range of pharmacological activities including expectorant, anti-inflammatory, vasoprotective, hypocholesterolemic, immunomodulatory, hypoglycaemic,

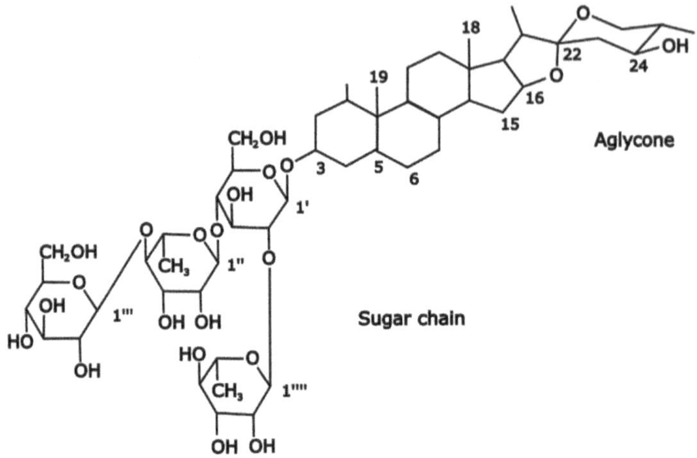

Fig. 5 The structure of plant saponins.

antifungal and antiparasitic. They also reduce blood cholesterol levels and the risk of cancer (Cheeke et al., 2006; Liu et al., 2016).

Saponins are also used in food and cosmetic industry, where they are utilized as emulsifiers or sweeteners. Another important application of saponins is their use as adjuvants in the production of vaccines (Sun, Xie, & Ye, 2009). They have the unique capacity to stimulate both the Th1 immune response and the production of cytotoxic T lymphocytes (CTLs) against exogenous antigens (Sun et al., 2009). Steroidal sapogenins for many years have served as economically important materials for the pharmaceutical industry in the production of steroidal hormones (Jesus et al., 2016).

2.1.6 *Tannins*

Tannins are natural polyphenols found in the latex of many plants (Khanbabaee & van Ree, 2001). They are complex chemical substances derived from phenolic acids, which are found in many species of plants, from all parts of the globe (Fig. 6).

They are traditionally thought to form an important factor of plant defense against herbivorous insects. Tannins are typically defined as water-soluble phenolic compounds that are able to bind and precipitate proteins and other macromolecules within aqueous solutions. Other plant latex phenolics, such as lignins and diphenols (catechols), play a defensive roles against herbivorous insects. Many other phenolic components of the latex, together with flavonoids, are effective antioxidants, as well as anticancer, antibacterial, cardioprotective and anti-inflammatory agents. They may promote the

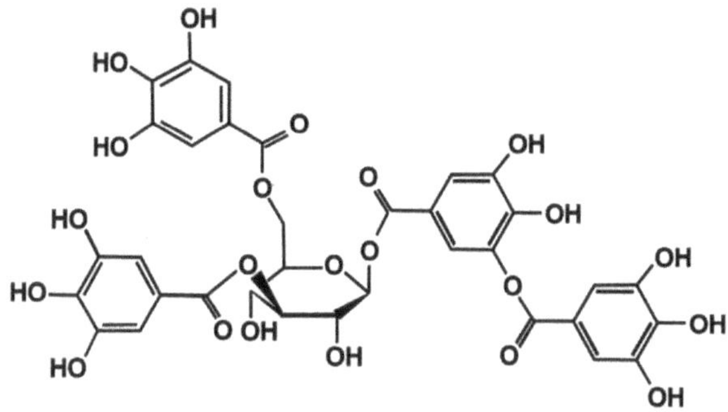

Fig. 6 Representative structure of tannin.

immune system, protect the skin from UV radiation, and are interesting candidates for numerous pharmaceutical and medical applications (Tungmunnithum, Thongboonyou, Pholboon, & Yangsabai, 2018).

2.1.7 Acetogenins

Acetogenins are a class of potent bioactive compounds in various plant species from the Annonaceae family (Fig. 7).

Acetogenins are a relatively new class of fatty acid-derived natural products that have a wide range of biological activities, such as cytotoxic, antitumor, antimalarial, and immunosuppressive, as well as pesticidal and antifeedant (González-Esquinca, De-La-Cruz-Chacón, Castro-Moreno, Orozco-Castillo, & Riley- Saldaña, 2014). The annonaceous acetogenins are also inhibitors of bovine heart mitochondrial complex I (Degli Esposti

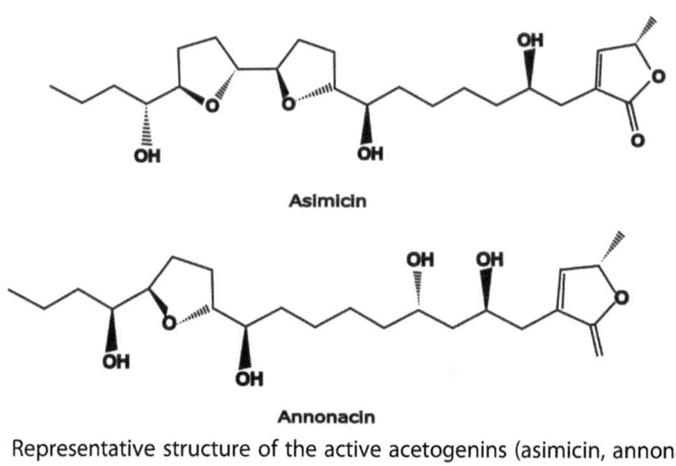

Fig. 7 Representative structure of the active acetogenins (asimicin, annonacin).

M, 1994). Acetogenins are characterized by the presence of one to three tetrahydrofuran (THF) rings in the center of a long alkyl chain with a butenolide moiety at the end. They are typified by a C32 or C34 fatty acid chains with a terminal γ-lactone. Besides mentioned classical types, acetogenins with a tetrahydropyran (THP) ring in the long chain, such as mucocin, jimenezin, muconin, pyranicin and pyragonicin, have also been discovered (Matsui et al., 2010).

Different pharmacological activities of the latex of different plants and their selected compounds are present in Table 1.

2.2 Latex proteins

Latex contains a wide range of proteins, which play major roles in defense mechanisms of plants. Several proteins and proteolytic enzymes have been identified in the latex of a variety of plants. For example, cysteine proteases were found in the latex of Caricaceae, Moraceae and Apocynaceae (Freitas et al., 2019; Matsuura & Fett-neto, 2015). Chitinases, that degrade chitin, were found in the latex of plants belonging to Caricaceae, Moraceae, Apocynaceae, and Euphorbiaceae families (Freitas et al., 2016a; Tomè & Colombo, 1995). The latex of *H. brasiliensis* was found to contain *cis*-prenyltransferase, the enzyme involved in rubber biosynthesis (Konno et al., 2004; Yamashita et al., 2016). Oxidases were also reported in the latex of many plants, which are responsible for browning, coagulation and wound healing after latex exposures to air (Freitas et al., 2016b; Nawrot et al., 2017). Latex of *Calotropis procera* was found to contain lysozymes (Yogi, Gupta, & Mishra, 2016).

2.2.1 Defense-related proteins of plant latex

Most of proteins identified in plant latex are classified as pathogenesis-related (PR) proteins. Expression of PR proteins is induced by different viral, bacterial and fungal pathogens (Sels, Mathys, De Coninck, et al., 2008). Despite sharing a similar defense function, they have various structures and activities and are categorized into 17 families: antifungal (PR-1), β-1,3-glucanases (PR-2), endochitinases (PR-3), chitinases types I and II (PR-4), thaumatin-like proteins (PR-5, osmotins), proteinase inhibitors (PR-6), endoproteinases (PR-7), chitinases type III (PR-8), peroxidases (PR-9), ribonuclease-like (PR-10), chitinases (PR-11), defensins (PR-12), thionins (PR-13), lipid transfer protein (PR-14), oxalate oxidases (PR-15), oxalate oxidase-like (PR-16), unknown (PR-17) (Fister et al., 2016; Sels, Mathys, De Coninck et al., 2008; van Loon et al., 2006). Recent data, based on

Table 1 Pharmacological activities of some latex-bearing plants or their compounds.

Plant (family)	Latex or its compounds	Pharmacological activities	References
Anticancer activity			
Euphorbia tirucalli (Euphorbiaceae)	Diterpenes, ingenol esters	Ingenol esters interact with microtubule proteins	(Khaleghian et al., 2010)
Croton celtidifolius (Euphorbiaceae)	Flavonols, flavan-3-ols	Flavonols are responsible for DNA fragmentation and apoptosis	(Biscaro et al., 2013)
Papaver sominiferum (Papaveraceae)	Noscapine (narcotine)	Induces apoptosis and metaphase arrest in dividing cells	(Kumar, 2018)
Cannabis sativa (Cannabaceae)	Cannabidiol (CBD)	Antiproliferative and proapoptotic effects	(Sharafi, He, & Nikfarjam, 2019)
Antimicrobial activity			
Jatropha curcas (Euphorbiaceae)	Petroleum ether fraction of latex	Against: *Staphylococcus aureus*	(Upadhyay, 2011)
Artocarpus heterophyllus (Moraceae)	Ether fraction of latex	*Pseudomonas aeruginosa, Candida albicans, Staphylococcus aureus*	(Siritapetawee,Thammasirirak, & Samosornsuk, 2012)
Ficus botryocarpa Miq. (Moraceae)	Phenantroindolizidine alkaloids (ficuseptine)	*Escherichia coli*	(Wau, Timi, & Harakuwe, 2016)
Taraxacum officinale (Asteraceae) (common name Dandelion)	n-hexane and ethyl acetate extracts	*Staphylococcus aureus, Escherichia coli, Klebsiella pneumoniae*	

(Continued)

Table 1 Pharmacological activities of some latex-bearing plants or their compounds.—cont'd

Plant (family)	Latex or its compounds	Pharmacological activities	References
Analgesic activity			
Calotropis procera (Asclepiadaceae)	Organic fraction of latex	Analgesic activity	(Jucá et al., 2013)
Papaver sominiferum (Papaveraceae)	Morphine Codeine		(Hagel & Facchini, 2013)
Antidiabetic activity			
Mulberry (Moraceae)	Sugar-mimic allaloid glycosidas inhibitors such as1,4 dideoxy-1, 4-imino-D-arabinitol, 4-imino-D-ribitol	Antidiabetic activity	(Konno et al., 2006)
Antioxidant activity			
Ferrula assa-foetida (Apiaceae)	Essential oil from the latex	Reactive oxygen species (ROS), reactive nitrogen species (RNS) and hydrogen peroxide scavenger	(Kavoosi, Tafsiry, Ebdam, & Rowshan, 2013)
Aloe harlana (Asphodelaceae)	Anthrone (aloin), chromone	Antioxidant activity	(Asamenew, Bisrat, Mazumder, & Asres, 2011; Upadhyay, 2011)
Croton lechleri (Euphorbiaceae)	Megastigmane, lignan, clerodane derivatives, Flavan-3-ols derivatives		(De Marino et al., 2008)
Jatropha curcas (Euphorbiaceae)	Polyphenolic		(Upadhyay, 2011)

Hancornia speciosa (Apocynaceae)	Latex		(Marinho, Alviano, Matheus, Alviano, & Fernandes, 2011)
Calotropis procera L. (Asclepiadacae)	Flavonoids of latex		(Yogi et al., 2016)
Angiogenic activity			
Synadenium umbellatum (Eophorbiaceae)	Latex	Angiogenic activity	(Melo-Reis et al., 2010)
Antifungal			
Hevea brasiliensis (Euphorbiaceae)	Latex	Antifungal	(Daruliza et al., 2011)
Psychoactive properties			
Calotropis procera (Asclepiadaceae)	Alkaloids: Alfa-amyrin Beta-amyrin	Psychoactive properties	(Kinda et al., 2017)
Papaver sominiferum (Papaveracea)	Opium alkaloids (e.g. morphine)		
Antiviral			
Ficus carica (Moracea)	Latex	HPV infection and cancer	(Ghanbari et al., 2019)
Ficus nitida (Moreaceae)		Antiviral properties	(Mahmoud, Gad-Rab, Hussein, & Shoreit, 2010)

(Continued)

Table 1 Pharmacological activities of some latex-bearing plants or their compounds.—cont'd

Plant (family)	Latex or its compounds	Pharmacological activities	References
Chelidonium majus (Papaveraceae)		Antiviral properties	(Colombo & Bosisio, 1996)
Aloe harlana (Asphodelaceae)	Anthrone (aloin)	Anti-HIV	(Upadhyay, 2011)
Euphorbia lactea (Eophorbiacae)	Diterpenoides	Anti-HIV	
Antialergic			
Papaver sominiferum (Papaveraceae)	Opium alkaloids Papaverine	The cough suppressant, antispasmodic	(Upadhyay, 2011)
Wound – healing activities			
Calotropis gigantea (Asclepiadaceae)		Wound-healing	(Nalwaya et al., 2009)
Croton lechleri (Euphorbiaceae)			(De Marino et al., 2008)

C. majus transcriptome search, have shown that *C. majus* contains the members from 12 pathogenesis-related protein families out of 17, including 2 predominant families in terms of protein quantities: PR-9 (peroxidases) and major latex proteins (MLPs) with high homology to PR-10 proteins (ribonuclease-like/Bet v 1 protein family). Moreover, *C. majus* latex contains also several other defense-related proteins, which do not belong to PR proteins – polyphenol oxidase (PPO), dirigent-like protein, lipoxygenase, glycine-rich RNA-binding protein, 14-3-3 proteins and others (Nawrot et al., 2017). Similarly, it was shown that the opium poppy (*P. somniferum*) latex contains 98 proteins, from enzymes, nucleases, chaperons, putative transcription factors and signaling, storage to defense-related proteins, like MLP (Decker, Wanner, Zenk, & Lottspeich, 2000). Subsequent analysis of opium poppy cell cultures by 2-DE-LC-MS/MS helped to identify the 219 of 340 proteins (Zulak, Khan, Alcantara, Schriemer, & Facchini, 2009). PR proteins and lectins have also been reported in the latex of other species, like *E. tirucalli* (Kitajima et al., 2016).

Some of the proteins of plant latex are ribosome-inactivating proteins (RIPs), like lectins from *E. characias* L. and *Hura crepitans* L. latex (Schrot, Weng, & Melzig, 2015). They are prevalent in different plant species and can suppress translation by enzymatic damage of the ribosomes (Shi, Mak, Wong, & Shaw, 2016; Musidlak, Nawrot, & Goździcka-Józefiak, 2017). They also have adenosine and guanosine glycosidase, chitinase, lipase, superoxide dismutase, DNase and RNase activities (Schrot et al., 2015). Their physiological roles could be connected with defense responses against predators, fungi and mainly viruses due to N-glycosidase activity on viral RNA. Due to the same N-glycosidase activity on genomic plant DNA, RIPs could also have a role in plant senescence (Schrot et al., 2015; Walsh, Dodd, & Hautbergue, 2013).

2.2.2 Plant latex proteins as allergens

The latex of many plants contains proteins which are allergens (Upadhyay, 2013). Many of them belong to PR proteins and their homologs, because of their size, stability, resistance to proteases, hydrolytic and membrane-permeabilizing properties. Many of them present high cross-reactivity and cause syndromes like latex-fruit or pollen-related food syndromes (Sinha et al., 2014). Expression of PR proteins growing in different conditions varies, so their level of allergenicity is different. Based on the sequence comparisons, the typical allergens belong to PR families 1, 2, 3, 4, 5, 8, 10 and 14 (Table 2). The most common are allergens from *H. brasiliensis*, the tree

Table 2 The most common PR protein allergens identified in different plants comparing to latex-bearing plants (according to Sinha et al., 2014, modified, unless additionally referenced).

PR family	Proteins	Common allergens identified in different plants	Allergens identified in latex-bearing plants
PR-1	PR-1a, PR-1b, PR-1c	Cuc m 3 (muskmelon)	–
PR-2	B-1,3-Glucanases	Mus a 5 (banana)	Hev b 2 (*H. brasiliensis*)
PR-3	Chitinase types I, II, IV, V, VI, VII	Pers a 1 (avocado) Mus a 2 (banana)	–
PR-4	Chitinase types I and II	–	Hev b 6.01 (prohevein), Hev b 6.02 (hevein), Hev b 6.03 (wound-inducible protein) (*H. brasiliensis*)
PR-5	Thaumatin-like proteins	Pru av 2 (cherry) Mal d 2 (apple) Act d 2 (kiwi) Mus a 4 (banana)	–
PR-8	Chitinase type III	Cof a 1 (coffee)	Hev b 14 (hevamine, *H. brasiliensis*)
PR-10	Parsley "PR-1" Bet v 1, Mal d 1, Dau c 1	Bet v 1 (birch pollen) Pru av 1 (cherry) Mal d 1 (apple) Dau c 1 (carrot)	PR-10 allergens with a strong cross-reactivity to Bet v 1 in Moraceae fruits (Hemmer et al., 2010)
PR-14	Lipid transfer proteins	Pru p 3 (peach) Mal d 3 (apple) Pru av 3 (cherry) Cor a 8 (hazelnut) Zea m 14 (maize)	Hev b 12 (*H. brasiliensis*) (Beezhold et al., 2003)

which is used to produce commercial latex. β-1,3 glucanases from *Hevea* (Hev b 2, Table 2) are the reason of the latex-fruit syndrome (hypersensitivity to some foods, like banana, avocado, fig, kiwi) caused by cross-reactivity due to the high overall conformation similarity of Hev b 2 to β-1,3-glucanases from banana (Mus a 5) and other fruits (Sinha et al., 2014). Prohevein (Hev b 6.01, Table 2) is a cysteine-rich 20 kDa chitin-binding protein (PR-4) from *H. brasiliensis* latex, responsible for natural rubber latex allergy, e.g. in the group of health care personnel. Products of its cleavage are hevein (Hev b 6.02) and wound-inducible protein (Hev b 6.03), which are also allergenic (Sinha et al., 2014). The other allergenic protein from *Hevea* latex is hevamine (Hev b 14), a PR-8 family protein with lysosome and chitinase activity against pathogenic fungi (Lee et al., 2010).

PR-10 are a family of small (MW around 15—16 kDa), acidic, intracellular proteins, important for the plant's physiology and development, however with unknown function. They are mainly induced under various stress conditions and in some cases expressed constitutively (Fernandes, Michalska, Sikorski, & Jaskolski, 2013). Together with MLPs, PR-10 belong to the large Bet v 1 superfamily. Bet v 1 proteins are common allergens responsible for birch pollen allergy, however the others were identified from celery, apple, cherry or carrot (Sinha et al., 2014). Proteins with the Bet v 1 fold have structural similarities and possess large internal hydrophobic cavity in its structure, which can bind low-molecular hydrophobic compounds, like fatty acids, flavonoids and cytokinins. The cavity could be therefore important for Bet v 1 physiological functions (Fernandes et al., 2013). Cross-reactivity between Bet v 1 and other food allergens manifests with clinical oral allergy syndrome. (Mogensen, Wimmer, Larsen, Spangfort, & Otzen, 2002). The study by Hemmer et al. (2010) demonstrated the presence of PR-10 allergens with a strong cross-reactivity to Bet v 1 in the fruits of fig (*F. carica*), mulberry (*Morus alba*) and jackfruit (*Artocarpus heterophyllus*), which belong to the latex-bearing Moraceae family (Hemmer et al., 2010).

PR-14 are small (9—10 kDa) lipid transfer proteins (LTPs), as they could transfer phospholipids and other fatty acids across membranes. They are involved mainly in plant defense against bacterial and fungal pathogens. They are reported in allergenic reactions to hazelnuts (Cor a 8), maize (Zea m 14) or peach (Pru p 3) (Finkina, Melnikova, Bogdanov, & Ovchinnikova, 2016; Sinha et al., 2014). Several families of cysteine-rich antimicrobial peptides (AMPs) have been described, including defensins (PR-12), lipid transfer proteins (LTPs) (PR-14) (both also classified as pathogenesis-related proteins) and hevein type peptides (Nawrot et al., 2014). Lipid transfer protein was also reported in *H. brasiliensis* latex, as a cross-reactive pan-allergen (Hev b 12) (Table 2) (Beezhold et al., 2003).

3. The potential use of active latex compounds in nanotechnology

Nanotechnology is a field of science with a wide range of applications, from energy production and material engineering, to biomedical area. The subject of nanotechnology are nanoparticles (NPs), which have unique properties. They comprise ultrafine particles with one or more dimensions in the size range between 1 and 100 nm (less than 100 nm) (European Commission Annual Report 2011). They possess different sizes, shapes, compositions and functionalities. Thus, NPs can be classified according to their material properties, whether carbon-based, ceramic, polymeric or way of their application. Nowadays, due to their unique physico-chemical characteristics (i.e. high specific surface area, high porosity) nanomaterials have received tremendous attention in biology and biomedical research. Functionalized nanomaterials, such as carbon nanotubes (CNTs), iron oxide, polymeric NPs, quantum dots (QDs), can be used as a contrast agents in medical diagnostic, imaging tools for cell labeling, drug carriers or specific scaffolds in tissue engineering and regenerative medicine. Moreover, nanomaterials can be applied in the field of plant biology. Recently, polymeric NPs can be utilized as a vehicle to deliver specific cargo (siRNA) in plant cells (Silva, Nguyen, Ye, Verchot, & Moon, 2010).

Generally, methods for the synthesis of nanomaterials can be divided into physical, chemical and biological. The green synthesis is a biological plant-based and eco-friendly method for the synthesis of NPs, in which small natural plant compounds are applied. Currently, the green synthesis is successfully used for the synthesis of metallic nanoparticles, such as silver, copper, gold or magnetite. Many studies have shown that those biosynthesized metallic nanoparticles possess remarkable properties and can be applied in anticancer therapies, or pharmacology and cosmetology, as antimicrobial agents, wound dressing materials, as well as in food industry (e.g. food packaging). Moreover, some studies have reported that biosynthesized nanoparticles have better defined morphology and sizes comparing to those produced using physical and chemical methods (Khandel, Yadaw, Soni, Kanwar, & Shahi, 2018). The key benefits of alternative green synthesis over conventional physico-chemical methods is the reduction of use of hazardous and toxic chemicals, as well as its low-cost, what makes the NPs production "eco-friendly". Other studies have reported that green NPs synthesis by plant compounds can be comparable to traditional methods, but detailed reduction of the mechanism, stability and nanoparticle yields are not known in this approach (Cruz et al., 2010; Peralta-Videa et al., 2016). Thus, plant-mediated synthesis of NPs has received considerable attention in the nanoscience area. In green synthesis of metallic NPs, plant natural compounds and their derivatives act as environmentally safe ion reducing agents. For green synthesis of nanoparticles, whole plant extract, crude plant latex or solution of pure plant-delivered bioactive components can be used. Essentially, plant latex contains numerous bioactive secondary metabolites, such as alkaloids, flavonoids, tannins and phenolic acids. These compounds can act as reducing and stabilizing agents during biosynthesis of NPs. For instance, *C. procera* latex was used for the synthesis of silver NPs with antibacterial properties (Mohamed, Ismail, Abdel-Mageed, & Mohamed Shoreit, 2014). Latex from *Aloe barbadensis* and *Aloe vera* was effectively used for the biosynthesis of antibacterial zinc oxide (ZnO) nanoparticles with optical properties (spherical shape, 5–20 nm), and antibacterial silver and gold nanoparticles (spherical shape, 50–350 nm), respectively (Sangeetha, Rajeshwari, & Venckatesh, 2011; Chandran, Chaudhary, Pasricha, Ahmad, & Sastry, 2006). Crude latex from medicinal plant *J. curcas* L. was used for the successful synthesis of stable antibacterial silver nanoparticles (AgNPs) from $AgNO_3$ (Bar et al., 2009). In another study, crude aqueous latex extract of *Calotropis gigantea* L. plant was used for the green synthesis of AgNPs for biomedical therapeutic applications (Rajkuberan, Sudha, Sathishkumar, & Sivaramakrishnan, 2015). The scheme of the green synthesis approach of nanoparticles is presented on Fig. 8.

Latex-bearing plants are a rich source of pharmacologically active natural compounds. Due to their antimicrobial, antioxidant, antitumor properties they are also used in the process of biofunctionalization of nanomaterials. Polymeric electro-spun nanofibers with medicinal plant extracts can be applied in medicine (particularly as scaffolds in regenerative medicine and tissue engineering), pharmaceutical and cosmeceutical industries. Many of

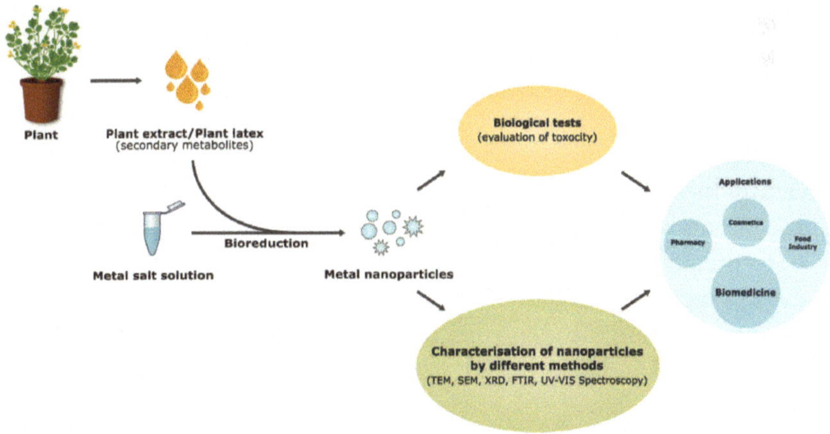

Fig. 8 The scheme of the green synthesis approach of nanoparticles. Plant-mediated nanoparticles can be used in biomedicine and as potential pharmaceutical agents. Synthesis of metal nanoparticles by using plant extracts or plant latex.

the natural, bioactive plant components possess poor bioavailability, i.e. curcumin or berberine. The use of nanocarriers and incorporating the alkaloids into biocompatible and polymer-based nanofibers provides an option to improve its stability, bioavailability and release in controlled rate (Khan, Xiangyang, Ahmad, & Mo, 2018). Electro-spun polymer-based nanofibers can be functionalized with natural products and used as a potential material for growth and differentiation of neuronal cells (Requejo-Aguilar et al., 2017). Incorporated natural compounds from medicinal plants can improve antibacterial ability, antioxidant properties and biocompatibility of obtained nanomaterials. Furthermore, they can modify their physical properties through structural and morphological changes. Additionally, using nanocarriers approach, like plant metabolites, functionalized nanodiamonts open a new way to use the plant bioactive molecules against cancer (Gismondi et al., 2015).

To sum up, the biosynthesis of nanoparticles using green resources, like the plant latex, is an alternative approach comparing to chemical synthesis of nanomaterials. Pharmacologically active compounds from extracts or latex of plants can be potentially utilized for functionalization of nanomaterials. Biofunctionalization improves availability and stability of natural compounds and their derivatives. Thus, it could improve the application potential of nanomaterials and plant natural products in medicine and pharmaceutical industry (Ocsoy et al., 2018).

4. Conclusions

In recent times the use of plant natural products has increased tremendously, and there is a very high demand of herbal products for therapeutic, clinical and agricultural purposes. Plant latex is a natural source of diverse secondary metabolites with unique pharmacological properties. Thus, plant latex compounds are potential candidates for bioactive agents in pharmaceutical and medical practice, to promote prevention and cure of various human diseases. Many of bioactive compounds from plant latex act not only as antiherbivorous and plant antiviral factors, but show wide spectrum of the multiple pharmacologic activities, such as antimicrobial, anti-inflammatory and anticancer. Different classes of active compounds from latex such as flavonoids, alkaloids, saponins, tannins, acetogenins and cardiac glycosides have been described as natural plant agents with positive effects on human health. Alkaloids from the latex-bearing plants show antidiabetic activity (berberine and its derivatives) and possess potential therapeutic effects against neurodegenerative diseases (berberine, morphine).

It is now recognized that isoquinoline alkaloids, like papaverine or morphine, which are stored in latex from *Papaver* sp., have strong pharmacological properties, long medicinal history and have been used for various medicinal purposes. Furthermore, based on recent scientific reports, the complex composition of different active molecules, like alkaloids and proteins in *C. majus* latex, contributes significantly to broad spectrum of its bioactivity (antimicrobial, antiviral, anticancer) and underlines the long history of its use in folk medicine. Similarly, *E. peplus* L. white latex sap has been used traditionally to treat skin warts and corns. The phytochemical study on the latex from this plant identified one of the most active anticancer natural compounds — ingenol mebutate. This diterpene acts effective against actinic keratosis, the precursor of squamous cell carcinoma. Vincristine, a naturally-occurring vinca alkaloid obtained from another latex bearing plant *Cataranthus roseus* L., is used in oncology to treat leukemia, neuroblastoma and lung cancer (Seca and Pinto, 2018).

Based on these facts, latex-bearing medicinal plants are interesting and useful for the pharmaceutical industry. Herein we conclude, that latex-bearing plants are a rich source of pharmacologically active compounds, which should be considered as natural therapeutic agents for combination therapy. Moreover, their modification using nanotechnological solution approaches, e.g. liposomes as nanocarriers, is still the possible strategic way to increase bioavailability, selectivity, distribution and improve the therapeutic applications of pharmaceutically active, natural latex compounds.

Acknowledgments

The authors acknowledge financial support from Polish National Science Center grant no. 2012/05/N/NZ9/01337.

References

Abotaleb, M., Samuel, S. M., Varghese, E., Varghese, S., Kubatka, P., Liskova, A., et al. (2019). Flavonoids in cancer and apoptosis. *Cancers, 11*(1). https://doi.org/10.3390/cancers11010028.

Agrawal, A. A., & Konno, K. (2016). All use subject to JSTOR terms and conditions latex : A model for understanding mechanisms. *Ecology, and Evolution of Plant Defense Against Herbivory, 40*(2009), 311–331. https://doi.org/10.1146/annurev.ecolsys.l.

Arora, D., & Sharma, A. (2013). A review on phytochemical and pharmacological potential of genus *Chelidonium*. *Pharmacognosy Journal, 5*(4), 184–190. https://doi.org/10.1016/j.phcgj.2013.07.006.

Asamenew, G., Bisrat, D., Mazumder, A., & Asres, K. (2011). In vitro antimicrobial and antioxidant activities of anthrone and chromone from the latex of *Aloe harlana* reynolds. *Phytotherapy Research, 25*(12), 1756–1760. https://doi.org/10.1002/ptr.3482.

Bar, H., Bhui, D. K., Sahoo, G. P., Sarkar, P., De, S. P., & Misra, A. (2009). Green synthesis of silver nanoparticles using latex of *Jatropha curcas*. *Colloids and Surfaces A: Physicochemical and Engineering Aspects, 339*(1–3), 134–139. https://doi.org/10.1016/j.colsurfa.2009.02.008.

Barnes, J., Anderson, L., & Phillipson, J. (2007). *Herbal medicines* (3rd ed.). London: London : Pharmaceutical Press.

Beaudoin, G. A., & Facchini, P. J. (July 2014). Benzylisoquinoline alkaloid biosynthesis in opium poppy. *Planta, 240*(1), 19–32.

Beezhold, D. H., Hickey, V. L., Kostyal, D. A., Puhl, H., Zuidmeer, L., van Ree, R., et al. (April 2003). Lipid transfer protein from *Hevea brasiliensis* (Hev b 12), a cross-reactive latex protein. *Annals of Allergy, Asthma, & Immunology, 90*(4), 439–445.

Biscaro, F., Parisotto, E. B., Zanette, V. C., Günther, T. M., Ferreira, E. A., Gris, E. F., et al. (2013). Anticancer activity of flavonol and flavan-3-ol rich extracts from Croton celtidifolius latex. *Pharmaceutical Biology, 51*(6), 737–743.

Biswas, S. J. (2013). Chelidonium majus L. - a review on pharmacological activities and clinical effects. *Global Journal of Research on Medicinal Plants and Indigenous Medicine, 2*, 238–245.

Cazzaniga, M. E., Camerini, A., Addeo, R., Nolè, F., Munzone, E., Collovà, E., et al. (February 2016). Metronomic oral vinorelbine in advanced breast cancer and non-small-cell lung cancer: Current status and future development. *Future Oncology, 12*(3), 373–387. https://doi.org/10.2217/fon.15.306.

Chandran, S., Chaudhary, M., Pasricha, R., Ahmad, A., & Sastry, M. (2006). Synthesis of gold nanotriangles and silver nanoparticles using *Aloe vera* plant extract. *Biotechnology Progress, 22*(2), 577–583.

Chavan Bhagyashri, A., Hundiwale Jogendra, C., & Patil Avinash, V. (2015). Plant latex: An inherent spring of pharmaceuticals. *World Journal of Pharmacy and Pharmaceutical Sciences, 4*(04), 1781–1796.

Cheeke, P. R., Piacente, S., & Oleszek, W. (2006). Antiinflammatory and antiarthritic effects of Yucca schidigera: A review. *Journal of Inflammation, 3*, 6. https://doi.org/10.1186/1476925536.

Colombo, M. L., & Bosisio, E. (1996). Pharmacological activities of *Chelidonium majus* L. (papaveraceae). *Pharmacological Research, 33*(2), 127–134. https://doi.org/10.1006/phrs.1996.0019.

Cruz, D., Falé, P. L., Mourato, A., Vaz, P. D., Luisa Serralheiro, M., & Lino, A. R. L. (2010). Preparation and physicochemical characterization of Ag nanoparticles biosynthesized by *Lippia citriodora* (Lemon Verbena). *Colloids and Surfaces B: Biointerfaces, 81*(1), 67–73. https://doi.org/10.1016/j.colsurfb.2010.06.025.

Daruliza, K. M. A., Lam, K. L., Yang, K. L., Priscilla, J. T., Sunderasan, E., & Ong, M. T. (2011). Anti-fungal effect of *Hevea brasiliensis* latex C-serum on *Aspergillus niger*. *European Review for Medical and Pharmacological Sciences, 15*(9), 1027–1033.

De Marino, S., Gala, F., Zollo, F., Vitalini, S., Fico, G., Visioli, F., et al. (2008). Identification of minor secondary metabolites from the latex of *Croton lechleri* (Muell-Arg) and evaluation of their antioxidant activity. *Molecules, 13*(6), 1219–1229. https://doi.org/10.3390/molecules13061219.

Decker, G., Wanner, G., Zenk, M. H., & Lottspeich, F. (2000). Characterization of proteins in latex of the opium poppy (Papaver somniferum) using two-dimensional gel electrophoresis and microsequencing. *Electrophoresis, 21*(16), 3500–3516. https://doi.org/10.1002/1522-2683(20001001)21:16<3500::AID-ELPS3500>3.0.CO;2-O.

Dixon, R. A., & Pasinetti, G. M. (2010). Flavonoids and isoflavonoids: From plant biology to agriculture and neuroscience. *Plant Physiology, 154*(2), 453–457. https://doi.org/10.1104/pp.110.161430.

Duo, J., Ying, G. G., Wang, G. W., & Zhang, L. (2012). Quercetin inhibits human breast cancer cell proliferation and induces apoptosis via Bcl-2 and Bax regulation. *Molecular Medicine Reports, 5*(6), 1453–1456. https://doi.org/10.3892/mmr.2012.845.

El-Mallakh, R. S., Brar, K. S., & Yeruva, R. R. (2019). Cardiac glycosides in human physiology and disease: Update for entomologists. *Insects, 10*(4). https://doi.org/10.3390/insects10040102.

Elfawal, M. A., Towler, M. J., Reich, N. G., Golenbock, D., Weathers, P. J., & Rich, S. M. (2012). Dried whole plant *Artemisia annua* as an antimalarial therapy. *PLoS One,* 7(12), 1–7. https://doi.org/10.1371/journal.pone.0052746.

Etxenagusia, M., Anda, M., González-Mahave, I., Fernández, E., & Fernández de Corrès, L. (2000). No Contact dermatitis from *Chelidonium majus* (greater celandine). *Contact Dermatitis, 43*(1), 47.

European Commission Joint Research Centre. (2011). *Annual Report 2011 – EUR 25231 EN.* Luxembourg: Publications Office of European Union. https://ec.europa.eu/jrc/sites/jrcsh/files/jrc_ar_2011.pdf.

Felpin, F. X., & Lebreton, J. (2004). History, chemistry and biology of alkaloids from *Lobelia inflata. Tetrahedron, 60*(45), 10127–10153. https://doi.org/10.1016/S0040-4020(04)01294-3.

Fernandes, H., Michalska, K., Sikorski, M., & Jaskolski, M. (2013). Structural and functional aspects of PR-10 proteins. *FEBS Journal, 280*(5), 1169–1199. https://doi.org/10.1111/febs.12114.

Finkina, E. I., Melnikova, D. N., Bogdanov, I. V., & Ovchinnikova, T. V. (2016). Plant pathogenesis-related proteins PR-10 and PR-14 as components of innate immunity system and Ubiquitous allergens. *Current Medicinal Chemistry, 24*(17), 1772–1787. https://doi.org/10.2174/0929867323666161026154111.

Fister, A. S., Mejia, L. C., Zhang, Y., Herre, E. A., Maximova, S. N., & Guiltinan, M. J. (2016). *Theobroma cacao* L. pathogenesis-related gene tandem array members show diverse expression dynamics in response to pathogen colonization. *BMC Genomics, 17*(1), 1–16. https://doi.org/10.1186/s12864-016-2693-3.

Freitas, C. D. T., Leite, H. B., Oliveira, J. P. B., Amaral, J. L., Egito, A. S., Vairo-Cavalli, S., et al. (2016b). Insights into milk-clotting activity of latex peptidases from *Calotropis procera* and *Cryptostegia grandiflora. Food Research International, 87*, 50–59.

Freitas, C. D. T., Silva, R. O., Ramos, M. V., Porfirio, C. T. M. N., Farias, D. F., Sousa, J. S., et al. (October 9, 2019). Identification, characterization, and antifungal activity of cysteine peptidases from *Calotropis procera* latex. *Phytochemistry, 169*, 112163.

Freitas, C. D. T., Viana, C. A., Vasconcelos, I. M., Moreno, F. B., Lima-Filho, J. V., Oliveira, H. D., et al. (2016a). First insights into the diversity and functional properties of chitinases of the latex of *Calotropis procera. Plant Physiology and Biochemistry, 108*, 361–371.

Ghanbari, A., Le Gresley, A., Naughton, D., Kuhnert, N., Sirbu, D., & Ashrafi, G. H. (2019). Biological activities of *Ficus carica* latex for potential therapeutics in Human Papillomavirus (HPV) related cervical cancers. *Scientific Reports, 9*(1), 1–11. https://doi.org/10.1038/s41598-018-37665-6.

Ghasemzadeh, A. (2011). Flavonoids and phenolic acids: Role and biochemical activity in plants and human. *Journal of Medicinal Plants Research, 5*(31), 6697–6703. https://doi.org/10.5897/jmpr11.1404.

Gilca, M., Gaman, L., Panait, E., Stoian, I., & Atanasiu, V. (2010). *Chelidonium majus* - an integrative review: Traditional knowledge versus modern findings. *Forschende Komplementärmedizin, 17*(5), 241–248. https://doi.org/10.1159/000321397.

Gismondi, A., Reina, G., Orlanducci, S., Mizzoni, F., Gay, S., Terranova, M. L., et al. (2015). Nanodiamonds coupled with plant bioactive metabolites: A nanotech approach for cancer therapy. *Biomaterials, 38*, 22–35. https://doi.org/10.1016/j.biomaterials.2014.10.057.

González-Esquinca, A. R., De-La-Cruz-Chacón, I., Castro-Moreno, M., Orozco-Castillo, J. A., & Riley- Saldaña, C. A. (2014). Alkaloids and acetogenins in Annonaceae development: Biological considerations. *Revista Brasileira de Fruticultura, 36*(spe1), 01–16. https://doi.org/10.1590/s0100-29452014000500001.

Groeneveld, H. W. (1982). Cardenolide synthesis of in and the triterpene laticifers Asclepias curassavica glycosides quantities Calotropis (Hesse Hoya Euphorbia mainly synthesised appeared (Groeneveld only (Groeneveld 1982). *Up CoA Reichstein Synthesis Triterpenes Asclepias, 31*(February), 5–10.

Grosso, C., Ferreres, F., Gil-Izquierdo, A., Valentão, P., Sampaio, M., Lima, J., et al. (2014). Box-Behnken factorial design to obtain a phenolic-rich extract from the aerial parts of *Chelidonium majus* L. *Talanta, 130*, 128–136. https://doi.org/10.1016/j.talanta.2014.06.043.

Guclu-Ustundag, Ö., & Mazza, G. (2007). Saponins: Properties, applications and processing. *Critical Reviews in Food Science and Nutrition, 47*(3), 231–258. https://doi.org/10.1080/10408390600698197.

Gu, R., Wang, Y., Long, B., Kennelly, E., Wu, S., Liu, B., et al. (2014). Pharmacologically active constituents from plants used in traditional medicine prospecting for bioactive constituents from traditional medicinal plants through ethnobotanical approaches. *Biological and Pharmaceutical Bulletin, 37*(June), 903–915.

Hagel, J. M., & Facchini, P. J. (2013). Benzylisoquinoline alkaloid metabolism: A century of discovery and a brave new world. *Plant and Cell Physiology, 54*(5), 647–672. https://doi.org/10.1093/pcp/pct020.

Hemmer, W., Focke, M., Marzban, G., Swoboda, I., Jarisch, R., & Laimer, M. (April 2010). Identification of Bet v 1 related allergens in fig and other Moraceae fruits. *Clinical and Experimental Allergy, 40*(4), 679–687.

Hussain, G., Rasul, A., Anwar, H., Aziz, N., Razzaq, A., & Wei, W. (2018). *Role of plant derived alkaloids and their mechanism in neurodegenerative disorders* (Vol. 14). https://doi.org/10.7150/ijbs.23247.

Hussain, M., Debnath, B., Qasim, M., Bamisile, B. S., Islam, W., Hameed, M. S., et al. (2019). Role of Saponins in Plant Defense Against Specialist Herbivores. *Molecules, 24*, 2067. https://doi.org/10.3390/molecules24112067.

Jakovljević, Z. D., Stanković, S. M., & Topuzović, D. M. (2013). Seasonal variability of *Chelidonium majus* L. secondary metabolites content and antioxidant activity. *EXCLI Journal, 12*, 260–268.

Jantova, S., Cipak, L., Cernakova, M., & Kostalova, D. (2003). *Effect of berberine on proliferation, cell cycle and apoptosis in HeLa and L1210 cells.* https://doi.org/10.1211/0022357021422.

Jesus, M., Martins, A. P., Gallardo, E., & Silvestre, S. (2016). Diosgenin: recent highlights on pharmacology and analytical methodology. *Journal of Analytical Methods in Chemistry, 2016*, 4156293. https://doi.org/10.1155/2016/4156293.

Jucá, M. M., Cysne Filho, F. M. S., de Almeida, J. C., Mesquita, D. D. S., Barriga, J. R. M., Dias, K. C. F., et al. (November 2018). Flavonoids: Biological activities and therapeutic potential. *Natural Product Research, 16*, 1–14. https://doi.org/10.1080/14786419.2018.1493588.

Jucá, T. L., Ramos, M. V., Moreno, F. B. M. B., Viana De Matos, M. P., Marinho-Filho, J. D. B., Moreira, R. A., et al. (2013). Insights on the phytochemical profile (cyclopeptides) and biological activities of calotropis procera latex organic fractions. *The Scientific World Journal, 2013*. https://doi.org/10.1155/2013/615454.

Kavoosi, G., Tafsiry, A., Ebdam, A. A., & Rowshan, V. (2013). Evaluation of antioxidant and antimicrobial activities of essential oils from *Carum copticum* seed and *Ferula assafoetida* latex. *Journal of Food Science, 78*(2). https://doi.org/10.1111/1750-3841.12020.

Kazemi, S., Esmaeili, H., Abachi, F., Khiali, S., Islam, B., & Haider, S. (April 2017). Selectivity of major isoquinoline alkaloids from *Chelidonium majus* towards telomeric G-quadruplex : A study using a transition-FRET (t-FRET) assay. *Biochimica et Biophysica Acta (BBA) - General Subjects, 1861*, 2020–2030. https://doi.org/10.1016/j.bbagen.2017.05.002.

Khaleghian, A., Rezaie, M., Nazari, H., Riazi, G. H., Ghafari, M., Takahashi, A., et al. (2010). Effect of inganen anticancer properties on microtobule organization. *Pakistan journal of pharmaceutical sciences, 23*(3), 273–278.

Khanbabaee, K., & van Ree, T. (2001). Tannins: Classification and definition. *Natural Product Reports, 18*(6), 641–649. https://doi.org/10.1039/b101061l.

Khandel, P., Yadaw, R. K., Soni, D. K., Kanwar, L., & Shahi, S. K. (2018). Biogenesis of metal nanoparticles and their pharmacological applications: Present status and application prospects. *Journal of Nanostructure in Chemistry, 8*. https://doi.org/10.1007/s40097-018-0267-4.

Khan, A., Xiangyang, S., Ahmad, A., & Mo, X. (2018). Electrospinning of crude plant extracts for antibacterial and wound healing applications: A review. *SM Journal of Biomedical Engineering, 4*(1), 1–8. Retrieved from https://smjournals.com/biomedical-engineering/fulltext/smjbe-v4-1024.pdf.

Kinda, P., Zerbo, P., Guenné, S., Compaoré, M., Ciobica, A., & Kiendrebeogo, M. (2017). Medicinal plants used for neuropsychiatric disorders treatment in the Hauts Bassins region of Burkina Faso. *Medicines, 4*(2), 32. https://doi.org/10.3390/medicines4020032.

Kitajima, S., Miura, K., Aoki, W., Yamato, K. T., Taira, T., Murakami, R., et al. (2016). Transcriptome and proteome analyses provide insight into laticifer's defense of *Euphorbia tirucalli* against pests. *Plant Physiology and Biochemistry, 108*, 434–446. https://doi.org/10.1016/j.plaphy.2016.08.008.

Konno, K. (2011). Plant latex and other exudates as plant defense systems: Roles of various defense chemicals and proteins contained therein. *Phytochemistry, 72*(13), 1510–1530. https://doi.org/10.1016/j.phytochem.2011.02.016.

Konno, K., Hirayama, C., Nakamura, M., Tateishi, K., Tamura, Y., Hattori, M., et al. (2004). Papain protects papaya trees from herbivorous insects: Role of cysteine proteases in latex. *The Plant Journal, 37*(3), 370–378. https://doi.org/10.1046/j.1365-313X.2003.01968.x.

Konno, K., Ono, H., Nakamura, M., Tateishi, K., Hirayama, C., Tamura, Y., et al. (2006). Mulberry latex rich in antidiabetic sugar-mimic alkaloids forces dieting on caterpillars. *Proceedings of the National Academy of Sciences, 103*(5), 1337–1341. https://doi.org/10.1073/pnas.0506944103.

Kumar, S. S. (2018). Noscapine - a novel anticancer drug for the treatment of primary/secondary brain tumor. *Asian Journal of Pharmaceutics, 12*(2), S428–S432. https://doi.org/10.22377/ajp.v12i02.2370.

Lee, M. F., Wang, N. M., Han, J. L., Lin, S. J., Tsai, J. J., & Chen, Y. H. (2010). Estimating allergenicity of latex gloves using Hev b 1 and hevamine. *Journal of Investigational Allergology and Clinical Immunology, 20*(6), 499–505.

Lewinsohn, T. M. (1991). The geographical distribution of plant latex. *Chemoecology, 2*, 64–68. https://doi.org/10.1016/j.flora.2010.10.002.

Liu, J., Li, Y., Shi, H., Wang, T., Wu, X., Sun, X., et al. (2016). Components characterization of total tetraploid jiaogulan (*Gynostemma pentaphyllum*) saponin and its cholesterol-lowering properties. *Journal of Functional Foods, 23*, 542–555. https://doi.org/10.1016/j.jff.2016.03.013.

Liu, Q., Tang, J. S., Hu, M. J., Liu, J., Chen, H. F., Gao, H., et al. (2013). Antiproliferative cardiac glycosides from the latex of *Antiaris toxicaria*. *Journal of Natural Products, 76*(9), 1771–1780. https://doi.org/10.1021/np4005147.

van Loon, L. C., Rep, M., & Pieterse, C. M. J. (2006). Significance of inducible defense-related proteins in infected plants. *Annual Review of Phytopathology, 44*(1), 135–162. https://doi.org/10.1146/annurev.phyto.44.070505.143425.

Luiza Andreazza, N., Vevert-Bizet, C., Bourg-Heckly, G., Sureau, F., Jose Salvador, M., & Bonneau, S. (2016). Berberine as a photosensitizing agent for antitumoral photodynamic therapy : Insights into its association to low density lipoproteins. *International Journal of Pharmaceutics*. https://doi.org/10.1016/j.ijpharm.2016.06.009.

Maji, A. K., & Banerji, P. (2015). *Chelidonium majus* L. (greater celandine) – a review on its phytochemical and therapeutic perspectives. *3*(1), 10–27.

Mandal, S. M., Chakraborty, D., & Dey, S. (2010). Phenolic acids act as signaling molecules in plant-microbe symbioses. *Plant Signaling & Behavior, 5*(4), 359–368. https://doi.org/10.4161/psb.5.4.10871.

Manoorkar, V. B., & Gachande, B. D. (2015). Phytochemical analysis of some plant latex. *International Journal of Life Sciences, 3*(1), 108–110.

Marinho, D. G., Alviano, D. S., Matheus, M. E., Alviano, C. S., & Fernandes, P. D. (2011). The latex obtained from *Hancornia speciosa* Gomes possesses anti-inflammatory activity. *Journal of Ethnopharmacology, 135*(2), 530–537. https://doi.org/10.1016/j.jep.2011.03.059.

Matsui, Y., Takeuchi, T., Kumamoto-Yonezawa, Y., Takemura, M., Sugawara, F., Yoshida, H., et al. (2010). The relationship between the molecular structure of natural acetogenins and their inhibitory activities which affect DNA polymerase, DNA topoisomerase and human cancer cell growth. *Experimental and Therapeutic Medicine, 1*(1), 19–26.

Matsuura, H. N., & Fett-neto, A. G. (2015). Plant alkaloids : Main features, toxicity , and mechanisms of action. *Plant Toxins*, 1–15. https://doi.org/10.1007/978-94-007-6728-7.

Ma, W., Zhu, M., Zhang, D., Yang, L., Yang, T., Li, X., et al. (2017). Berberine inhibits the proliferation and migration of breast cancer ZR-75-30 cells by targeting Ephrin-B2. *Phytomedicine, 25*, 45–51. https://doi.org/10.1016/j.phymed.2016.12.013.

Melo-Reis, P., Andrade, L., Silva, C., Araújo, L., Pereira, M., Mrue, F., et al. (2010). Angiogenic activity of *Synadenium umbellatum* Pax latex. *Brazilian Journal of Biology, 70*(1), 189–194. https://doi.org/10.1590/s1519-69842010000100026.

Men, X., Wang, F., Chen, G., Zhang, H., & Xian, M. (2019). *Biosynthesis of natural Rubber : Current state and perspectives*. https://doi.org/10.3390/ijms20010050.

Mogensen, J. E., Wimmer, R., Larsen, J. N., Spangfort, M. D., & Otzen, D. E. (2002). The major birch allergen, Bet v 1, shows affinity for a broad spectrum of physiological ligands. *Journal of Biological Chemistry, 277*(26), 23684–23692. https://doi.org/10.1074/jbc.M202065200.

Mohamed, N. H., Ismail, M. A., Abdel-Mageed, W. M., & Mohamed Shoreit, A. A. (2014). Antimicrobial activity of latex silver nanoparticles using *Calotropis procera*. *Asian Pacific Journal of Tropical Biomedicine, 4*(11), 876–883. https://doi.org/10.12980/apjtb.4.201414b216.

Mahmoud, S. Y. M., Gad-Rab, S. M. F., Hussein, N., & Shoreit, A. A. M. (2010). Antiviral Activity of Latex from Ficus nitida Against Plant Viruses. *Global Journal of Biotechnology & Biochemistry, 5*(3), 198–205.

Moses, T., Papadopoulou, K. K., & Osbourn, A. (2014). Metabolic and functional diversity of saponins, biosynthetic intermediates and semi-synthetic derivatives. *Critical Reviews in Biochemistry and Molecular Biology, 49*(6), 439–462. https://doi.org/10.3109/10409238.2014.953628.

Moudi, M., Go, R., Yien, C., & Nazre, M. (2013). Vinca alkaloids. *International Journal of Preventive Medicine, 4*(11), 1231–1235.

Munro, B., Vuong, Q., Chalmers, A., Goldsmith, C., Bowyer, M., & Scarlett, C. (2015). Phytochemical, antioxidant and anti-cancer properties of *Euphorbia tirucalli* methanolic and aqueous extracts. *Antioxidants, 4*(4), 647–661. https://doi.org/10.3390/antiox4040647.

Musidlak, O., Nawrot, R., & Goździcka-Józefiak, A. (2017). Which plant proteins are involved in antiviral defense? Review on in vivo and in vitro activities of selected plant proteins against viruses. *International Journal of Molecular Sciences, 18*(11). https://doi.org/10.3390/ijms18112300.

Nalwaya, N., Pokhorna, G., Deb, L., & Jain, N. K. (2009). Wound healing activity of latex of Calotropis gigantea. *International Journal of Pharmacy and Pharmaceutical Sciences, 1*(1), 176–181.

Nawrot, R., Barylski, J., Lippmann, R., Altschmied, L., & Mock, H. P. (2016). Combination of transcriptomic and proteomic approaches helps to unravel the protein composition of *Chelidonium majus* L. milky sap. *Planta, 244*(5), 1055–1064. https://doi.org/10.1007/s00425-016-2566-7.

Nawrot, R., Kalinowski, A., & Gozdzicka-Jozefiak, A. (2007). Proteomic analysis of *Chelidonium majus* milky sap using two-dimensional gel electrophoresis and tandem mass spectrometry. *Phytochemistry, 68*(12), 1612–1622. https://doi.org/10.1016/j.phytochem.2007.03.039.

Nawrot, R., Lesniewicz, K., Pienkowska, J., & Gozdzicka-Jozefiak, A. (2007). A novel extracellular peroxidase and nucleases from a milky sap of *Chelidonium majus*. *Fitoterapia, 78*(7–8), 496–501. https://doi.org/10.1016/j.fitote.2007.04.012.

Nawrot, R., Lippmann, R., Matros, A., Musidlak, O., Nowicki, G., & Mock, H. P. (2017). Proteomic comparison of *Chelidonium majus* L. latex in different phases of plant development. *Plant Physiology and Biochemistry, 112*, 312–325. https://doi.org/10.1016/j.plaphy.2017.01.010.

Nawrot, R., Zauber, H., & Schulze, W. X. (2014). Global proteomic analysis of *Chelidonium majus* and *Corydalis cava* (Papaveraceae) extracts revealed similar defense-related protein compositions. *Fitoterapia, 94*, 77–87. https://doi.org/10.1016/j.fitote.2014.01.015.

Ocsoy, I., Tasdemir, D., Mazicioglu, S., & Tan, W. (2018). Nanotechnology in plants. *Advances in Biochemical Engineering, 164*, 263–275. https://doi.org/10.1007/10_2017_53.

Ono, K., Yoshiike, Y., Takashima, A., Hasegawa, K., Naiki, H., & Yamada, M. (2003). Potent anti-amyloidogenic and fibril-destabilizing effects of polyphenols in vitro: Implications for the prevention and therapeutics of Alzheimer's disease. *Journal of Neurochemistry, 87*(1), 172–181. https://doi.org/10.1046/j.1471-4159.2003.01976.x.

Özbilgin, S., Acıkara, Ö. B., Akkol, E. K., Süntar, I., Keleş, H., & İşcan, G. S. (2018). In vivo wound-healing activity of *Euphorbia characias* subsp. wulfenii: Isolation and quantification of quercetin glycosides as bioactive compounds. *Journal of Ethnopharmacology, 224*, 400–408. https://doi.org/10.1016/j.jep.2018.06.015.

Panche, A. N., Diwan, A. D., & Chandra, S. R. (2016). Flavonoids: An overview. *Journal of Nutritional Science, 5*. https://doi.org/10.1017/jns.2016.41.

Paris, D., Mathura, V., Ait-Ghezala, G., Beaulieu-Abdelahad, D., Patel, N., Bachmeier, C., et al. (2012). Flavonoids lower Alzheimer's Aß production via an NFkB dependent mechanism. *Bioinformation, 6*(6), 229–236. https://doi.org/10.6026/97320630006229.

Peralta-Videa, J. R., Huang, Y., Parsons, J. G., Zhao, L., Lopez-Moreno, L., Hernandez-Viezcas, J. A., et al. (2016). Plant-based green synthesis of metallic nanoparticles: Scientific curiosity or a realistic alternative to chemical synthesis? *Nanotechnology for Environmental Engineering, 1*(1), 1–29. https://doi.org/10.1007/s41204-016-0004-5.

Priya, P., Yadav, A., Chand, J., & Yadav, G. (January 10, 2018). Terzyme: A tool for identification and analysis of the plant terpenome. *Plant Methods, 14*, 4. https://doi.org/10.1186/s13007-017-0269-0.

Rajkuberan, C., Sudha, K., Sathishkumar, G., & Sivaramakrishnan, S. (2015). Antibacterial and cytotoxic potential of silver nanoparticles synthesized using latex of *Calotropis gigantea*

L. *Spectrochimica Acta Part A: Molecular and Biomolecular Spectroscopy, 136*(PB), 924–930. https://doi.org/10.1016/j.saa.2014.09.115.

Ramos, M. V., Demarco, D., & Cristina, I. (2019). Laticifers, latex, and their role in plant defense. *Trends in Plant Science, 24*(6), 553–567. https://doi.org/10.1016/j.tplants.2019.03.006.

Requejo-Aguilar, R., Alastrue-Agudo, A., Cases-Villar, M., Lopez-Mocholi, E., England, R., Vicent, M. J., & Moreno-Manzano, V. (2017). Combined polymer-curcumin conjugate and ependymal progenitor/stem cell treatment enhances spinal cord injury functional recovery. *Biomaterials, 113*, 18–30. https://doi.org/10.1016/j.biomaterials.2016.10.032.

Sangeetha, G., Rajeshwari, S., & Venckatesh, R. (2011). Green synthesis of zinc oxide nanoparticles by aloe barbadensis miller leaf extract: Structure and optical properties. *Materials Research Bulletin, 46*(12), 2560–2566. https://doi.org/10.1016/j.materresbull.2011.07.046.

Schrot, J., Weng, A., & Melzig, M. F. (2015). Ribosome-inactivating and related proteins. *Toxins,* 7(5), 1556–1615. https://doi.org/10.3390/toxins7051556.

Seca, A. M. L., & Pinto, D. C. G. A. (January 16, 2018). Plant secondary metabolites as anticancer agents: Successes in clinical trials and therapeutic application. *International Journal of Molecular Sciences, 19*(1), E263. https://doi.org/10.3390/ijms19010263.

Sels, J., Mathys, J., De Coninck, B. M. A., Cammue, B. P. A., & De Bolle, M. F. C. (2008). Plant pathogenesis-related (PR) proteins: A focus on PR peptides. *Plant Physiology and Biochemistry, 46*(11), 941–950. https://doi.org/10.1016/j.plaphy.2008.06.011.

Sharafi, G., He, H., & Nikfarjam, M. (2019). Potential use of cannabinoids for the treatment of pancreatic cancer. *Journal of Pancreatic Cancer, 5*, 1–7. https://doi.org/10.1089/pancan.2018.0019.

Shi, W. W., Mak, A. N. S., Wong, K. B., & Shaw, P. C. (2016). Structures and ribosomal interaction of ribosome-inactivating proteins. *Molecules, 21*(11). https://doi.org/10.3390/molecules21111588.

Shuntang, G. (2018). Current topics in saponins and the bitter taste. *Research in Medical & Engineering Sciences, 5*(1), 390–391. https://doi.org/10.31031/rmes.2018.05.000601.

Sikander, M., Hafeez, B. B., Malik, S., Alsayari, A., Halaweish, F. T., Yallapu, M. M., et al. (2016). Cucurbitacin D exhibits potent anti-cancer activity in cervical cancer. *Scientific Reports, 6*. https://doi.org/10.1038/srep36594.

Silva, A. T., Nguyen, A., Ye, C., Verchot, J., & Moon, J. H. (2010). Conjugated polymer nanoparticles for effective siRNA delivery to tobacco BY-2 protoplasts. *BMC Plant Biology, 10*(1), 291. https://doi.org/10.1186/1471-2229-10-291.

Singh, B., & Sharma, R. A. (April 2015). Plant terpenes: Defense responses, phylogenetic analysis, regulation and clinical applications. *3 Biotech, 5*(2), 129–151. https://doi.org/10.1007/s13205-014-0220-2.

Sinha, M., Singh, R. P., Kushwaha, G. S., Iqbal, N., Singh, A., Kaushik, S., et al. (2014). Current overview of allergens of plant pathogenesis related protein families. *The Scientific World Journal, 2014*. https://doi.org/10.1155/2014/543195.

Siritapetawee, J., Thammasirirak, S., & Samosornsuk, W. (2012). Antimicrobial activity of a 48-kDa protease (AMP48) from *Artocarpus heterophyllus* latex. *European Review for Medical and Pharmacological Sciences, 16*(1), 132–137.

Stancic-Rotaru, M., Mititelu, M., Crasmaru, M., & Balaban, D. (2003). Spectroanalytical profile of flavonoids from *Chelidonium majus* L. *Romanian Biotechnological Letters, 8*, 1093–1100.

Sun, H. X., Xie, Y., & Ye, Y. P. (March 13, 2009). Advances in saponin-based adjuvants. *Vaccine, 27*(12), 1787–1796. https://doi.org/10.1016/j.vaccine.2009.01.091.

Takshak, S. (2018). Bioactive compounds in medicinal plants: A condensed review. *SEJ Pharmacognosy, 1*(1).

Tomè, F., & Colombo, M. L. (1995). Distribution of alkaloids in *Chelidonium majus* and factors affecting their accumulation. *Phytochemistry, 40*(1), 37–39. https://doi.org/10.1016/0031-9422(95)00055-C.

Tungmunnithum, D., Thongboonyou, A., Pholboon, A., & Yangsabai, A. (2018). Flavonoids and other phenolic compounds from medicinal plants for pharmaceutical and medical aspects: An overview. *Medicines, 5*(3), 93. https://doi.org/10.3390/medicines5030093.

Upadhyay, R. (2011). Plant latex: A natural source of pharmaceuticals and pesticides. *International Journal of Green Pharmacy, 5*(3), 169. https://doi.org/10.4103/0973-8258.91222.

Upadhyay, R. K. (2013). Latex and plant food allergens , its clinical effects and diagnosis : A review. *3*, 120–131.

Veeramuthu, D., Raja, W., Raja, T., & Al-dhabi, N. A. (2017). *Flavonoids : Anticancer properties, IntechOpen, Chapter 13.* https://doi.org/10.5772/68095.

Walsh, M. J., Dodd, J. E., & Hautbergue, G. M. (2013). Ribosome-inactivating proteins potent poisons and molecular tools. *Virulence, 4*(8). https://doi.org/10.4161/viru.26399.

Wang, H., Li, K., Ma, L., Wu, S., Hu, J., Yan, H., et al. (2017). Berberine inhibits enterovirus 71 replication by downregulating the MEK/ERK signaling pathway and autophagy. *Virology Journal*, 1–8. https://doi.org/10.1186/s12985-016-0674-4.

Warowicka, A., Łukasz, P., Grażyna, B., Oskar, M., Jagoda, L., Dorota, K., et al. (2019). US CR. *Phytomedicine*, 152919. https://doi.org/10.1016/j.phymed.2019.152919.

Wau, J., Timi, D., & Harakuwe, A. (2016). Isolation of three bioactive phenantroindolizidine alkaloids from the fruit latex of *Ficus botryocarpa* Miq. *Natural Products Chemistry & Research, 03*(06), 6–10. https://doi.org/10.4172/2329-6836.1000197.

Xiao, Y., Tian, C., Huang, T., Han, B., Wang, M., Ma, H., et al. (2018). 8-Cetylberberine inhibits growth of lung cancer in vitro and in vivo. *Life Sciences, 192*(1), 259–269. https://doi.org/10.1016/j.lfs.2017.11.012.

Yamashita, S., Yamaguchi, H., Waki, T., Aoki, Y., Mizuno, M., Yanbe, F., et al. (October 28, 2016). Identification and reconstitution of the rubber biosynthetic machinery on rubber particles from *Hevea brasiliensis*. *Elife, 5*, e19022.

Yogi, B., Gupta, S. K., & Mishra, A. (2016). Calotropis procera (Madar): A medicinal plant of various therapeutic uses-A review. *Bulletin of Environment, Pharmacology and Life Sciences, 5*(57), 74–81. Retrieved from https://pdfs.semanticscholar.org/ca82/a0c618880673723a32de3433c9e5b9f73035.pdf.

Zielinska, S., Jezierska-Domaradzka, A., Wójciak-Kosior, M., Sowa, I., Junka, A., & Matkowski, A. M. (2018). Greater celandine's ups and downs-21 centuries of medicinal uses of *Chelidonium majus* from the viewpoint of today's pharmacology. *Frontiers in Pharmacology, 9*(APR). https://doi.org/10.3389/fphar.2018.00299.

Zielinska, S., Wójciak-Kosior, M., Płachno, B. J., Sowa, I., Włodarczyk, M., & Matkowski, A. (2018). Quaternary alkaloids in *Chelidonium majus* in vitro cultures. *Industrial Crops and Products, 123*(June), 17–24. https://doi.org/10.1016/j.indcrop.2018.06.062.

Zulak, K. G., Khan, M. F., Alcantara, J., Schriemer, D. C., & Facchini, P. J. (2009). Plant defense responses in opium poppy cell cultures revealed by liquid chromatography-tandem mass spectrometry proteomics. *Molecular & Cellular Proteomics, 8*(1), 86–98. https://doi.org/10.1074/mcp.m800211-mcp200.

CHAPTER SIX

Latex proteins and enzymes involved in terpenoid biosynthesis of selected *Euphorbia* species: *Euphorbia kansui* Liou and *Euphorbia helioscopia* L.

Xueyan Zhao[a,b] and Xia Cai[a,*]
[a]Key Laboratory of Resource Biology and Biotechnology in Western China, Northwest University, Ministry of Education, Xi'an, China
[b]Xi'an Botanical Garden of Shaanxi Province, Xi'an, China
*Corresponding author: E-mail: caix@nwu.edu.cn

Contents

Advances in Botanical Research, Volume 93
ISSN 0065-2296
https://doi.org/10.1016/bs.abr.2019.09.005

Abstract

Euphorbia is the largest genus in Euphorbiaceae. The most species have non-articulated laticifers. During the development of non-articulated laticifers, some cytoplasm parts degenerate and lots of latex particles are synthesized and accumulated in the central vacuole. Latex proteins play significant roles in the laticifer development and latex particle synthesis. Therefore, latex proteomics of selected *Euphorbia* species: *Euphorbia kansui* Liou and *Euphorbia helioscopia* L. were introduced, especially latex proteins involved in laticifer development and terpenoid biosynthesis. Immunoblot and immunocytochemistry analyses on cysteine protease, clathrin heavy chain and AP-2 complex subunit beta-1 belonging to lysosome pathway showed that they were mainly distributed in vesicles derived from dicytosomes, which are called lysosome-like vesicles, and ubiquitin belonging to the proteasome pathway was widely distributed in the cytosol. As a result, it was speculated that lysosome pathway and proteasome pathway might be involved in the degradation of organelles and some cytoplasmic matrix in *E. helioscopia* laticifers. In addition, acetyl-CoA acetyltransferase, hydroxymethylglutaryl-CoA synthase, hydroxymethylglutaryl-CoA reductase, hydroxymethylglutaryl-CoA lyase, mevalonate disphosphate decarboxylase, isopentenyl diphosphate isomerase and farnesyl pyrophosphate synthase in terpenoid biosynthesis were detected in latex of *Euphorbia* plants. Immunocytochemistry analyses showed that acetyl-CoA acetyltransferase and mevalonate disphosphate decarboxylase were located in cisternae of endoplasmic reticulum, small vacuoles from endoplasmic reticulum and also distributed randomly in protoplasm of laticifer cells. Consequently, the isopentenyl pyrophosphate were speculated to be synthesized in the endoplasmic reticulum, and then transported to parts which dilated into small vacuoles, and the terpenoid was further synthesized into latex particles in these small vacuoles in *E. helioscopia* laticifers.

1. Introduction of latex proteins of *Euphorbia*

The genus *Euphorbia* is one of the largest genera of angiosperms in the world, with the excess in Africa and Central and South America. There are about 2000 species and most of them contain non-articulated laticifers.

The laticifers are specialized secretory structures in plants. They exist in more than 20 families of angiosperms. Latex, the milky-like material contained in laticifers, is an important plant resource used by human beings. Latex contains many unique metabolites, such as natural rubber, which is widely used in human economy and industry; cardiac steroids, alkaloids and terpenoids, which can be used to treat human diseases. Proteins, as a part of protoplasts, also exist in the latex and play significant roles in laticifer differentiation and development as bioactive regulators, as well as in the synthesis of various metabolites in the latex. Therefore, the study of latex proteomics is the theoretical basis for the in-depth study of the occurrence and development of laticifers and the regulation of biosynthesis of bioactive substances.

Fig. 1 Plants of *Euphorbia helioscopia* and *E. kansui.*

With the development of mass spectrometry and proteomics technology, it is possible to elucidate the proteins in laticifers. However, latex protein identification of different *Euphorbia* species is limited because of the lack of genome-sequence information, the extraction techniques of latex proteins and the identification techniques of low abundance proteins.

Euphorbia helioscopia is a biennial herb belonging to the *Euphorbia* genus (Euphorbiaceae), is native to the whole of Europe, Turkey, the Eastern Mediterranean, the Caucasus, Iran, Central Asia, the Himalaya, West India, North Africa and is naturalized in Japan and China (Fig. 1). *E. helioscopia* is also a traditional medicinal plant. The whole plant has anticancer properties. The latex is applied externally to skin eruptions. The seeds, mixed with roasted pepper, have been used in the treatment of cholera. The oil from the seeds has purgative properties (Cai, Wang, & Liang, 1999; Wang et al. 2012). In Euphorbiaceae, greater than 31,595 reference sequences have been constructed. As a result, using NCBI Euphorbiaceae RefSeq database, 202 proteins were confidently identified in *E. helioscopia* latex (Table 1). Many of these proteins are present in low abundance and participate in various biological processes and metabolic pathways.

Gene Ontology (GO) analysis revealed that 179 proteins had relevant entries. GO includes three ontologies: molecular function, cellular component, and biological process. In molecular functional classes, more than half of the identified proteins consisted of two major protein classes: metabolic process (28.57%) and cellular process (23.55%). These results suggested that laticifer cells are metabolically active. In addition, 14.86% were proteins involved in responses to stimuli. This finding coincided with previous latex proteome analyses, which contained many viral and defense-related

Table 1 Proteins identified in *Euphorbia helioscopia* latex with iTRAQ proteomics.[a]

Group ID	Accession	Description	Score	Cov	Peptide	Ratio2/1	Ratio4/3
1	gi|255573627|	Heat shock protein, putative	497	16.1	8	4.736	4.846
1	gi|255575054|	Heat shock protein, putative	497	16	8		
2	gi|255549601|	5-Methyltetrahydropteroyltriglutamate–homocysteine methyltransferase, putative	442	6.3	4	2.513	2.039
3	gi|255544516|	ATP synthase alpha subunit vacuolar, putative	361	12.6	5	1.155	0.873
4	gi|255578112|	Aspartic proteinase precursor, putative	342	3.5	1	0.974	0.467
5	gi|255562375|	Actin, putative	312	24.4	7	2.39	2.385
5	gi|255568080|	Actin, putative	312	24.4	7		
6	gi|255557008|	60S ribosomal protein L12, putative	311	5.4	1	0.539	0.488
6	gi|255578850|	60S ribosomal protein L12, putative	311	5.4	1		
7	gi|255554815|	Aspartic proteinase precursor, putative	303	7.6	4	0.682	0.537
8	gi|255569484|	5-Methyltetrahydropteroyltriglutamate–homocysteine methyltransferase, putative	292	3.9	2	0.845	1.583
9	gi|255575355|	Enolase, putative	277	9.4	3	1.705	2.045
10	gi|255560497|	ATP synthase subunit beta vacuolar, putative	266	18.5	6	1.082	0.924
11	gi|255585546|	Malate dehydrogenase, putative	261	12	3	0.961	1.799
12	gi|255540341|	Glyceraldehyde 3-phosphate dehydrogenase, putative	220	8.3	2	2.844	2.423
12	gi|255591569|	Glyceraldehyde 3-phosphate dehydrogenase, putative	220	13.6	2		
13	gi|255579745|	Actin, putative	218	15.9	5	—	—
13	gi|255580037|	Actin, putative	218	19.4	5		
14	gi|255555659|	Heat shock protein, putative	211	9.5	5	1.9	2.095
15	gi|255544582|	Phosphoglycerate kinase, putative	207	8.5	2	3.016	2.863
16	gi|255547385|	Malate dehydrogenase, putative	197	9.5	3	3.105	2.992
17	gi|255565631|	60S ribosomal protein L10a, putative	192	5.6	1	0.519	0.449
18	gi|255575226|	NADP-specific isocitrate dehydrogenase, putative	185	7.7	2	8.316	6.139

19	gi|255563606|	Adenosylhomocysteinase, putative	171	10.7	4	3.067	2.155
20	gi|255584426|	Pectinesterase-3 precursor, putative	169	5.9	2	1.023	0.448
21	gi|255582911|	ATP synthase beta subunit, putative	162	14.3	5	5.451	3.725
22	gi|255568281|	Ketol-acid reductoisomerase, chloroplast precursor, putative	159	3.3	1	2.372	2.189
22	gi|255563128|	Ketol-acid reductoisomerase, chloroplast precursor, putative	159	3.2	1		
23	gi|255566672|	Peroxidase 12 precursor, putative	157	6.5	2	0.623	0.553
24	gi|255545176|	Heat shock protein, putative	153	6.5	4	4.055	5.691
25	gi|255556390|	ATP-citrate synthase, putative	150	8.7	4	3.461	4.971
26	gi|255537011|	Glyceraldehyde 3-phosphate dehydrogenase, putative	145	9.5	2	—	—
27	gi|255537671|	6-Phosphogluconate dehydrogenase, putative	144	7.3	3	2.271	2.597
28	gi|255558017|	Mta/sah nucleosidase, putative	143	2.8	1	0.71	0.392
28	gi|255558029|	Mta/sah nucleosidase, putative	143	2.7	1		
29	gi|255572331|	Protein with unknown function, putative	131	15.1	3	1.886	1.574
30	gi|255554865|	Lactoylglutathione lyase, putative	128	15.4	3	1.223	1.478
31	gi|255563084|	Malate dehydrogenase, putative	122	6.7	2	2.513	3.312
31	gi|255566555|	Malate dehydrogenase, putative	122	6.7	2		
32	gi|255540177|	Alpha-1,4-glucan-protein synthase [UDP-forming], putative	121	9.7	2	0.957	1.684
32	gi|255547137|	Alpha-1,4-glucan-protein synthase [UDP-forming], putative	121	9.7	2		
33	gi|255560311|	Diphosphomevalonate decarboxylase, putative	120	2.9	1	1.315	2.449
34	gi|255579273|	Succinate dehydrogenase, putative	116	3.4	2	6.8	6.805
35	gi|255562564|	Zinc finger protein, putative	115	1.7	1	0.713	0.668
36	gi|255544634|	Ubiquitin, putative	114	26	4	0.655	0.576
36	gi|255548313|	Ubiquitin, putative	114	25.6	4		

(Continued)

Table 1 Proteins identified in *Euphorbia helioscopia* latex with iTRAQ proteomics.[a]—cont'd

Group ID	Accession	Description	Score	Cov	Peptide	Ratio2/1	Ratio4/3
36	gi|255578975|	Ubiquitin, putative	114	31.2	4		
36	gi|255568846|	Ubiquitin, putative	114	25.6	4		
36	gi|255578906|	Ubiquitin, putative	114	31.2	4		
37	gi|255542380|	Cysteine synthase, putative	103	8	2	4.372	3.658
38	gi|255568502|	Protein with unknown function, putative	101	7.9	1	0.955	0.865
38	gi|255570175|	Protein with unknown function, putative	101	6.2	1		
39	gi|255558795|	NADP-dependent glyceraldehyde-3-phosphate dehydrogenase, putative	101	9.4	3	—	—
39	gi|255565005|	NADP-dependent glyceraldehyde-3-phosphate dehydrogenase, putative	101	8.3	3		
40	gi|255548998|	Ran, putative	97	8.6	2	0.469	0.364
41	gi|255580774|	Syntaxin, putative	97	10.2	1	0.404	0.538
42	gi|255537890|	Tubulin alpha chain, putative	93	3.3	1	—	—
42	gi|255582570|	Tubulin alpha chain, putative	93	3.3	1		
42	gi|255562192|	Tubulin alpha chain, putative	93	3.3	1		
43	gi|255543555|	Lysosomal alpha-mannosidase, putative	93	4.3	4	0.74	0.498
44	gi|255580564|	Nutrient reservoir, putative	87	7	2	2.045	1.251
45	gi|255536917|	ATP binding protein, putative	87	1.8	1	0.48	1.144
46	gi|255547492|	Phosphatidylcholine-sterol O-acyltransferase, putative	82	4.6	2	0.56	0.614
47	gi|255587664|	Rubisco subunit binding-protein alpha subunit, ruba, putative	80	8.2	3	1.19	1.233
48	gi|255537239|	Histone h4, putative	80	21.4	2	—	—
48	gi|255568195|	Histone h4, putative	80	21.4	2		
48	gi|255581707|	Histone h4, putative	80	21.4	2		
48	gi|255584136|	Histone h4, putative	80	21.4	2		
48	gi|255555809|	Histone h4, putative	80	21.4	2		
48	gi|255581703|	Histone h4, putative	80	21.4	2		

49	gi|255548762|	NADP-dependent glyceraldehyde-3-phosphate dehydrogenase, putative	76	8.5	3	—	—
50	gi|255546339|	Hyoscyamine 6-dioxygenase, putative	73	3.8	1	—	—
51	gi|255540059|	Lysosomal alpha-mannosidase, putative	73	2.9	2	0.473	0.495
52	gi|255572953|	Endosomal P24A protein precursor, putative	72	4.2	2	0.63	0.64
53	gi|255581578|	40S ribosomal protein S5, putative	71	10.4	2	2.596	5.522
53	gi|255587562|	40S ribosomal protein S5, putative	71	10.5	2		
54	gi|255558626|	Proteasome subunit beta type, putative	70	6.9	1	—	—
55	gi|255576991|	Leucine aminopeptidase, putative	69	7.7	3	2.177	2.491
56	gi|255544686|	Eukaryotic translation elongation factor, putative	66	2.4	2	5.429	3.942
56	gi|255565836|	Eukaryotic translation elongation factor, putative	66	2.4	2		
57	gi|255539911|	Protein with unknown function, putative	65	10.8	2	0.548	0.655
57	gi|255584780|	Protein with unknown function, putative	65	10.2	2		
57	gi|255587494|	Protein with unknown function, putative	65	10.2	2		
57	gi|255581964|	Protein with unknown function, putative	65	10.8	2		
57	gi|255549670|	Protein with unknown function, putative	65	10	2		
58	gi|255577538|	Conserved hypothetical protein, putative	64	2.7	1	0.082	0.343
58	gi|255587380|	Conserved hypothetical protein, putative	64	3.1	1		
59	gi|255573008|	Dtdp-glucose 4-6-dehydratase, putative	64	4.3	1	—	—
60	gi|255577165|	Aspartate carbamoyltransferase 2, chloroplast precursor, putative	63	7.8	3	4.211	9.399
61	gi|255550694|	Triacylglycerol lipase, putative	63	1.9	1	0.17	0.65
61	gi|255602447|	Hypothetical protein RCOM_1869450, putative	63	6	1		
61	gi|255585239|	Triacylglycerol lipase, putative	63	1.9	1		
61	gi|255586340|	Triacylglycerol lipase, putative	63	1.9	1		

(Continued)

Table 1 Proteins identified in *Euphorbia helioscopia* latex with iTRAQ proteomics.[a]—cont'd

Group ID	Accession	Description	Score	Cov	Peptide	Ratio2/1	Ratio4/3
62	gi|255573724|	Phosphoglucomutase, putative	63	2.1	1	8.345	5.563
63	gi|255579212|	3-Isopropylmalate dehydrogenase, putative	63	2.7	1	1.95	3.099
64	gi|255555150|	Chloroplast alpha-glucan water dikinase, putative	62	1.8	2	5.433	2.112
65	gi|255575381|	Fructose-bisphosphate aldolase, putative	62	5.9	3	1.957	3.002
66	gi|255538970|	Endoplasmin, putative	61	3.3	2	2.554	2.173
67	gi|255545916|	Glucose-6-phosphate isomerase, putative	60	5.3	2	1.448	1.508
68	gi|255556334|	Dihydrolipoamide dehydrogenase, putative	60	2.3	1	—	—
69	gi|255542636|	Beta-mannosidase, putative	60	2.3	2	0.805	0.541
70	gi|255539845|	Quinone oxidoreductase, putative	59	3.7	1	—	—
71	gi|255544658|	Cyclin-L2, putative	59	1.4	1	0.346	0.419
72	gi|255587307|	Myo-inositol-1 phosphate synthase, putative	59	2.2	1	1.45	1.814
73	gi|255540249|	Hydrolase, putative	58	3.4	2	1.136	0.621
74	gi|255546541|	Aldehyde dehydrogenase, putative	58	4	2	0.859	1.563
75	gi|255554066|	Elongation factor 1-alpha, putative	58	6.5	3	0.521	0.637
75	gi|255554068|	Elongation factor 1-alpha, putative	58	6.5	3		
75	gi|255574227|	Elongation factor 1-alpha, putative	58	6.5	3		
76	gi|255557562|	Biotin caboxylase subunit (BC) of Het-ACCase, putative	58	6.8	2	1.205	1.343
77	gi|255575353|	Peroxiredoxin, putative	56	4.9	1	—	—
78	gi|255539891|	Phosphofructokinase, putative	55	2.5	1	—	—
78	gi|255549790|	Phosphofructokinase, putative	55	2.4	1		
78	gi|255546259|	Phosphofructokinase, putative	55	2.1	1		
79	gi|255538612|	Glucose-6-phosphate isomerase, putative	54	2.1	1	2.436	3.264
80	gi|255553165|	Phopholipase d alpha, putative	53	1.5	1	—	—
81	gi|255558946|	Serine/threonine protein phosphatase, putative	53	4.2	1	0.656	0.993
81	gi|255581451|	Serine/threonine protein phosphatase, putative	53	3.3	1		

81	gi|255561919|	Serine/threonine protein phosphatase, putative	53	3.1	1		
82	gi|255564820|	Rubisco subunit binding-protein beta subunit, rubb, putative	53	2.1	1	—	—
83	gi|255542876|	Calreticulin, putative	52	2.2	1	2.712	3.786
84	gi|255548734|	Endosomal P24A protein precursor, putative	52	2	1	—	—
84	gi|255576995|	Endosomal P24A protein precursor, putative	52	2	1		
85	gi|255552287|	Peroxiredoxin, putative	51	5.5	1	—	—
86	gi|255579064|	Phosphoenolpyruvate carboxylase, putative	51	3.4	3	3.751	5.677
87	gi|255571035|	Nucleoside diphosphate kinase, putative	51	12.2	2	1.949	1.721
88	gi|255579936|	6-Phosphogluconate dehydrogenase, putative	51	5.3	3	—	—
89	gi|255587693|	fk506-binding protein, putative	51	10.1	1	—	—
90	gi|255557353|	50 kDa ketoavyl-ACP synthase, putative	51	2.6	1	—	—
91	gi|255548798|	Aspartate aminotransferase, putative	50	4.7	2	3.911	4.744
92	gi|255543757|	Diaminopimelate decarboxylase, putative	50	2.9	1	—	—
93	gi|255543008|	ATP-citrate synthase, putative	49	2.4	1	—	—
94	gi|255583291|	Xylose isomerase, putative	48	4.4	1	—	—
95	gi|255580086|	Conserved hypothetical protein, putative	48	7.3	1	—	—
96	gi|255538410|	Acylamino-acid-releasing enzyme, putative	48	1.8	1	—	—
97	gi|255555109|	Flavoprotein wrbA, putative	47	5.4	1	1.781	1.597
98	gi|255559509|	Peptide-N4-(N-acetyl-beta-glucosaminyl) asparagine amidase A, putative	46	2.2	1	—	—
99	gi|255539346|	Phosphofructokinase, putative	46	1.6	1	—	—
100	gi|255556721|	endo-1,4-Beta-glucanase, putative	45	2.3	1	—	—
101	gi|255540703|	Polyamine oxidase, putative	45	3.7	2	0.534	0.489
102	gi|255545264|	Tripeptidyl peptidase II, putative	44	0.9	1	—	—
103	gi|255552027|	Sulfate adenylyltransferase, putative	44	4.3	2	1.792	1.996
104	gi|255539759|	Beta-fructofuranosidase, soluble isoenzyme I precursor, putative	44	4.1	2	0.834	0.501
105	gi|255555501|	Zinc metalloprotease, putative	43	1.7	1	—	—
106	gi|255577916|	Succinyl-CoA synthetase beta chain, putative	43	2.4	1	3.471	4.577

(Continued)

Table 1 Proteins identified in *Euphorbia helioscopia* latex with iTRAQ proteomics.[a]—cont'd

Group ID	Accession	Description	Score	Cov	Peptide	Ratio2/1	Ratio4/3
107	gi|255563750|	Acetyl-CoA acetyltransferase, mitochondrial, putative	43	4.7	2	2.091	3.681
108	gi|255550319|	Sucrose synthase, putative	42	1.6	1	—	—
109	gi|255578174|	Short-chain dehydrogenase, putative	40	3.1	1	—	—
110	gi|255561094|	Vacuolar ATP synthase subunit ac39, putative	40	5.7	1	—	—
111	gi|255558087|	Vacuolar ATP synthase subunit h, putative	39	2.5	1	0.945	0.804
112	gi|255537805|	Nucleoside diphosphate kinase, putative	39	5.1	1	0.542	0.571
113	gi|255538186|	D-3-phosphoglycerate dehydrogenase, putative	39	1.8	1	—	—
113	gi|255568067|	D-3-phosphoglycerate dehydrogenase, putative	39	1.7	1		
113	gi|255555301|	D-3-phosphoglycerate dehydrogenase, putative	39	1.8	1		
114	gi|255559165|	Cell elongation protein diminuto, putative	39	1.9	1	—	—
115	gi|255574808|	Clathrin heavy chain, putative	39	0.7	1	—	—
116	gi|255596152|	Aldehyde dehydrogenase, putative	39	11.3	1	—	—
117	gi|255557267|	Pyruvate dehydrogenase, putative	38	2.4	1	—	—
118	gi|255564579|	GMP synthase, putative	38	2.1	1	—	—
119	gi|255563171|	Hypothetical protein RCOM_0971700, putative	38	7	1	—	—
119	gi|323649907|	ATP synthase F1 subunit 1 (mitochondrion), putative	38	2	1		
120	gi|255540003|	Calnexin, putative	38	2	1	—	—
121	gi|255542092|	Nucleotide pyrophosphatase/ phosphodiesterase, putative	38	1.7	1	—	—
122	gi|255581402|	Catalytic, putative	38	2.1	1	—	—
123	gi|255540299|	Adenylate kinase 1, putative	38	3.7	1	—	—
124	gi|255546205|	Hydrogen-transporting ATP synthase, rotational mechanism, putative	38	7.4	1	0.494	0.57

125	gi|255546279|	Profilin, putative	37	17.6	2	3.407	4.033
125	gi|255549802|	Profilin, putative	37	17.4	2		
126	gi|255573147|	ATP-dependent Clp protease proteolytic subunit, putative	37	4.5	1	—	—
127	gi|255570990|	Heat shock protein, putative	37	3	2	2.613	3.615
128	gi|255580772|	Protein phsophatase-2a, putative	37	3.3	1	—	—
129	gi|255581258|	Triosephosphate isomerase, putative	37	6.8	1	0.689	0.598
129	gi|255584863|	Triosephosphate isomerase, putative	37	3.1	1		
130	gi|255560408|	40S ribosomal protein S19, putative	36	6.3	1	—	—
130	gi|255570270|	40S ribosomal protein S19-3, putative	36	6.3	1		
130	gi|255570268|	40S ribosomal protein S19, putative	36	6.3	1		
131	gi|255557510|	Conserved hypothetical protein, putative	36	1.3	1	—	—
131	gi|255590338|	Carboxyl-tansferase, subunit of Het-ACCase, putative	36	3.3	1		
131	gi|372450148|	accD gene product (chloroplast), putative	36	1.8	1		
132	gi|255577354|	Receptor for activated protein kinase C, putative	36	3.7	1	—	—
133	gi|255554022|	Beta-galactosidase, putative	36	0.9	1	0.857	0.488
134	gi|255567869|	Cysteine protease, putative	36	6	1	—	—
135	gi|255543230|	Transaldolase, putative	36	2.5	1	—	—
136	gi|255552241|	Basic 7S globulin 2 precursor small subunit, putative	35	2.3	1	—	—
137	gi|255548666|	Eukaryotic translation initiation factor 3 subunit, putative	34	1.2	1	—	—
138	gi|255540271|	Elongation factor 1-beta, putative	34	3.9	1	0.491	0.478
138	gi|255564428|	Elongation factor 1-beta, putative	34	4	1		
139	gi|255571322|	Mgatp-energized glutathione s-conjugate pump, putative	34	0.6	1	—	—

(Continued)

Table 1 Proteins identified in *Euphorbia helioscopia* latex with iTRAQ proteomics.[a]—cont'd

<table>
<tr><th>Group ID</th><th>Accession</th><th>Description</th><th>Score</th><th>Cov</th><th>Peptide</th><th>Ratio2/1</th><th>Ratio4/3</th></tr>
<tr><td>140</td><td>gi|255555403|</td><td>Ribosomal protein S9, putative</td><td>34</td><td>6.6</td><td>1</td><td>—</td><td>—</td></tr>
<tr><td>140</td><td>gi|255573872|</td><td>Ribosomal protein S9, putative</td><td>34</td><td>6.6</td><td>1</td><td></td><td></td></tr>
<tr><td>141</td><td>gi|255580047|</td><td>Dolichyl-phosphate beta-glucosyltransferase, putative</td><td>34</td><td>3.6</td><td>1</td><td>—</td><td>—</td></tr>
<tr><td>142</td><td>gi|255546141|</td><td>Casein kinase, putative</td><td>34</td><td>2</td><td>1</td><td>—</td><td>—</td></tr>
<tr><td>142</td><td>gi|255560729|</td><td>Casein kinase, putative, putative</td><td>34</td><td>1.9</td><td>1</td><td></td><td></td></tr>
<tr><td>143</td><td>gi|255563186|</td><td>GTP cyclohydrolase II, putative</td><td>34</td><td>4.7</td><td>1</td><td>—</td><td>—</td></tr>
<tr><td>143</td><td>gi|255587021|</td><td>GTP cyclohydrolase II, putative</td><td>34</td><td>1.7</td><td>1</td><td></td><td></td></tr>
<tr><td>144</td><td>gi|255558602|</td><td>Conserved hypothetical protein, putative</td><td>34</td><td>1.1</td><td>1</td><td>0.888</td><td>0.457</td></tr>
<tr><td>145</td><td>gi|255542070|</td><td>UDP-glucose 6-dehydrogenase, putative</td><td>33</td><td>1.9</td><td>1</td><td>—</td><td>—</td></tr>
<tr><td>145</td><td>gi|255542080|</td><td>UDP-glucose 6-dehydrogenase, putative</td><td>33</td><td>1.9</td><td>1</td><td></td><td></td></tr>
<tr><td>145</td><td>gi|255542078|</td><td>UDP-glucose 6-dehydrogenase, putative</td><td>33</td><td>1.9</td><td>1</td><td></td><td></td></tr>
<tr><td>146</td><td>gi|255571216|</td><td>Acid phosphatase 1 precursor, putative</td><td>33</td><td>3.3</td><td>1</td><td>—</td><td>—</td></tr>
<tr><td>147</td><td>gi|255546147|</td><td>Dead box ATP-dependent RNA helicase, putative</td><td>33</td><td>1.3</td><td>1</td><td>—</td><td>—</td></tr>
<tr><td>147</td><td>gi|255560725|</td><td>Dead box ATP-dependent RNA helicase, putative</td><td>33</td><td>2.4</td><td>1</td><td></td><td></td></tr>
<tr><td>148</td><td>gi|255572307|</td><td>Peroxidase 12 precursor, putative</td><td>33</td><td>2.5</td><td>1</td><td>—</td><td>—</td></tr>
<tr><td>149</td><td>gi|255540107|</td><td>Proteasome subunit alpha type, putative</td><td>33</td><td>3.6</td><td>1</td><td>—</td><td>—</td></tr>
<tr><td>149</td><td>gi|255540109|</td><td>Proteasome subunit alpha type, putative</td><td>33</td><td>3.6</td><td>1</td><td></td><td></td></tr>
<tr><td>150</td><td>gi|255559084|</td><td>Fumarylacetoacetate hydrolase, putative</td><td>32</td><td>2.9</td><td>1</td><td>—</td><td>—</td></tr>
<tr><td>151</td><td>gi|255543771|</td><td>60S ribosomal protein L44, putative</td><td>32</td><td>7.6</td><td>1</td><td>—</td><td>—</td></tr>
<tr><td>151</td><td>gi|255549179|</td><td>60S ribosomal protein L44, putative</td><td>32</td><td>7.7</td><td>1</td><td></td><td></td></tr>
<tr><td>151</td><td>gi|255571924|</td><td>60S ribosomal protein L44, putative</td><td>32</td><td>7</td><td>1</td><td></td><td></td></tr>
<tr><td>152</td><td>gi|255588145|</td><td>Spliceosomal protein sap, putative</td><td>32</td><td>0.7</td><td>1</td><td>—</td><td>—</td></tr>
</table>

153	gi|255555351|	Hypothetical protein RCOM_0812220, putative	32	4.1	1	—	—
154	gi|255542058|	14-3-3 protein, putative	31	3.8	1	—	—
154	gi|255579722|	14-3-3 protein sigma, gamma, zeta, beta/alpha, putative	31	10.5	1		
154	gi|255545912|	14-3-3 protein, putative	31	3.9	1		
154	gi|255545792|	14-3-3 protein, putative	31	3.9	1		
154	gi|255545736|	14-3-3 protein, putative	31	4	1		
154	gi|255564764|	14-3-3 protein, putative	31	3.8	1		
154	gi|255551400|	14-3-3 protein, putative	31	7.4	1		
155	gi|255537603|	Serine/threonine protein phosphatase, putative	31	4.4	1	—	—
156	gi|255571263|	Coatomer delta subunit, putative	31	2.3	1	—	—
157	gi|255547446|	Acyl-[acyl-carrier-protein] desaturase, chloroplast precursor, putative	31	2	1	—	—
157	gi|255582193|	Stearoy-ACP desaturase, putative	31	2	1		
157	gi|255547448|	Stearoyl-ACP desaturase, putative	31	2.4	1		
158	gi|255579021|	Hsp90 co-chaperone AHA1, putative	31	3.7	2	0.601	0.734
159	gi|255586003|	Heparanase, putative	31	2	1	—	—
160	gi|255586766|	Monodehydroascorbate reductase, putative	30	3.2	1	—	—
161	gi|255563608|	Serine hydroxymethyltransferase, putative	30	1.5	1	—	—
162	gi|255572579|	Oligopeptidase A, putative	30	1.3	1	—	—
163	gi|255581736|	Peroxidase 17 precursor, putative	30	3.7	1	—	—
164	gi|255547584|	Regulator of ribonuclease activity A, putative	30	6.6	1	—	—
165	gi|255539945|	Glutamate dehydrogenase, putative	30	2.4	1	—	—
165	gi|255568914|	Glutamate dehydrogenase, putative	30	2.4	1		
166	gi|255581813|	Beta-hexosaminidase, putative	29	1.4	1	—	—
167	gi|255558550|	40S ribosomal protein S26, putative	29	6.8	1	—	—

(Continued)

Table 1 Proteins identified in *Euphorbia helioscopia* latex with iTRAQ proteomics.[a]—cont'd

<table>
<tr><th>Group ID</th><th>Accession</th><th>Description</th><th>Score</th><th>Cov</th><th>Peptide</th><th>Ratio2/1</th><th>Ratio4/3</th></tr>
<tr><td>167</td><td>gi|255587062|</td><td>40S ribosomal protein S26, putative</td><td>29</td><td>6.8</td><td>1</td><td></td><td></td></tr>
<tr><td>167</td><td>gi|255559669|</td><td>40S ribosomal protein S26, putative</td><td>29</td><td>6.8</td><td>1</td><td></td><td></td></tr>
<tr><td>168</td><td>gi|255563092|</td><td>Triacylglycerol lipase, putative</td><td>29</td><td>2.4</td><td>1</td><td>—</td><td>—</td></tr>
<tr><td>168</td><td>gi|255563096|</td><td>Triacylglycerol lipase, putative</td><td>29</td><td>2</td><td>1</td><td></td><td></td></tr>
<tr><td>168</td><td>gi|255563098|</td><td>Triacylglycerol lipase, putative</td><td>29</td><td>1.8</td><td>1</td><td></td><td></td></tr>
<tr><td>169</td><td>gi|255567297|</td><td>Chaperonin containing t-complex protein 1, alpha subunit, tcpa, putative</td><td>29</td><td>1.8</td><td>1</td><td>—</td><td>—</td></tr>
<tr><td>170</td><td>gi|255553153|</td><td>Dead box ATP-dependent RNA helicase, putative</td><td>29</td><td>2.3</td><td>1</td><td>—</td><td>—</td></tr>
<tr><td>170</td><td>gi|255592393|</td><td>Dead box ATP-dependent RNA helicase, putative</td><td>29</td><td>9.5</td><td>1</td><td></td><td></td></tr>
<tr><td>171</td><td>gi|255553522|</td><td>DNA topoisomerase II, putative</td><td>29</td><td>0.9</td><td>1</td><td>—</td><td>—</td></tr>
<tr><td>172</td><td>gi|255574007|</td><td>Conserved hypothetical protein, putative</td><td>29</td><td>2.1</td><td>1</td><td>—</td><td>—</td></tr>
<tr><td>173</td><td>gi|255546389|</td><td>Lactoylglutathione lyase, putative</td><td>29</td><td>2.2</td><td>1</td><td>—</td><td>—</td></tr>
<tr><td>174</td><td>gi|255554204|</td><td>Potassium channel beta, putative</td><td>29</td><td>3</td><td>1</td><td>—</td><td>—</td></tr>
<tr><td>175</td><td>gi|255608231|</td><td>Conserved hypothetical protein, putative</td><td>29</td><td>2.1</td><td>1</td><td>—</td><td>—</td></tr>
<tr><td>176</td><td>gi|255553805|</td><td>Epoxide hydrolase, putative</td><td>29</td><td>2.2</td><td>1</td><td>—</td><td>—</td></tr>
<tr><td>176</td><td>gi|255553807|</td><td>Epoxide hydrolase, putative</td><td>29</td><td>2.2</td><td>1</td><td></td><td></td></tr>
<tr><td>177</td><td>gi|255548858|</td><td>Iron (III)-zinc(II) purple acid phosphatase precursor, putative</td><td>29</td><td>2.1</td><td>1</td><td>—</td><td>—</td></tr>
<tr><td>178</td><td>gi|255546003|</td><td>40S ribosomal protein S3, putative</td><td>28</td><td>3.4</td><td>1</td><td>—</td><td>—</td></tr>
<tr><td>178</td><td>gi|255550145|</td><td>40S ribosomal protein S3, putative</td><td>28</td><td>3.4</td><td>1</td><td></td><td></td></tr>
<tr><td>179</td><td>gi|255583642|</td><td>Acid beta-fructofuranosidase precursor, putative</td><td>28</td><td>1.5</td><td>1</td><td>—</td><td>—</td></tr>
<tr><td>180</td><td>gi|255544922|</td><td>Conserved hypothetical protein, putative</td><td>28</td><td>0.8</td><td>1</td><td>—</td><td>—</td></tr>
<tr><td>181</td><td>gi|255573091|</td><td>Conserved hypothetical protein, putative</td><td>28</td><td>1.2</td><td>1</td><td>—</td><td>—</td></tr>
</table>

181	gi|255573093|	Conserved hypothetical protein, putative	28	1.2	1		
182	gi|255570875|	Conserved hypothetical protein, putative	28	4.4	1	—	—
183	gi|255542088|	Glycogen phosphorylase, putative	28	0.6	1	—	—
184	gi|255541320|	Proteasome subunit beta type, putative	28	4	1	—	—
185	gi|255585331|	Fructokinase, putative	28	2.7	1	—	—
185	gi|255541404|	Fructokinase, putative	28	2.4	1		
186	gi|255564498|	AP-2 complex subunit beta-1, putative	28	0.9	1	—	—
187	gi|255555807|	Cucumisin precursor, putative	27	0.9	1	—	—
188	gi|255541894|	Enolase, putative	27	1.8	1	—	—
189	gi|255558896|	Conserved hypothetical protein, putative	27	4.3	1	—	—
190	gi|255549476|	Pentatricopeptide repeat-containing protein, putative	27	2.3	1	—	—
191	gi|255567027|	12-Oxophytodienoate reductase opr, putative	27	2.8	1	—	—
192	gi|255541534|	Dead box ATP-dependent RNA helicase, putative	27	1.2	1	—	—
193	gi|255548339|	Nitrogen regulatory protein P-II, putative	27	10	1	—	—
194	gi|255575724|	Isocitrate dehydrogenase, putative	26	1.6	1	—	—
195	gi|255546650|	Galactokinase, putative	26	1.6	1	—	—
196	gi|255554535|	Methionine sulfoxide reductase, putative	25	2.4	1	—	—
196	gi|255568464|	Protein alx, putative	25	1.5	1		
197	gi|255542878|	Pyrophosphate-energized vacuolar membrane proton pump, putative	25	1.1	1	—	—
197	gi|255553279|	Pyrophosphate-energized vacuolar membrane proton pump, putative	25	0.8	1		
197	gi|255579837|	Pyrophosphate-energized vacuolar membrane proton pump, putative	25	1	1		
198	gi|255546311|	Ceramide glucosyltransferase, putative	25	1.1	1	—	—

(Continued)

Table 1 Proteins identified in *Euphorbia helioscopia* latex with iTRAQ proteomics.[a]—cont'd

<table>
<tr><th>Group ID</th><th>Accession</th><th>Description</th><th>Score</th><th>Cov</th><th>Peptide</th><th>Ratio2/1</th><th>Ratio4/3</th></tr>
<tr><td>199</td><td>gi|255559961|</td><td>Pentatricopeptide repeat-containing protein, putative</td><td>23</td><td>1.1</td><td>1</td><td>—</td><td>—</td></tr>
<tr><td>200</td><td>gi|255588191|</td><td>Serine carboxypeptidase, putative</td><td>21</td><td>6.5</td><td>1</td><td>—</td><td>—</td></tr>
<tr><td>201</td><td>gi|255580797|</td><td>Histone-lysine n-methyltransferase, suvh, putative</td><td>14</td><td>1.8</td><td>1</td><td>—</td><td>—</td></tr>
<tr><td>202</td><td>gi|255546684|</td><td>Cellulose synthase, putative</td><td>14</td><td>0.5</td><td>1</td><td>10.796</td><td>6.739</td></tr>
</table>

[a]Matched organism of 202 proteins are *Ricinus communis*.
Group ID: The protein number (I—) obtained from the iTRAQ experiments.
Accession: Accession number of sequences from NCBI Euphorbiaceae RefSeq database.
Description: The assigned protein name is given along with the *Ricinus communis* from which the annotation was derived and its GenBank accession number.
Score: The identification Score.
Cov: The percent of sequence coverage value for the matched peptides with higher than 95% confidence.
Peptide: The number of matched peptides with higher than 95% confidence.
Ratio: The changed ration of target protein in the sample 2/1 and 4/3 identified by iTRAQ.

proteins. In biological process classes, 48.63% proteins had catalytic activity, and 42.81% are binding proteins.

Pathway analysis was performed using the KEGG database. A total of 179 proteins were involved in 138 metabolic pathways (Table 2), such as biosynthesis of secondary metabolites, terpenoid backbone biosynthesis, amino sugar and nucleotide sugar metabolism, ribosome, protein processing in endoplasmic reticulum (ER), lysosome and ubiquitin-mediated proteasomes.

Table 2 KEGG pathway annotation of different proteins in laticifers of *Euphorbia helioscopia* L.

	Pathway	Different proteins with pathway annotation (179)	Pathway ID
1	Metabolic pathways	90 (50.28%)	ko01100
2	Biosynthesis of secondary metabolites	47 (26.26%)	ko01110
3	Microbial metabolism in diverse environments	41 (22.91%)	ko01120
4	Glycolysis/Gluconeogenesis	16 (8.94%)	ko00010
5	Pyruvate metabolism	12 (6.7%)	ko00620
6	Starch and sucrose metabolism	12 (6.7%)	ko00500
7	Phagosome	11 (6.15%)	ko04145
8	Citrate cycle (TCA cycle)	11 (6.15%)	ko00020
9	Lysosome	11 (6.15%)	ko04142
10	Methane metabolism	10 (5.59%)	ko00680
11	Amino sugar and nucleotide sugar metabolism	9 (5.03%)	ko00520
12	Ribosome	9 (5.03%)	ko03010
13	Oxidative phosphorylation	9 (5.03%)	ko00190
14	Carbon fixation in photosynthetic organisms	9 (5.03%)	ko00710
15	Pentose phosphate pathway	8 (4.47%)	ko00030
16	Purine metabolism	7 (3.91%)	ko00230
17	Protein processing in endoplasmic reticulum	7 (3.91%)	ko04141
18	Galactose metabolism	6 (3.35%)	ko00052
19	*Vibrio cholerae* infection	6 (3.35%)	ko05110
20	Fructose and mannose metabolism	6 (3.35%)	ko00051
21	Antigen processing and presentation	6 (3.35%)	ko04612
22	RNA degradation	6 (3.35%)	ko03018
23	Endocytosis	5 (2.79%)	ko04144
24	Cysteine and methionine metabolism	5 (2.79%)	ko00270

(Continued)

Table 2 KEGG pathway annotation of different proteins in laticifers of *Euphorbia helioscopia* L.—cont'd

	Pathway	Different proteins with pathway annotation (179)	Pathway ID
25	Regulation of actin cytoskeleton	5 (2.79%)	ko04810
26	Spliceosome	5 (2.79%)	ko03040
27	RNA transport	5 (2.79%)	ko03013
28	Other glycan degradation	5 (2.79%)	ko00511
29	Alzheimer's disease	5 (2.79%)	ko05010
30	Reductive carboxylate cycle (CO2 fixation)	5 (2.79%)	ko00720
31	Fatty acid biosynthesis	4 (2.23%)	ko00061
32	Propanoate metabolism	4 (2.23%)	ko00640
33	Glutathione metabolism	4 (2.23%)	ko00480
34	Focal adhesion	4 (2.23%)	ko04510
35	Two-component system	4 (2.23%)	ko02020
36	Glycine, serine and threonine metabolism	4 (2.23%)	ko00260
37	Glyoxylate and dicarboxylate metabolism	4 (2.23%)	ko00630
38	Huntington's disease	4 (2.23%)	ko05016
39	Ascorbate and aldarate metabolism	4 (2.23%)	ko00053
40	Phenylpropanoid biosynthesis	4 (2.23%)	ko00940
41	Epithelial cell signaling in *Helicobacter pylori* infection	4 (2.23%)	ko05120
42	Phenylalanine metabolism	4 (2.23%)	ko00360
43	Peroxisome	3 (1.68%)	ko04146
44	Selenoamino acid metabolism	3 (1.68%)	ko00450
45	Bacterial invasion of epithelial cells	3 (1.68%)	ko05100
46	Glycosaminoglycan degradation	3 (1.68%)	ko00531
47	Aminobenzoate degradation	3 (1.68%)	ko00627
48	Valine, leucine and isoleucine degradation	3 (1.68%)	ko00280
49	Plant-pathogen interaction	3 (1.68%)	ko04626
50	Collecting duct acid secretion	3 (1.68%)	ko04966
51	Pentose and glucuronate interconversions	3 (1.68%)	ko00040
52	Parkinson's disease	3 (1.68%)	ko05012
53	Insulin signaling pathway	3 (1.68%)	ko04910
54	Lysine degradation	3 (1.68%)	ko00310
55	Folate biosynthesis	3 (1.68%)	ko00790
56	Shigellosis	3 (1.68%)	ko05131
57	Alanine, aspartate and glutamate metabolism	3 (1.68%)	ko00250

Table 2 KEGG pathway annotation of different proteins in laticifers of *Euphorbia helioscopia* L.—cont'd

	Pathway	Different proteins with pathway annotation (179)	Pathway ID
58	Arginine and proline metabolism	3 (1.68%)	ko00330
59	Pyrimidine metabolism	3 (1.68%)	ko00240
60	Proteasome	3 (1.68%)	ko03050
61	Oocyte meiosis	3 (1.68%)	ko04114
62	Pathogenic Escherichia coli infection	3 (1.68%)	ko05130
63	Valine, leucine and isoleucine biosynthesis	3 (1.68%)	ko00290
64	Glycosphingolipid biosynthesis - ganglio series	2 (1.12%)	ko00604
65	Vascular smooth muscle contraction	2 (1.12%)	ko04270
66	Leukocyte transendothelial migration	2 (1.12%)	ko04670
67	NOD-like receptor signaling pathway	2 (1.12%)	ko04621
68	Toxoplasmosis	2 (1.12%)	ko05145
69	Tryptophan metabolism	2 (1.12%)	ko00380
70	Lysine biosynthesis	2 (1.12%)	ko00300
71	Glycerophospholipid metabolism	2 (1.12%)	ko00564
72	Pathways in cancer	2 (1.12%)	ko05200
73	Sulfur metabolism	2 (1.12%)	ko00920
74	Streptomycin biosynthesis	2 (1.12%)	ko00521
75	Tyrosine metabolism	2 (1.12%)	ko00350
76	Prostate cancer	2 (1.12%)	ko05215
77	Viral myocarditis	2 (1.12%)	ko05416
78	MAPK signaling pathway	2 (1.12%)	ko04010
79	Inositol phosphate metabolism	2 (1.12%)	ko00562
80	Terpenoid backbone biosynthesis	2 (1.12%)	ko00900
81	MAPK signaling pathway - yeast	2 (1.12%)	ko04011
82	Fatty acid metabolism	2 (1.12%)	ko00071
83	Butanoate metabolism	2 (1.12%)	ko00650
84	Arrhythmogenic right ventricular cardiomyopathy (ARVC)	2 (1.12%)	ko05412
85	Meiosis - yeast	2 (1.12%)	ko04113
86	Dilated cardiomyopathy	2 (1.12%)	ko05414
87	Type I diabetes mellitus	2 (1.12%)	ko04940
88	Tight junction	2 (1.12%)	ko04530
89	Phototransduction - fly	2 (1.12%)	ko04745
90	Proximal tubule bicarbonate reclamation	2 (1.12%)	ko04964
91	Long-term potentiation	2 (1.12%)	ko04720
92	Cyanoamino acid metabolism	2 (1.12%)	ko00460
93	Hypertrophic cardiomyopathy (HCM)	2 (1.12%)	ko05410
94	Adherens junction	2 (1.12%)	ko04520

(Continued)

Table 2 KEGG pathway annotation of different proteins in laticifers of *Euphorbia helioscopia* L.—cont'd

	Pathway	Different proteins with pathway annotation (179)	Pathway ID
95	Gap junction	1 (0.56%)	ko04540
96	Drug metabolism - other enzymes	1 (0.56%)	ko00983
97	PPAR signaling pathway	1 (0.56%)	ko03320
98	Systemic lupus erythematosus	1 (0.56%)	ko05322
99	Bile secretion	1 (0.56%)	ko04976
100	N-Glycan biosynthesis	1 (0.56%)	ko00510
101	Protein export	1 (0.56%)	ko03060
102	Chagas disease	1 (0.56%)	ko05142
103	Amebiasis	1 (0.56%)	ko05146
104	Histidine metabolism	1 (0.56%)	ko00340
105	Biosynthesis of unsaturated fatty acids	1 (0.56%)	ko01040
106	mRNA surveillance pathway	1 (0.56%)	ko03015
107	Hedgehog signaling pathway	1 (0.56%)	ko04340
108	Protein digestion and absorption	1 (0.56%)	ko04974
109	Sphingolipid metabolism	1 (0.56%)	ko00600
110	Styrene degradation	1 (0.56%)	ko00643
111	Isoquinoline alkaloid biosynthesis	1 (0.56%)	ko00950
112	D-Glutamine and D-glutamate metabolism	1 (0.56%)	ko00471
113	Steroid biosynthesis	1 (0.56%)	ko00100
114	Cell cycle - Caulobacter	1 (0.56%)	ko04112
115	Beta-alanine metabolism	1 (0.56%)	ko00410
116	Glycosphingolipid biosynthesis - globo series	1 (0.56%)	ko00603
117	Benzoate degradation	1 (0.56%)	ko00362
118	Fc gamma R-mediated phagocytosis	1 (0.56%)	ko04666
119	ABC transporters	1 (0.56%)	ko02010
120	Alpha-Linolenic acid metabolism	1 (0.56%)	ko00592
121	Tropane, piperidine and pyridine alkaloid biosynthesis	1 (0.56%)	ko00960
122	GnRH signaling pathway	1 (0.56%)	ko04912
123	Circadian rhythm - fly	1 (0.56%)	ko04711
124	Neurotrophin signaling pathway	1 (0.56%)	ko04722
125	Pantothenate and CoA biosynthesis	1 (0.56%)	ko00770
126	Carotenoid biosynthesis	1 (0.56%)	ko00906
127	Ether lipid metabolism	1 (0.56%)	ko00565
128	Phenylalanine, tyrosine and tryptophan biosynthesis	1 (0.56%)	ko00400
129	Nitrogen metabolism	1 (0.56%)	ko00910
130	Progesterone-mediated oocyte maturation	1 (0.56%)	ko04914

Table 2 KEGG pathway annotation of different proteins in laticifers of *Euphorbia helioscopia* L.—cont'd

	Pathway	Different proteins with pathway annotation (179)	Pathway ID
131	Prion diseases	1 (0.56%)	ko05020
132	Glycerolipid metabolism	1 (0.56%)	ko00561
133	Synthesis and degradation of ketone bodies	1 (0.56%)	ko00072
134	Cell cycle	1 (0.56%)	ko04110
135	One carbon pool by folate	1 (0.56%)	ko00670
136	SNARE interactions in vesicular transport	1 (0.56%)	ko04130
137	Novobiocin biosynthesis	1 (0.56%)	ko00401
138	Tetracycline biosynthesis	1 (0.56%)	ko00253

Euphorbia kansui is an endemic *Euphorbia* species in China and mainly located in Gansu, Shanxi, Shaanxi, Ningxia and Henan provinces of Chinese mainland. *E. kansui* is a perennial herb which is mostly found on low slopes, barren slopes, sandy lands, fields and roadsides (Fig. 1). Early, a proteomic approach was applied to analyze the *E. kansui* latex proteins by using sodium dodecyl sulfate-polyacrylamide gel electrophoresis and two-dimensional polyacrylamide gel electrophoresis. A total of 125 milky sap proteins associated with development of laticifers, disease and defense, and general metabolism were identified, and 19 differentially expressed proteins at two different developmental stages of laticifers were successfully detected. Peroxidase, cytochrome P450 mono-oxygenase superfamily, lipoxygenase, and multidrug resistance protein ABC transporter family may be involved in laticifer development, secondary metabolite synthesis and transport, and plant physiology. Lately, an iTRAQ-based quantitative proteome characterization of laticifers at different development stages and with UV-B radiation was investigated to reveal laticifer development and synthesis metabolism of secondary metabolites. A protein database for searching was deduced from a transcriptome assembly database of *E. kansui* (accession number: SRP067381) and NCBI Euphorbiaceae RefSeq protein database. A global profiling of quantitative proteome was obtained on 4 samples (latex from the laticifer at early stage, CK1; latex from the laticifer at early stage with UV-B radiation, T1; latex from the laticifer at later stage, CK2; latex from the laticifer at later stage with UV-B radiation, T2). Overall, more than 1501 unique peptides, corresponding to 322 proteins were identified with most of molecular weights ranging from 20 kDa to 80 kDa, which contained at least two peptides, and 69.26% protein sequence coverage is more than 10%.

2. Latex proteins involved in laticifer development

2.1 Introduction

It has been reported that a mass of cytoplasm including mitochondria, plastids and dicytosomes is gradually degraded in the process of laticifer development. As a result, the mature laticifers only have a thin layer of cytoplasm and the central vacuole is filled with latex particles (Cai, Li, & Yin, 2009; Inamdar, Murugan, & Subramanian, 1988; Stockstill & Nessler, 1986; Zhang et al., 2018). For a long time, numerous studies have focused on how the cytoplasm was degraded in laticifers. Various proteases, acidic phosphatases and hydrolases were found in the extracted latex and were hypothesized to participate in laticifer formation (Cai et al. 2009; Giordani, 1978).

In *E. helioscopia* laticifers, the cytoplasm was relatively uniform, and the organelles, including nucleus, dicytosomes, ER and plastids, exhibited normal structures at the early developmental stage (Fig. 2A). Noticeably, the ER at this stage was tube-shaped (Fig. 2B). As development proceeds, some ER was dilated and formed small vesicles. These resultant vesicles enlarged and began to accumulate many gray particles of various sizes called latex particles (Fig. 2C—E, arrows). Latex particles were finally released into the central vacuole by vesicles fusing with vacuoles (Fig. 2F, arrows). In the late developmental laticifers, a large number of dictyosome vesicles were distributed around degrading organelles (Fig. 2G). Some black osmium particles were noted in mitochondria (Fig. 2H). Then, mitochondria and plastids began to collapse. The outer membranes of these organelles were blurred. The ridge disappeared, and a large vacuole appeared inside (Fig. 2I). The rough ER swelled into various shapes or crooked into cup-shaped structure (shallow gray box) and ring-like structure (white box; Fig. 2J). Furthermore, many vesicles were distributed around the degrading mitochondria (gray box; Fig. 2K).

Nevertheless, the degradation mechanism of laticifer cytoplasm remains unclear. At present, we know that the lysosome pathway (LP) and proteasome pathway (PP) are two major degradation pathways in eukaryotic cells (Luo, Wong, & Wong, 2010). LP was originally considered to rid the cells of

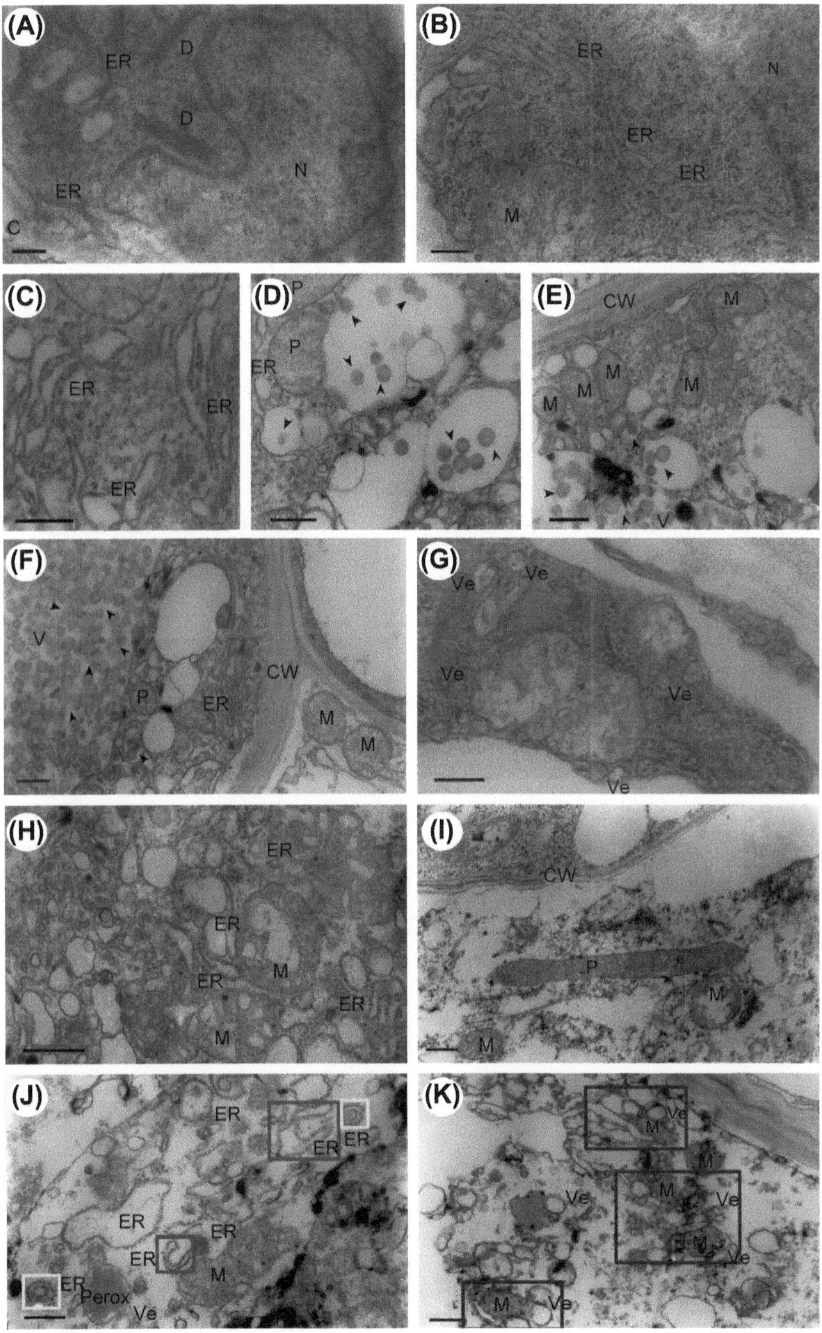

Fig. 2 Electron microscopic observations of changes of different organelles during *Euphorbia helioscopia* laticifer development. At the early developmental stage of laticifers, nucleus, rich dictyosomes and ER were clearly visible in protoplasm (A) as well as ER and mitochondria (B). At the mid-developmental stage of laticifers, the expanding

waste products, but it also has broader functions in metabolism and degradation of toxic proteins and damaged organelles (Huber & Teis 2016, Xu et al. 2017). PP is highly selective and mediated by ubiquitin and believed to be involved in the degradation of misfolded proteins and short-lived regulatory proteins (Collins, Wang, Caldwell, & Chopra, 2017; Glickman & Ciechanover, 2002). As a result, we investigated the distribution of LP and PP in non-articulated laticifers of *E. helioscopia.*

2.2 Latex proteins of lysosome pathway

In *E. helioscopia* latex, 10 lysosomal acid hydrolases were detected: cysteine protease (CP), beta-galactosidase, beta-hexosaminidase, beta-mannosidase, lysosomal alpha-mannosidase, phosphatidylcholine-sterol O-acyltransferase, vacuolar adenosine triphosphate (ATP) synthase subunit h, vacuolar ATP synthase subunit ac39, clathrin heavy chain (CHC) and AP-2 complex subunit beta-1 (AP2B1). CP is a marker enzyme of lysosomes, and CHC and AP2B1 are involved in the formation of lysosomes (Kuriyama & Fukuda, 2002). Immunoblotting analysis of *E. helioscopia* latex proteins showed that obvious bands appeared at 40 kDa, 170 kDa, and 100 kDa corresponded to the relative molecular mass of CP, CHC and AP2B1 respectively (Fig. 3C,D,E). For further analysis of the localization of three proteins in *E. helioscopia* laticifers, IgG-colloidal gold (10 nm) was used as probe. The distributions of the three proteins were examined using primary anti-bodies and electron microscopy. In *E. helioscopia* laticifers at mid-development stage, we observed some immuno-gold particles located around degenerative organelles (Fig. 4A, labeled with the AP2B1 antibodies). Gold particles with anti-CHC antibodies and anti-CP antibodies were distributed in dictyosome vesicles, which we referred to as lysosome-like vesicles (Fig. 4B,C). In laticifers at the late development stage, gold particles with anti-AP2B1, anti-CHC and anti-CP were located predominantly in

ER (C) and vesicles enlarged from ER (D) began to accumulate the latex particles (arrows). The small vesicles fused and latex particles (arrows) were released into the central vacuole after the vesicles fused with the vacuole (E) until the central vacuole was filled with latex particles (F; arrows) and some vesicles were distributed around the degrading organelles (G). At the late developmental stage of laticifers, mitochondria were degraded (H), the cytoplasm became thin, and plastids and mitochondria were degraded (I), the rough ER became swollen and crooked and the peroxisome is degraded (J), mitochondria were surrounded by some vesicles (K). Bars: A, C-K = 200 nm; B = 100 nm. CW, cell wall; D, dictyosomes; ER, endoplasmic reticulum; M, mitochondrion; P, Plastid; Perox, peroxysome; Ve, lysosome-like vesicle.

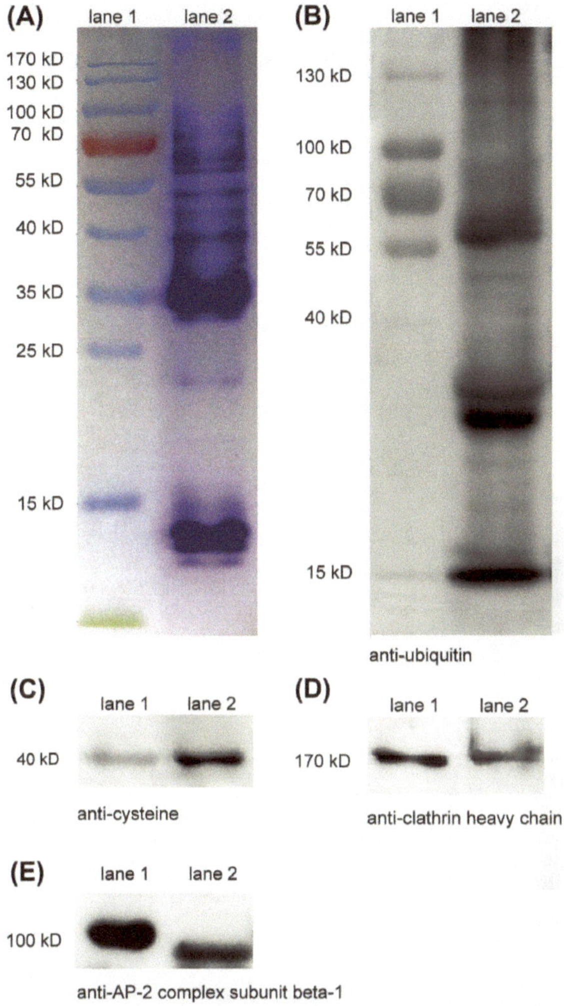

Fig. 3 Western blot analyses for detection of four different proteins in *Euphorbia helioscopia* latex. Laticifer total protein samples were extracted from latex; subjected to SDS-PAGE and analyzed with immunoblotting using anti-ubiquitin, anti-CP, anti-CHC and anti-AP2B1 antibodies. (A) Gel stained with Coomassie brilliant blue to ensure the loading of equivalent amounts of protein. Lane 1: Marker; lane 2: Total protein. Anti-ubiquitin (B), anti-CP (C), anti-CHC (D) and anti-AP2B1 (E) antibodies recognize latex protein in (E*) helioscopia.*

lysosome-like vesicles (Fig. 4D–F), and degenerative organelles were observed (Fig. 4E,F). The results revealed that CP, CHC and AP2B1 were endogenous proteins in laticifers and mainly distributed in dictyosome vesicles. These findings demonstrate that the dictyosome vesicles

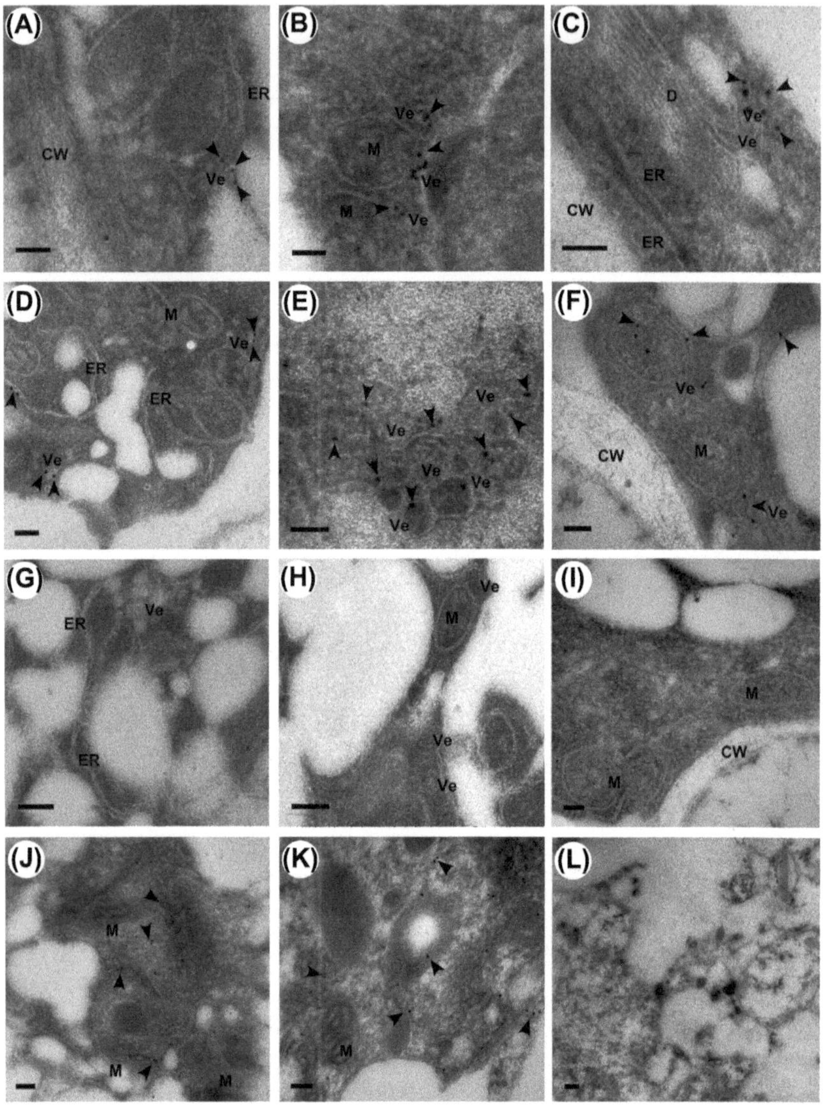

Fig. 4 Immunocytochemistry analyses of four types of antibodies in *Euphorbia helioscopia* laticifers. Golden particles conjugated with anti-AP2B1 antibodies were located in degraded organelles, endoplasmic reticulum, vesicles, and the membrane of the vesicles (A; arrows) and around mitochondria and membranes of vesicles (D; arrows). Golden particles conjugated with anti-CHC antibodies were located on membranes of the vesicles (B and E; arrows). Golden particles conjugated to anti-CP antibodies were located in some vesicles originating from dictyosomes (C; arrows), organelles to be degraded and some vesicles near the cell walls (F; arrows). (G) Control group of A and D. (H) Control group of B and E. (I) Control group of C and F. (J and K) Golden particles conjugated with anti-ubiquitin antibodies were widely distributed in the cytoplasm (arrows). (L) Control group of J and K. Bars: A-L, 100 nm. CW, cell wall; D, dictyosomes; ER, endoplasmic reticulum; M, mitochondrion; Ve, lysosome-like vesicle.

surrounding the degrading organelle in laticifers observed under electron microscope were indeed lysosome-like vesicles.

Therefore, it was hypothesized that LP is involved in the cytoplasmic degeneration of laticifers via the fusion of lysosome-like vesicles with degrading organelles and proteins. Hence, further studies were performed to discover the mechanism of these lysosome-like vesicles fuse with degrading organelles and proteins. It was found that autophagy is involved in cytoplasmic degradation through directly engulfing cytosol and organelles by autophagosomes and then fusing with lysosome-like vesicles during the development of non-articulated laticifers in *E. kansui* (Zhang et al. 2018).

CP was also identified in the latex of *Carica papaya* (Konno et al. 2004), *Calotropis procera* (Rawdkuen, Chaiwut, Pintathong, & Benjakul, 2010), *Asclepias curassavica* (Obregón et al. 2009), *Morrenia brachystephana* (Cavalli, Arribére, Caffini, & Priolo, 2003) *Ervatamia coronaria* (Nallamsetty, Kundu, & Jagannadham, 2003) and *Hevea brasiliensis* (Peng, Zhu, Li, & Tian, 2008). The presence of CP in latex fluids may be related to the defensive role of latex (Ra & Jones, 2004) and protein degradation during laticifer development (Agrawal, Lajeunesse, & Mark, 2008; Obregón et al. 2009).

2.3 Latex proteins of ubiquitin/proteasome pathway

In *E. helioscopia* latex, ubiquitin, proteasome subunit-alpha type and proteasome subunit-beta type were also identified. Ubiquitin is a small polypeptide composed of 76 amino acids. Ubiquitin has also been reported in *Chelidonium majus* latex (Nawrot, Kalinowski, & Gozdzicka-Jozefiak, 2007). Ubiquitin expression was identified by immunoblotting in *E. helioscopia* laticifers. Ubiquitin mainly existed in the form of ubiquitinated protein, and minimal free ubiquitin was observed. As a result, anti-ubiquitin antibodies detected many bands on the lane of total latex proteins and the bands were of various thicknesses. This result indicated that ubiquitinated proteins were widely distributed in laticifers and had different molecular weights (Fig. 3B). Immuno-electron microscopy analysis showed that numerous immunogold particles conjugated with anti-ubiquitin antibodies were widely distributed in the cytoplasm (Fig. 3J,K).

The proteasome activity in the laticifers of *E. helioscopia* was detected using various concentrations of the inhibitor MG132 for proteasome activity assays by monitoring the release of the fluorophore AMC from the synthetic peptide suc-Leu-Leu-Val-Tyr (sLLVY)-AMC that is widely used to detect proteasome activity *in vitro*. It was found that the fluorescence decreased with the addition of various concentrations of MG132, and the cleavage

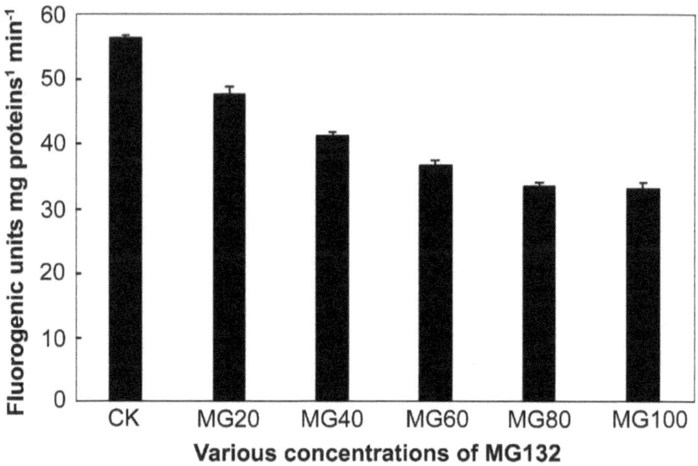

Fig. 5 Proteasome activity measurements. Protein extracts of latex of *Euphorbia helioscopia* were incubated in 100-μM sLLVY-AMC and various concentrations of MG132 at room temperature. Control check (CK), MG 20, MG40, MG60, MG80 and MG100 represented 0-, 10-, 20-, 40-, 60-, 80- and 100-μM MG132, respectively, and were added to the mixture above.

of sLLVY-AMC was significantly reduced upon treatment with 20, 40 and 60 μM MG132 (84.5%, 73.34%, and 65.19% compared with control, respectively). The fluorescence was not notably altered between MG80 and MG100 (Fig. 5).

The results above showed that ubiquitinated proteins were widely distributed in the cytosol, and proteasome activity was significantly reduced when various concentrations of the inhibitor MG132 were added to the latex total protein. Therefore, it could be deduced that proteins in laticifer cytoplasm that need degradation were first ubiquitin-modified and then degraded by proteasome. As a result, PP probably played a role in the degradation of cytoplasm protein during the development of *E. helioscopia* laticifers.

3. Latex proteins involved in terpenoid biosynthesis pathway in *Euphorbia*

3.1 Introduction

Terpenoids comprise isoprene polymers and its derivatives. They are important natural medicinal ingredients with broad physiological activity and varied structures.

3.2 The structure type of the terpenoids

Terpenoids are mainly distributed in plants. So far, more than 22,000 terpenoids have been found. Terpenoids are usually divided into eight categories according to the number of isoprene units in their molecular structure: hemiterpene, monoterpene, sesquiterpenes, diterpene, sesterterpenes, triterpenes, tetraterpenes, and polymer terpene. In addition, Terpenoids can also be divided into chain terpenes, monocyclic terpenes, icyclic terpenes and tetracyclic terpenes according to the presence or absence of carbon ring and the number of carbon ring in their molecular structure.

Monoterpenes are made up of two isoprene units, and they contain 10 carbon atoms. Monoterpenes are the main component of volatile oil, and distributed in gland, vitta, and resin canal of seed plants. Some monoterpenes exist in the form of glycosides, but not volatilitye. Monoterpenes included chain type and ring type, and kinds of compounds, such as lavender alkanes, citronella alkanes, menthol, and so on.

Sesquiterpenes are composed of three isoprene units, built of 15 carbon atoms. They are distributed in plants and microorganisms. They exist in the form of volatile oil, and mostly in forms of alcohol, ketone, glycosides and lactone in plants. Sesquiterpenes are divided into no ring, one ring to 12 ring type. There are thousands of compounds, such as artemisinin, farnesene and so on.

Diterpenoids are composed of four isoprene units, and they contained 20 carbon atoms. Diterpenoids mostly belong to crystalline solid. They are widely distributed in plant kingdom. Both latex and resin are diterpene derivatives. Kinds of important drugs, such as taxol, tanshinone, and lactones ginkgo all belong to diterpenoids. Diterpenoids can divide into chain type and ring type. The chemical structure types of diterpenoids have been found mainly included abietane, atisane, kaurane, pimarane, tigliane, ingenane, daphane, lathyrane, Jatrophane, casbabe, lathrane, myrsinol, rosane, euphoractine A, euphoractine B, and euphoractine C types. Abietane type diterpenoids were main compounds of *Euphorbia* plants, and lathyrane type diterpenoids were one of main stimulating compounds (Baole & Na, 2018).

Sesterterpene are composed of five isoprene units, and they contained 25 carbon atoms. They are distributed in ferns, lichens, and the secretion of some insects. The first sesterterpene was found in 1965. Now, there are about 30 compounds, such as serpentine and so on.

Triterpenes are composed of six isoprene units, and include 30 carbon atoms. Triterpene compounds exist in free form, or in form of glycosides which are known as triterpenoid saponins. Triterpene and triterpenoid saponins are widely found in fungi, ferns and seed plants and animals, and most of them are distributed in dicotyledonous plants. It is reported that triterpene compounds in free form exist mainly in Asteraceae, Leguminosae, Euphorbiaceae, Meliaceae, and Labiatae plants. Additionally, triterpenoid saponins exist mainly in Leguminosae, Araliaceae, Cucurbitaceae, and Umbelliferae plants. There are many structure types of triterpene compounds. Most of the triterpenoid compounds are tetracyclic and pentacyclic triterpenes. Tetracyclic triterpenes mainly include dammarane, lanostane, tirucallanes, and cucurbitacins types. Pentacyclic triterpenes mainly include oleanolic acid, ursanes, lupane and suberane and other types. The chemical structure types of triterpenes have been found including mainly cycloartane, lanostane, euphane, tirucallane, oleanane, ursane, and lupane types; and cycloartane type of triterpenes were main compounds of *Euphorbia* plants (Baole & Na, 2018).

Tetraterpenes mainly form a part of carrot hydroxyl pigment. The polymers of terpenes are mainly considered as rubber and hard rubber (Zhou, 1993).

3.3 Roles and uses of terpenoids from *Euphorbia*

Euphorbia is the largest genera in Euphorbiaceae. There are more than 2000 plants, and a variety of *Euphorbia* plants are used for medicinal plants. *Euphorbia pekinensis*, *E. pallasii*, *E. bracteolate*, *E. kansui*, *Euphorbia lathyris*, *E. hirta*, and *E. humifusae* are recorded in Phamacopoeia of the People's Republic of China (Chinese Pharmacopeia Committe, 2015). Some plants are often used as folk herbs in China. The main active components of *Euphorbia* plants were terpenoids, and showed many biological activities.

The dry root of *E. kansui* is a well known traditional Chinese medicine (Chinese Pharmacopeia Committee, 2015). Diterpene esters are the major active substances that show strong antiviral, anticancer, and inhibiting cell proliferation properties (Cheng et al. 2015; Hou et al. 2017; Liu, Yu, et al., 2018; Pan & Min, 2003; Zheng, 2004). These biologically active compounds mainly accumulate in the latex (Dickenson & Fairbairn, 1975). As an ancient folk herbal medicine, *E. helioscopia* has the efficacy of reducing phlegm, eliminating edema, reducing swelling and itching, and so on. Its active ingredients are terpenoids which showed antitumor activity (Cai, Wang, & Liang, 1999; Wang et al. 2012). *E. lathyris* is a biennial herbaceous species of *Euphorbia* (Euphorbiaceae). The seeds can be used as a medicine. It was reported that *E. lathyris* has diarrheogenic effect and antitumor activity (Chinese Pharmacopeia Committee, 2015).

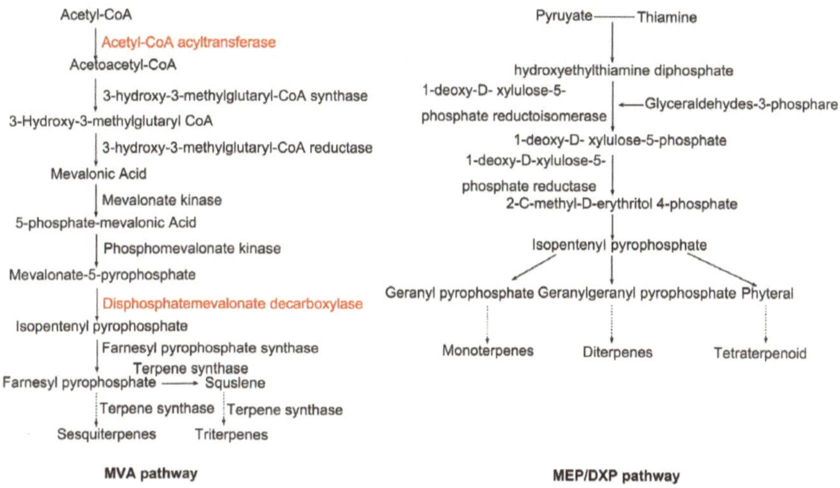

Fig. 6 A scheme showing mevalonic acid (MVA) and methylerythritol 4-phosphate (MEP) pathways in terpenoid biosynthesis.

3.4 Terpenoid biosynthesis pathways

There are two ways in terpenoid biosynthesis, namely the mevalonic acid (MVA) pathway and the methylerythritol 4-phosphate (MEP) pathway (Fig. 6, Rohmer, Knani, Simonin, Sutter, & Sahm, 1993). MVA pathway mainly occurs in the cytoplasm, so it is also called cytosolic pathway. This metabolic pathway was first found in animals and yeast by Conrad Bloch and Feodor Lynen in 1958 (Lichtenthaler, 1999). MEP pathway, which is also called 1-deoxy-D-xylulose-5-phosphate pathway, was found in bacteria by Rohmer in 1993, and it occurred in plasmid in plants. Acetyl-coenzyme A was used to synthesize isopentenyl diphosphate (IPP) by the MVA pathway. Glyceraldehyde-3-phosphate was used to synthesize dimethylallyl diphosphate (DMAPP). IPP was then isomerized to DMAPP. IPP and its isomer DMADP are the basic building blocks for all of the isoprenoids in living organisms (Hemmerlin, Harwood, & Bach, 2012).

3.5 The enzymes related to terpenoid biosynthesis pathway in latex of *Euphorbia*

Terpenoids accumulated mostly in laticifers are the main active constituents in *Euphorbia* plants, which have been reported in *E. kansui* by histochemical technology (Li, Cai, & Wang, 2009). The enzymes involved in terpenoid backbone and terpenoid biosynthesis pathways found in latex of *Euphorbia* plants were introduced in the following paragraphs (from 3.6 to 3.11).

3.6 Acetyl-CoA acetyltransferase

Acetyl-CoA acyltransferase is the initial enzyme in MVA pathway, which belongs to thiolase. Acetyl-CoA acyltransferase catalyzes two acetyl CoAs, which were condensed into acetoacetyl-CoA. Acetyl-CoA acyltransferase was identified in *E. helioscopia* latex (Fang et al. 2018).

The gene of acetyl-CoA acetyltransferase was cloned successfully from *E. helioscopia* using RT-PCR and RACE methods, named as *EhAACT* (accession number: KP995935). The full length cDNA of *EhAACT* was 1847 bp, and it contained an open reading frame of 1173 bp, a 5′ untranslated region of 207 bp and a 3′ untranslated region of 401 bp. Acetyl-CoA acetyltransferase contained 413 amino acids. Bioinformatic analyses showed that its predicted molecular weight was 42.606 kDa, theoretical pI was 8.7. Acetyl-CoA acetyltransferase belonged to hydrophobic protein without transmembrane domain and signal peptide, but with 22 phosphorylation sites. Acetyl-CoA acetyltransferase was located in the cytoplasm and belonged to the thiolase superfamily. Phylogenetic analysis revealed that *E. helioscopia* acetyl-CoA acetyltransferase had the highest identity and the closest genetic relationship with acetyl-CoA acetyltransferase in *Ricinus communis*. Recombinant protein of *E. helioscopia* acetyl-CoA acetyltransferase was expressed by procaryotic expression system. The molecular weight of expressed protein was about 43 kDa, which was consistent with the predicted result by software (Wang et al., 2017).

The mRNA transcripts of *EhAACT* were detected by real-time RT-PCR. The *EhAACT* expression level was similar in roots and stems, and little higher in leaves, and there was no significant difference ($P < 0.05$). In addition, western blotting results illustrated that the *E. helioscopia* acetyl-CoA acetyltransferase antibodies specifically recognized the endogenous proteins in *E. helioscopia* laticifers (Fig. 7A). Colloidal gold particles of acetyl-CoA acetyltransferase were observed to locate in the cytosol, endoplasmic reticulum cisternae, the cytosol around endoplasmic reticulum, and small vacuoles from endoplasmic reticulum (Fig. 8A–D). With the development of laticifers, it could be seen that endoplasmic reticulum cisternae swelled to form lots of vacuoles, and these vacuoles were always separated by cytosol. Colloidal gold particles were distributed around endoplasmic reticulum and mainly in these vacuoles (Fig. 8E,F). Some colloidal gold particles were also observed to locate in cytosol and endoplasmic

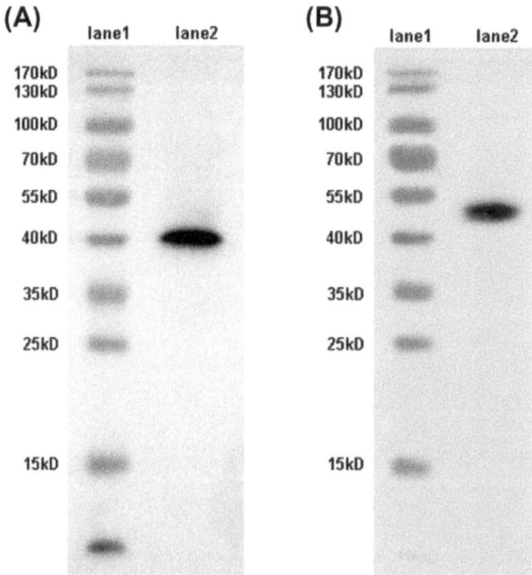

Fig. 7 Immunoblot analyses with acetyl-CoA acyltransferase and mevalonate disphosphate decarboxylase antibody from the protein extracts of *Euphorbia helioscopia* laticifers. (A) Immunoblot analysis of acetyl-CoA acyltransferase. (B) Immunoblot analysis of mevalonate disphosphate decarboxylase. Lane 1: marker; Lane 2: the corresponding protein gel blot probed with affinity purified antibodies.

reticulum between vacuoles (Fig. 8G). In the negative control which was not treated with the primary antibody incubation, no colloidal gold particles were found (Fig. 8H). As a result, we speculated that in *E. helioscopia* laticifers, acetyl-CoA acetyltransferase located in cytosol would be transferred to small vacuoles dilated from endoplasmic reticulum, and the precursors of terpenoids were synthesized in these small vacuoles, then terpenoids were further synthesized into latex particles (Wang et al. 2017).

At present, the acetyl-CoA acetyltransferase gene has been cloned in rubber tree (Sando, Takaoka, & Mukai, 2008), *Ganoderma lucidum* (Fang et al. 2013), *Salvia miltiorrhiza* (Cui, Wang, Feng, Zhao, & Huang, 2010), and *Aquilaria sinensis* (Liu, Xu, et al., 2014), *Camellia oleifera* (Zhang, Yan, et al., 2011) and *Ginkgo biloba* (Kim et al. 2008). Cui et al. (2010) found that expression level of acetyl-CoA acetyltransferase gene increased with the increase of tanshinones content under the induction of yeast extract and Ag^+ by cDNA expression microarray and real-time fluorescent quantitative PCR (Cui et al. 2010). In *G. lucidum,* there was a positive correlation between expression level of acetyl-CoA acetyltransferase gene and the accumulation of triterpenoid through semi-quantitative PCR analysis (Fang et al.

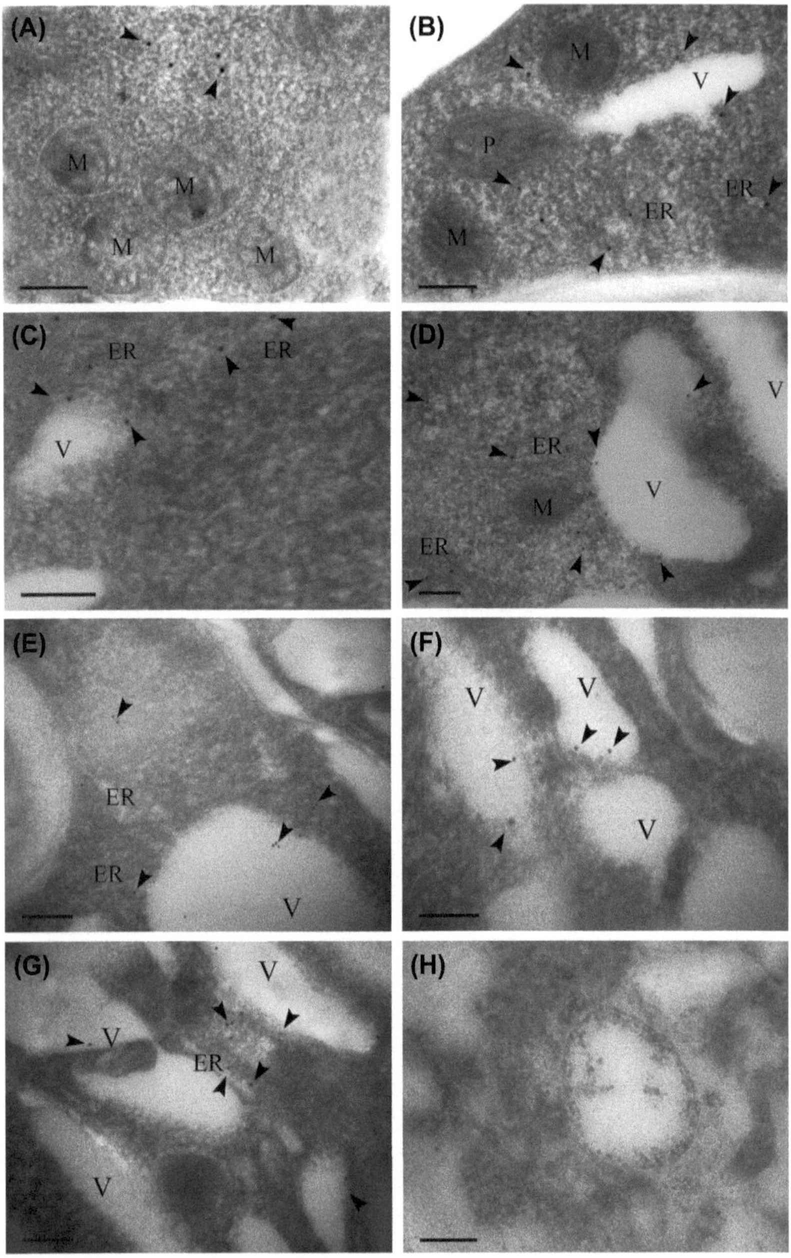

Fig. 8 Immunolocalization of the acetyl-CoA acyltransferase in *Euphorbia helioscopia* laticifers. (A) Acetyl-CoA acyltransferase signals appeared to locate in the cytosol randomly (arrows). (B) Acetyl-CoA acyltransferase signals appeared to locate mainly in endoplasmic reticulum and cytosol near small vacuoles from endoplasmic reticulum

2013). The result of qRT-PCR displayed that the highest expression level was at 4 h with jasmonic acid methyl ester treatment which indicated that it was possibly involved in early-stage response to wound in *A. sinensis* (Liu, Xu, et al., 2014).

3.7 Hydroxymethylglutaryl-CoA synthase

In the MVA pathway, the 3-hydroxy-3-methylglutaryl-coenzyme A synthase (HMG-CoA synthase) is an essential enzyme in mevalonate biosynthesis that catalyses the condensation of acetyl-CoA with acetoacetyl-CoA to yield 3-hydroxyl-3-methyl-glutaryl coenzyme A (HMG-CoA).

HMG-CoA synthase was detected in *E. kansui* latex. HMG-CoA synthase was an important enzyme of rubber synthesis, and it was mainly detected in *H. brasiliensis* latex, while the content in the leaves was very low (Suwanmanee, Sirinupong, & Suvachittanont, 2004). The cDNA sequence of HMG-CoA synthase has been cloned in rubber tree (Sirinupong, Suwanmanee, Doolittle, & Suvachitanont, 2005).

HMG-CoA synthase gene family in four genome sequenced Euphorbiaceae plants was analyzed. As a result, HMG-CoA synthase genes were identified from *R. communis*, *Jatropha curcas*, *Manihot esculenta* and *H. brasiliensis*, in a number of 1, 1, 2 and 2, respectively, and all of them contain 11 introns. Homology analysis indicated that HMG-CoA synthase genes are widely distributed in living organisms including archaea, eubacteria and eukaryotes, suggesting an early origin of this gene family. In most genome released plants including *Volvox*, *Arabidopsis*, *R. communis* and *J. curcas*, a single copy of HMG-CoA synthase gene was found, while in *M. tesculenta* and *H. brasiliensis* plants, gene duplication was found just like that in moss, rice and poplar. In *R. communis*, expression profiling suggested that in five examined tissues such as leaf, flower, II/III stage endosperm, V/VI stage endosperm and seed,

(arrows). (C) Acetyl-CoA acyltransferase signals appeared to locate in the endoplasmic reticulum, cytosol near small vacuoles from endoplasmic reticulum (arrows). (D) Golden particles conjugated acetyl-CoA acyltransferase antibodies were found to locate in the endoplasmic reticulum, cytosol near small vacuoles from endoplasmic reticulum (arrows). (E) Acetyl-CoA acyltransferase signals locate in small vacuoles from endoplasmic reticulum and cytosol (arrows). (F) Acetyl-CoA acyltransferase signals locate in small vacuoles from endoplasmic reticulum (arrows). (G) Golden particles conjugated acetyl-CoA acyltransferase antibodies were found to locate in the endoplasmic reticulum, small vacuoles from endoplasmic reticulum and cytosol between small vacuoles (arrows). (H) Negative controls showed no gold labeling in the laticifer cells. Bars = 200 nm. ER, Endoplasmic reticulum; M, Mitochondrion; P, Plastid; V, Vacuole.

HMG-CoA synthase gene was expressed strongly in flower and II/III stage endosperm, moderately in V/VI stage endosperm and leaf, weakly in seed (Zou, Xie, Mo, An, & Yang, 2014).

The cDNA sequence of HMG-CoA synthase has been cloned in *Arabidopsis thaliana* (Montamat, Guilloton, Karst, & Delrot, 1995), *Pinus sylvestris* (Wegener et al. 1997), *Brassica juncea* (Nagegowda, Bach, & Chye, 2004), rubber tree (Sirinupong et al. 2005), *Taxus media* (Kai et al. 2006), and *S. miltiorrhiza* (Zhang, Yan, et al., 2011). Alex, Bach, and Chye (2000) found that HMG-CoA synthase genes expressed in all organs, and the highest expression in flower of *Raphanus raphanistrum*. The expression level of HMG-CoA synthase gene was correlated with cell differentiation and growth. HMG-CoA synthase gene was down-regulated under abscisic acid, osmotic stress and dehydration (Alex et al. 2000). Nagegowda et al. (2004) detected HMG-CoA synthase in *R. raphanistrum*. The molecular mass was about 105.8 kDa, and the optimum pH and temperature of HMG-CoA synthase were 8.5 and 35 °C. HMG-CoA synthase could be activated by 5 mM disulfide erythritol, and inhibited by F-244 (Nagegowda et al., 2004). In *S. miltiorrhiza*, the full length of HMG-CoA synthase cDNA sequence was 1655 bp, and the open reading frame encoded 460 amino acids. Analysis on tissue expression pattern showed that the expression level of HMG-CoA synthase gene was the highest in leaf; stem second; the lowest in root. Moreover, methyl jasmonic acid and salicylic acid could induce HMG-CoA synthase gene expression (Zhang, Yan, et al., 2011). In *T. media*, HMG-CoA synthase gene expressed higher in the leaves and stems, and there was no expression in the root (Kai et al., 2006).

3.8 Hydroxymethylglutaryl-CoA reductase and Hydroxymethylglutaryl-CoA lyase

Hydroxymethylglutaryl-CoA reductase (HMG-CoA reductase) catalyzes HMG-CoA and 2 NADPH to form mevalonate. It is the first rate-limiting enzyme in MVA pathway. This step is also an important regulation site in terpenoids metabolism (Choi, Ward, & Bostock, 1992). Polypeptide chain structure of HMG-CoA reductase generally contained four parts: N terminal region, transmembrane region, linkage region and C terminal region. The sequences of N terminal region and linkage region were greatly different among various species in plant. Whereas, the transmembrane region and the C terminal region were highly conservative (Roberts, 2007). Amino acid sequence of HMG-CoA reductase contained four functionally important conservative regions, 2 HMG-CoA binding sites: EMP (V/I)

GY(V/I)QIP and TTEGCLVA; 2 NADP(H) binding sites: DAMGMNM and GTVGGGT (Dai, Cui, Zhou, Zhang, & Huang, 2011; Shen et al. 2006). Transmembrane region was composed of a hydrophilic group and two hydrophobic groups at both ends. Hydrophobic group has important role in membrane positioning for HMG-CoA reductase (Wang, Darnay, & Rodwell, 1990).

The seed of *E. lathyris* can be used as a traditional Chinese medicine. In addition, it was one of the first species proposed as an energy crop. Triterpenoids were the major constituents of the latex. Skrukrud (1987) investigated biosynthesis of triterpenoids (triterpene esters and triterpenols) by isolated latex of *E. lathyris*. Both HMG-CoA reductase (EC 1.1.1.34) and hydroxymethylglutaryl-CoA lyase (HMG-CoA lyase) (EC 4.1.3.4) were detected in isolated latex of *E. lathyris*. Electron micrographs of isolated *E. lathyris* latex showed the presence of latex particles and rod-shaped starch grains as well as a single-membrane-bounded structure which related to the activity of the mevalonate being converted to triterpenoids. HMG-CoA reductase was localized to a membrane-bound fraction of a 5000 g pellet of latex. The rate of conversion of HMG-CoA to mevalonate by this enzyme is comparable to the overall rate of acetate incorporation into the triterpenoids, what suggests that this enzyme is rate-determining in the biosynthesis of triterpenoids in *E. lathyris* latex. HMG-CoA reductase of *E. lathyris* vegetative tissue was localized to the membrane-bound portion of a particulate fraction (18,000 g). Differences in the optimal pH for activity of HMG-CoA reductases from the latex and vegetative tissue suggested that its isozymes might be present in the two tissue types. In addition, it is likely that different forms of the enzyme associated with the plastids, mitochondria, and endoplasmic reticulum exist (Skrukrud, 1987).

HMG-CoA reductase genes have been cloned from many speciess, such as *Ginkgo biloba* (Shen et al. 2006), *Eucommia ulmoides* (Jiang et al. 2006), *H. brasiliensis* (Sando et al. 2008), *E. pekinensis* (Cao et al. 2010), *S. miltiorrhiza* (Dai et al. 2011), *Glycyrrhiza uralensis* (Rong, Xu, Liu, & Huang, 2011). The mRNA length of HMG-CoA reductase was about 2000 bp in most plants. There were several HMG-CoA reductase genes in *A. thaliana*, *S. miltiorrhiza*, *H. brasiliensis* and other plants (Liu, Zhang, Cheng, Huang, & Gao, 2013). *E.* pekinensis is a perennial herbaceous plant of Euphorbia. The full length of E. pekinensis HMG-CoA reductase gene (EpHMGR, EF062569) was 2200 bp. The open reading frame length encoding 583 amino acids was 1752 bp. The theoretical molecular weight was 62.7 kDa, and the pH was 6.51. The full length of HMG-CoA reductase gene has two HMG CoA binding sites and NADP (H) binding sites. The expression level of EpHMGR genes was higher in roots than in stems. Complementary experiments were performed for functional analysis. The HMG-CoA reductase gene in *Escherichia coli* transformed strain could accelerate the synthesis of carotenoid. The results showed that HMG-CoA reductase gene could regulate the biosynthesis of terpenoids (Cao et al. 2010).

3.9 Mevalonate disphosphate decarboxylase

Mevalonate disphosphate decarboxylase is the last rate-limiting enzyme of generating the isopentenyl pyrophosphate precursor of terpenoid in MVA pathway. Mevalonate disphosphate decarboxylase catalyzes mevalonate disphosphate (MVAPP) to form IPP in C5 by decarboxylation reaction. This is the final step to generate IPP in MVA pathway, and ATP and Mg^{2+} participated in the reaction (Jabalquinto, Alvear, & Cardemil, 1988). In terms of protein structure, mevalonate disphosphate decarboxylase belongs to GHMP superfamily. It is a homologous dimer made up of two uniform subunits. Recently, two subunits have been found with molecular weight of 34–37 kDa and 45–46 kDa, respectively (Bonanno et al., 2001). GHMP superfamily is a kind of specific ATP dependent enzyme, and it contains four types of original members: galactokinase, homoserine kinase, mevalonate kinase and phosphomevalonate kinase (Bork, Sander, & Valencia, 1993).

Mevalonate disphosphate decarboxylase was respectively identified in *E. helioscopia* and *E. kansui* latex using iTRAQ and two-dimensional liquid chromatography mass spectrometry. Hereafter, the full length of mevalonate disphosphate decarboxylase gene (named as *EhMDC*, accession number: KP995936) was cloned successfully in *E. helioscopia* with rapid amplification of cDNA ends (RACE) method. The full length cDNA of *EhMDC* was 1653 bp, with an open reading frame of 1245 bp, a 5′untranslated region of 184 bp and a 3′untranslated region of 224 bp. *EhMDC* contained 415 amino acids. Bioinformatics analyses showed that predicted molecular weight was 45.971 kDa, and its theoretical pI was 5.77. Mevalonate disphosphate decarboxylase belonged to hydrophilic protein without transmembrane domain and signal peptide, but it had 19 predicted phosphorylation sites. Mevalonate disphosphate decarboxylase was located in the cytoplasm and belonged to GHMP kinases superfamily. Homologous sequence analysis showed that amino acid sequence of mevalonate disphosphate decarboxylase had the highest identity of 88% with *R. communis* mevalonate disphosphate decarboxylase. Phylogenetic analysis indicated *E. helioscopia*, *H. brasiliensis*, *R. communis*, and *J. carcas* mevalonate disphosphate decarboxylase were classified into one class.

Real-time PCR assay demonstrated that *EhMDC* was constitutively expressed in roots, stems and leaves with a similar transcription level. The protein of mevalonate disphosphate decarboxylase was expressed by procaryotic expression system. Immunoblotting demonstrated that the antibody was specific for a band estimated to be about 45.971 kDa being consistent with the calculated molecular mass of mevalonate disphosphate decarboxylase (Fig. 7B). In addition, immunogold assay was performed. It could be seen that golden particles with mevalonate disphosphate decarboxylase antibodies were located in cisternae of endoplasmic reticulum and small vacuoles from endoplasmic reticulum and also distributed randomly in protoplasm of laticifer cells. Whereas, preimmune controls showed there was no immunogold labeling. Therefore, the isopentenyl pyrophosphate (IPP) were speculated to be synthesized in the endoplasmic reticulum, and then transported into parts which dilated into small vacuoles, and the terpenoid was further synthesized into latex particles in these small vacuoles in *E. helioscopia* laticifers (Chai et al., 2017).

Cordier, Karst, and Bergès (1999) cloned the mevalonate disphosphate decarboxylase gene in *A. thaliana* by EST methods, and found that the MDC gene contained 9 exons and 8 introns. The length of the DNA sequence was 2837 bp, and the open reading frame length was 1239 bp. The mevalonate disphosphate decarboxylase gene encoded a protein of 45.58 kDa which was composed of 412 amino acid residues (Cordier et al., 1999).

At present, the cDNA sequence of mevalonate disphosphate decarboxylase has been cloned in *Homo sapiens* (Toth & Huwyler, 1996), *Rattus norvegicus* (Toth, Huwyler, & Park, 1996), *Saccharomyces cerevisiae, A. thaliana* (Cordier et al., 1999), *G. lucidum* (Qin, 2009) and so on. The mevalonate disphosphate decarboxylase gene in *G. lucidum* and *Eleutherococcus senticosus* were specifically expressed in different growth stages and tissues (Berthelot, Estevez, Deffieux, & Peruch, 2012). The full length cDNA of *G. lucidum* mevalonate disphosphate decarboxylase gene was 1203 bp, and the promoter sequence length was 1193 bp. The mevalonate disphosphate decarboxylase gene and protein expression levels in fruiting body primordia period were both significantly higher than that of mycelium period in *G. lucidum* (Qin, 2009).

3.10 Isopentenyl-diphosphate isomerase

Isopentenyl diphosphate isomerase is an important rate-limiting enzyme in terpenoid synthesis, which catalyzes the interconversion between isoprene pyrophosphate and dimethylpropylene pyrophosphate. Isoprene pyrophosphate and dimethylpropylene pyrophosphate are precursors of synthesis of geranyl diphosphate (GPP), farnesyl diphosphate (FPP) and geranylgeranyl diphosphate (GGPP). Isopentenyl diphosphate isomerase was identified in *E. kansui* latex. Isopentenyl diphosphate isomerase exists not only in the cytoplasm, and it also was found in the mitochondria, peroxide, and plastid (Cunningham, and Gantt, 2000). There are two isopentenyl diphosphate isomerases in most plants (Zhou, Li, Wiberley-Bradford, Weise, & Sharkey, 2013). Type I isopentenyl diphosphate isomerase gene sequences are very conservative and abundant in plants. There are two or more subtypes. The isopentenyl diphosphate isomerase gene subtype with longer sequence mostly exists in plasmids; while the isopentenyl diphosphate isomerase gene subtype with shorter sequence mostly exists in the cytoplasm (Dutoit et al., 2008). Type II isopentenyl diphosphate isomerase mainly exists in Gram positive bacterium, cyanobacteria and archaea. It has not been found in plants. Its gene sequence was longer, the number of coded amino acids were more than 300. The active sites were almost distributed in all amino acid sequence (Berthelot et al., 2012). *A. thaliana* double mutant plants with isopentenyl diphosphate isomerase gene defect were dwarfish and male sterility under long-day condition. Sterol and ubiquinone content were lower than the wild type. The result showed that IPP was originally produced through MVA way rather than the MEP pathway by isotope experiment in vivo (Okada et al., 2008). In tobacco, isopentenyl diphosphate isomerase gene expression level was higher under strong light, low temperature and salt stress conditions. The result showed that isopentenyl diphosphate isomerase was involved in plant stress responses (Nakamura, Shimada, Masuda, Ohta, & Takamiya, 2001). The genes sequences encoding isopentenyl diphosphate isomerase have been cloned in tobacco (Nakamura et al., 2008), *A. thaliana* (Campbell, Hahn, Poulter, & Leustek, 1998), *Daucus carota* (Just et al., 2007), and *E. ulmoides* (Liu, Xu, et al., 2014).

Isopentenyl diphosphate isomerase was a key enzyme of diterpene biosynthesis in *Rabdosia rubescens*. The cDNA encoding isopentenyl diphosphate isomerase was obtained from the leaves of aseptic seedlings of *R. rubescens* by reverse transcription PCR based on the results of transcriptome data. The results showed that the full length of isopentenyl diphosphate isomerase gene was 1050 bp, and contained an open reading frame of 906 bp and encoding 301 amino acids. The theoretical molecular weight was 27.4 kDa, and isoelectric point was predicted as 5.06. It was a kind of hydrophilic proteins. Real-time fluorescent quantitative PCR method was used to analyze the expression pattern of different tissues, and the result showed that the expression level of isopentenyl diphosphate isomerase gene was relatively high in the leaves, with the lowest expression in callus (Su, Yin, & Chen, 2016). Hu, Zhang, Tian, Xu, and Chen (2016) cloned a 1536 promoter fragment of isopentenyl diphosphate isomerase gene, and the bioinformatics analysis showed that the fragment contained conserved promoter sequence, such as TATA-box, CAAT-box. The promoter had promoter activities, and the activities were increased with the deleted length of promoters (Hu et al., 2016).

3.11 Farnesyl pyrophosphate synthase

IPP and DAMPP were important substrates which were formed to polyterpenes, triterpenes and sesquiterpenes by farnesyl pyrophosphate synthase (Okada, Saito, Nakagawa, Kawamukai, & Kamiya, 2000). Farnesyl pyrophosphate synthase was identified in *E. kansui* latex. It was purified (Archer and Cockbain, 1969; Light, Lazarus, & Dennis, 1989), and its cDNA sequence had been cloned and highly expressed in rubber tree laticifer cells (Adiwilaga and Kush, 1996; Takaya et al., 2003).

FPP synthase is a key enzyme in the synthetic pathway of terpenoids in plants. In addition, the gene sequence had been cloned in many plants, such as *A. thaliana* (Delourme, Lacroute, & Karst, 1994), *Acanthopanax senticosus* , *A. sinensis* (Yang, Wei, Liu, & Xu, 2013), *Lepidium apetalum* (Ma et al., 2017), and *Panax ginseng* (Liu, Yu, et al., 2018; Yang, Yang, Ma, & Han, 2017). Yang et al. (2017) cloned the full length sequence of farnesyl pyrophosphate synthase gene from *P. ginseng*, which was 1134 bp, and has an open reading frame, encoding a protein with 342 amino acids. Farnesyl pyrophosphate synthase has isoelectric point of 5.83 and a calculated relative molecular weight about 39,567 Da, without transmembrane regions and signal peptide. Protein characteristics, phylogenetic tree and tissue expression in different growth stages were analyzed. In the secondary structure, the main alpha helix accounted for 58.19%. The homologous analysis indicated that nucleotide sequence was 99% homologous with *P. quinquefolius* (ADJ68004) and *P. notoginseng* (AGS79228). Real-time quantitative PCR showed that farnesyl pyrophosphate synthase gene was expressed in all growth stages. Expression level of *P. ginseng* farnesyl pyrophosphate synthase was the highest in leaf-expanding period. Correlation analysis showed that expression of farnesyl pyrophosphate synthase gene was significantly positively correlated with content of Rb3 ($P < 0.05$) (Yang et al., 2017).

The farnesyl pyrophosphate synthase gene was cloned from *Taraxacum kok-saghyz* Rodin (*TkFPS*, accession Number: KJ558350.1) and transformed into the wild types. The open reading frame was 1029 bp which encoded 342 amino acid residues with the predicted molecular weight of 39.35 kD, and isoeletric point value of 5.41. Amino acid sequence analysis showed that farnesyl pyrophosphate synthase gene contained five conserved domains in higher plants. Phylogenetic tree analysis showed that the *T. kok-saghyz* farnesyl pyrophosphate synthase had the closest evolutionary relationship with *Herba Taraxaci* farnesyl pyrophosphate synthase. Subcellular localization analysis of TMHMM and TargetP indicated that it might be targeted to the cytoplasm without transmembrane region. Farnesyl pyrophosphate synthase gene was expressed in the root, petiole, leaf, peduncle, flower and seed of *T. kok-saghyz* by the tissue-specificity expression. The expressive content of farnesyl pyrophosphate synthase gene in different tissues of *T. kok-saghyz* was different: the highest was in the leaf, the next was root, and the lowest level was in the flower. The fluorescent quantitative RT-PCR showed that the relative expression levels of the farnesyl pyrophosphate synthase in transgenic lines were significantly higher, with the increases up to 2.8 times over the wild types. On average, rubber assay results showed that rubber content in roots of transgenic lines was increased by about 3.92% compared to the wild types. This result indicated that farnesyl pyrophosphate synthase played an important role in the biosynthetic pathway of the rubber in *T. kok-saghyz* (Cao et al., 2016).

4. Conclusions

Although the gene sequences of all enzymes which participate in the terpenoids synthesis were found in *E. kansui* transcriptome data, at the present, only acetyl-CoA acetyltransferase, hydroxymethylglutaryl-CoA synthase, hydroxymethylglutaryl-CoA reductase, hydroxymethylglutaryl-CoA lyase, mevalonate disphosphate decarboxylase, isopentenyl diphosphate isomerase and farnesyl pyrophosphate synthase were identified in latex of *Euphorbia* plants. Other enzymes in terpenoids synthesis pathway have not been detected in *Euphorbia* plants, because there is a lack of the genome information of *Euphorbia* plants, or the content of those enzymes in latex is low. With more and more research on *Euphorbia* plants and the development of proteomic technology, more enzymes will be found in the latex of *Euphorbia* plants in the future.

References

Adiwilaga, K., & Kush, A. (1996). Cloning and characterization of cDNA encoding farnesyl diphosphate synthase from rubber tree (*Hevea brasiliensis*). *Plant Molecular Biology, 30*, 935–946.

Agrawal, A. A., Lajeunesse, M. J., & Mark, F. (2008). Evolution of latex and its constituent defensive chemistry in milkweeds (*Asclepias*): A phylogenetic test of plant defense escalation. *Entomologia Experimentalis et Applicata, 128*, 126–138.

Alex, D., Bach, T. J., & Chye, M. (2000). Expression of *Brassica juncea* 3-hydroxy-3-methylglutaryl CoA synthase is developmentally regulated and stress-responsive. *The Plant Journal, 22*(5), 415–426.

Archer, B. L., & Cockbain, E. G. (1969). Rubber transferase from *Hevea brasiliensis* latex. *Methods in Enzymology, 15*, 476–480.

Baole, C. L., & Na, S. S. (2018). Advances on modern medicinal researches on plants of *Euphorbia* L. *World Chinese Medicine, 13*, 2090–2094.

Berthelot, K., Estevez, Y., Deffieux, A., & Peruch, F. (2012). Isopentenyl diphosphate isomerase: A checkpoint to isoprenoid biosynthesis. *Biochimie, 94*, 1621–1634.

Bonanno, J. B., Edo, C., Eswar, N., Pieper, U., Romanowski, M. J., Ilyin, V., et al. (2001). Structural genomics of enzymes involved in sterol/isoprenoid biosynthesis. *Proceedings of the National Academy of Sciences, 98*, 12129–12896.

Bork, P., Sander, C., & Valencia, A. (1993). Convergent evolution of similar enzymatic function on different protein folds: The hexokinase, ribokinase, and galactokinase families of sugar kinases. *Protein Science: A Publication of the Protein SOC* (Vol. 2, 31–40.

Cai, X., Li, W., & Yin, L. F. (2009). Ultrastructure and cytochemical localization of acid phosphatase of laticifers in Euphorbia kansui Liou. *Protoplasma, 238*, 3–10.

Cai, Y., Wang, J., & Liang, B. (1999). Antitumor activity of the root of Euphorbia helioscopia in vitro. *Journal of Chinese Medica Materia* (Vol. 22, 85–87.

Campbell, M., Hahn, F., Poulter, C. D., & Leustek, T. (1998). Analysis of the isopentenyl diphosphate isomerase gene family from *Arabidopsis thaliana*. *Plant Molecular Biology* (Vol. 36, 323–328.

Cao, X. W., Wang, X. Z., Li, Y. M., Zhao, L. J., Ma, H. X., & Yan, J. (2016). Molecular cloning and functional analysis of farnesyl pyrophosphate synthase genes from *Taraxacum kok-saghyz* Rodin. *Scientia Agricultura Sinica* (Vol. 49, 1034–1046.

Cao, X. Y., Zong, Z. M., Ju, X. Y., Sun, Y., Dai, C. C., Liu, Q., et al. (2010). Molecular cloning, characterization and function analysis of the gene encoding HMG-CoA reductase from *Euphorbia Pekinensis* Rupr. *Molecular Biology Reports, 37*, 1559–1567.

Cavalli, S. E. V., Arribére, M. A., Caffini, N. O., & Priolo, N. S. (2003). Morrenain b I, a papain-like endopeptidase from the latex of *Morrenia brachystephana* Griseb. (Asclepiadaceae). *Journal of Protein Chemistry, 22*, 15–22.

Chai, J., Wang, D., Peng, Y., Zhao, X. Y., Zhang, Q., Li, P., et al. (2017). Molecular cloning, expression and immunolocalization analysis of diphosphomevalonate decarboxylase involved in terpenoid biosynthesis from *Euphorbia helioscopia* L. *Biotechnology & Biotechnological Equipment, 31*, 1106–1115.

Cheng, F. F., Yang, Y. J., Zhang, L., Cao, Y. D., Yao, W. F., Tang, Y. P., et al. (2015). A natural triterpene derivative from *Euphorbia kansui* inhibits cell proliferation and induces apoptosis against rat intestinal epithelioid cell line in vitro. *International Journal of Molecular Sciences, 16*, 18956–18975.

Chinese Pharmacopeia Committee. (2015). *Phamacopoeia of the People's Republic of China (part I)*. Beijing: Chinese Medical Science Press.

Choi, D., Ward, B. L., & Bostock, R. M. (1992). Differential induction and suppression of potato 3-hydroxy-3-methylglutaryl coenzyme A reductase genes in response to *Phytophthora infestans* and to its elicitor arachidonic acid. *The Plant Cell Online, 4*, 1333–1344.

Collins, I., Wang, H., Caldwell, J. J., & Chopra, R. (2017). Chemical approaches to targeted protein degradation through modulation of the ubiquitin-proteasome pathway. *Biochemitry Journal, 474*, 1127–1147 (CrossRef partial stl atl).

Cordier, H., Karst, F., & Bergès, T. (1999). Heterologous expression in *Saccharomyces cerevisiae* of an *Arabidopsis thaliana* cDNA encoding mevalonate diphosphate decarboxylase. *Plant Molecular Biology, 39*, 953–967.

Cui, G. H., Wang, X. Y., Feng, H., Zhao, J. X., & Huang, L. Q. (2010). Molecular cloning and SNP analysis of a acetyl-CoA C-acetyltransferase gene (*SmAACT*) from *Salvia miltiorrhiza. Acta Pharmaceutica Sinica, 45*, 785–790.

Cunningham, F. X., & Gantt, E. (2000). Identification of multi-gene families encoding isopentenyl diphosphate isomerase in plants by heterologous complementation in *Escherichia coli. Plant and Cell Physiology, 41*, 119–123.

Dai, Z., Cui, G., Zhou, S. F., Zhang, X., & Huang, L. (2011). Cloning and characterization of a novel 3-hydroxy-3-methylglutaryl coenzyme A reductase gene from *Salvia miltiorrhiza* involved in diterpenoid tanshinone accumulation. *Journal of Plant Physiology, 168*, 148–157.

Delourme, D., Lacroute, F., & Karst, F. (1994). Cloning of an *Arabidopsis thaliana* cDNA coding for farnesyl diphosphate synthase by functional complementation in yeast. *Plant Molecular Biology, 26*, 1867–1873.

Dickenson, P. B., & Fairbairn, J. W. (1975). The ultrastructure of the alkaloidal vesicles of *Papaver somniferum* latex. *Annals of Botany, 39*, 707–712.

Dutoit, R., de Ruyck, J., Durisotti, V., Legrain, C., Jacobs, E., & Wouters, J. (2008). Overexpression, physicochemical characterization, and modeling of a hyperthermophilic pyrococcus furiosus type 2 IPP isomerase. *Proteins: Structure, Function, and Bioinformatics, 71*, 1699–1707.

Fang, X., Shi, L., Ren, A., Jiang, A. L., Wu, F. L., & Zhao, M. W. (2013). The cloning, characterization and functional analysis of a gene encoding an acetyl-CoA acetyltransferase involved in triterpene biosynthesis in *Ganoderma lucidum*. *Mycoscience, 54*, 100–105.

Fang, X. A., Zhang, Y., Wang, M., Li, P., Zhang, Q., Si, J. J., et al. (2018). Lysosome and proteasome pathways are distributed in laticifers of *Euphorbia helioscopia* L. *Physiologia Plantarum*. https://doi.org/10.1111/ppl.12869.

Giordani, R. (1978). Autophagie cellulaire et differenciation des laticiferes non articules chezune asclepiade. *Biology of the Cell, 33*, 253–260.

Glickman, M. H., & Ciechanover, A. (2002). The ubiquitin-proteasome proteolytic pathway: Destruction for the sake of construction. *Physiological Reviews, 82*, 373–428 (CrossRef partial stl atl).

Hemmerlin, A., Harwood, J. L., & Bach, T. J. (2012). A raison d'être for two distinct pathways in the early steps of plant isoprenoid biosynthesis? *Progress in Lipid Research, 51*, 95–148 (CrossRef partial atl).

Hou, J. J., Shen, Y., Yang, Z., Fang, L., Cai, L. Y., Yao, S., et al. (2017). Anti-proliferation activity of terpenoids isolated from *Euphorbia kansui* in human cancer cells and their structure-activity relationship. *Chinese Journal of Natural Medicines, 15*, 766–774.

Hu, X. L., Zhang, M., Tian, Y. S., Xu, Y., & Chen, F. (2016). Cloning and transient expression analysis of isopentenyl diphosphate isomerase gene (IPI) promoter from *Jatropha curcas*. *Journal of Sichuan University, 53*, 1409–1414.

Inamdar, J. A., Murugan, V., & Subramanian, R. B. (1988). Ultrastructure of non-articulated laticifers in *Allamanda violacea*. *Annals of Botany, 62*, 583–588.

Jabalquinto, A. M., Alvear, M., & Cardemil, E. (1988). Physiological aspects and mechanism of action of mevalonate 5-diphosphate decarboxylase. *Comparative Biochemistry and Physiology - Part B, 90*, 671–677 (CrossRef partial stl).

Jiang, J. H., Kai, G. Y., Cao, X. Y., Chen, F. M., He, D. N., & Liu, Q. (2006). Molecular cloning of a HMG-CoA reductase gene from *Eucommia ulmoides* Oliver. *Bioscience Reports, 26*, 171–181.

Just, B. J., Santos, C. A. F., Fonseca, M. E. N., Boiteux, L. S., Oloizia, B. B., & Simon, P. W. (2007). Carotenoid biosynthesis structural genes in carrot (*Daucus carota*): Isolation, sequence-characterization, single nucleotide polymorphism (SNP) markers and genome mapping. *Theoretical and Applied Genetics, 114*, 693–704.

Kai, G., Miao, Z., Zhang, L., Zhao, D., Liao, Z., Sun, X., et al. (2006). Molecular cloning and expression analyses of a new gene encoding 3-hydroxy-3-methylglutaryl-CoA synthase from *Taxus media. Biologia Plantarum, 50*, 359–366.

Kim, S., Kuzuyama, T., Kobayashi, A., Sando, T., Chang, Y. J., & Kim, S. U. (2008). 1-Hydroxy-2-methyl-2-(E)-butenyl 4-diphosphate reductase (IDS) is encoded by multicopy genes in gymnosperms *Ginkgo biloba* and *Pinus taeda. Planta, 227*, 287–298.

Konno, K., Hirayama, C., Nakamura, M., Tateishi, K., Tamura, Y., Hattori, M., et al. (2004). Papain protects papaya trees from herbivorous insects: Role of cysteine proteases in latex. *The Plant Journal: For Cell and Molecular Biology, 37*, 370–378 (CrossRef partial stl).

Kuriyama, H., & Fukuda, H. (2002). Developmental programmed cell death in plants. *Current Opinion in Plant Biology, 5*, 568–573.

Li, W., Cai, X., & Wang, Y. J. (2009). *Studies on anatomy of laticifers and localization of diterpene esters in vegetative organs in* Euphorbia kansui, *29* pp. 176–181).

Lichtenthaler, H. K. (1999). The 1-deoxydxylulose-5-phosphate pathway of isoprenoid biosynthesis in plant. *Annual Review of Plant Physiology and Plant Molecular Biology, 50*, 47–65.

Light, D. R., Lazarus, R. A., & Dennis, M. S. (1989). Rubber elongation by farnesyl pyrophosphate synthase involves a novel switch in enzyme stereospecificity. *Journal of Biological Chemistry, 264*, 18598–18607.

Liu, J., Xu, Y. H., Yang, Y., Liang, L., Han, X. M., & Gao, Z. H. (2014). Cloning and gene expression of acetyl-CoA C-acetyl transferase gene (AsAACT) from *Aquilaria sinensis. China Journal of Chinese Materia Medica, 6*, 972–980.

Liu, M. J., Yu, Y. L., Jiang, S., Zhang, X., Liu, D. Q., & Ge, F. (2018). Effect of farnesylpyrophosphate synthase (FPS) on the biosynthesis of saponins in *Panax japonicas. Bulletin of Botanical Research, 38*, 611–618.

Liu, Y. J., Zhang, X. N., Cheng, Q. Q., Huang, L. Q., & Gao, W. (2013). Advances in studies on 3-hydroxy-3-metllylglutaryl coenzyme A reductase in terpenoids biosynthesis of medicinal plants. *China Journal of Chinese Materia Medica, 38*, 3226–3233.

Luo, H. L., Wong, J., & Wong, B. (2010). Protein degradation systems in viral myocarditis leading to dilated cardiomyopathy. *Cardiovascular Research, 85*, 347–356 (CrossRef exact).

Ma, L. G., Zhao, L., Fu, X. D., Feng, W. S., Kuang, H. X., & Zheng, X. K. (2017). Cloning and prokaryotic expression of farnesyl pyrophosphate synthase gene from *Lepidium apetalum. Journal of Henan Agricultural, 46*, 92–97.

Montamat, F., Guilloton, M., Karst, F., & Delrot, S. (1995). Isolation and characterization of a cDNA encoding *Arabidopsis thaliana* 3-hydroxy-3-methylglutaryl-coenzyme A synthase. *Gene, 167*, 197–201.

Nagegowda, D., Bach, T., & Chye, M. (2004). *Brassica juncea* 3-hydroxy-3-methylglutaryl (HMG)-CoA synthase 1: Expression and characterization of recombinant wild-type and mutant enzymes. *Biochemical Journal, 383*, 517–527.

Nakamura, A., Shimada, H., Masuda, T., Ohta, H., & Takamiya, K. (2001). Two distinct isopentenyl diphosphate isomerases in cytosol and plastid are differentially induced by environmental stresses in tobacco. *FEBS Letters, 50*, 61–64 (CrossRef partial vol).

Nallamsetty, S., Kundu, S., & Jagannadham, M. V. (2003). Purification and biochemical characterization of a highly active cysteine protease ervatamin a from the latex of *Ervatamia coronaria*. *Journal of Protein Chemistry, 22*, 1–13.

Nawrot, R., Kalinowski, A., & Gozdzicka-Jozefiak, A. (2007). Proteomic analysis of *Chelidonium majus* milky sap using two-dimensional gel electrophoresis and tandem mass spectrometry. *Phytochemistry, 68*, 1612–1622.

Obregón, W. D., Liggieri, C. S., Trejo, S. A., Aviles, F. X., Vairocavalli, S. E., & Priolo, N. S. (2009). Characterization of papain-like isoenzymes from latex of *Asclepias curassavica* by molecular biology validated by proteomic approach. *Biochimie, 91*, 1457–1464.

Okada, K., Kasahara, H., Yamaguchi, S., Kawaide, H., Kamiya, Y., Nojiri, H., et al. (2008). Genetic evidence for the role of isopentenyl diphosphate isomerases in the mevalonate pathway and plant development in *Arabidopsis*. *Plant and Cell Physiology, 49*, 604–616.

Okada, K., Saito, T., Nakagawa, T., Kawamukai, M., & Kamiya, Y. (2000). Five geranylgeranyl diphosphate synthases expressed in different organs are localized into three subcellular compartments in *Arabidopsis*. *Plant Physiology, 122*, 1045–1056.

Pan, Q., & Min, Z. D. (2003). Studies on ingenol-type diterpene esters in root tuber of *Euphorbia kansui*. *Chinese Traditional and Herbal Drugs, 34*, 489–492.

Peng, S. Q., Zhu, J. H., Li, H. L., & Tian, W. M. (2008). Cloning and characterization of a novel cysteine protease gene (HbCP1) from *Hevea brasiliensis*. *Journal of Biosciences, 33*, 681–690.

Qin, L. (2009). *Molecular cloning and analysis of expression profiles of mevalonate diphosphate decarboxylase gene from* Ganoderma lucidum. *Master degree*. Nanjing: Nanjing Agricultural University.

Ra, V. A. D. H., & Jones, J. D. (2004). The plant proteolytic machinery and its role in defence. *Current Opinion in Plant Biology, 7*, 400–407.

Rawdkuen, S., Chaiwut, P., Pintathong, P., & Benjakul, S. (2010). Three-phase partitioning of protease from *Calotropis procera* latex. *Biochemical Engineering Journal, 50*, 145–149.

Roberts, S. C. (2007). Production and engineering of terpenoids in plant cell culture. *Nature Chemical Biology, 3*, 387–395.

Rohmer, M., Knani, M., Simonin, P., Sutter, B., & Sahm, H. (1993). Isoprenoid biosynthesis in bacteria: A novel pathway for the early steps leading to isopentenyl diphosphate. *Biochemical Journal, 295*, 517–524 (CrossRef exact).

Rong, Q. X., Xu, Q. X., Liu, C. S., & Huang, L. Q. (2011). Cloning and characterizationof 3-hydroxy-3-methylglutary CoA reductase cDNA of *Glycyrrhiza uralensis*. *China Journal of Chinese Materia Medica, 36*, 1275–1279.

Sando, T., Takaoka, C., & Mukai, Y. (2008). Cloning and characterization of mevalonate pathway genes in a natural rubber producing plant, *Hevea brasiliensis*. *Bioscience Biotechnology & Biochemistry, 72*, 2049–2060.

Shen, G., Pang, Y., Wu, W., Liao, Z., Zhao, L., Sun, X., et al. (2006). Cloning and characterization of a root-specific expressing gene encoding 3-hydroxy-3-methylglutaryl coenzyme a reductase from *Ginkgo biloba*. *Molecular Biology Reports, 33*, 117–127.

Sirinupong, N., Suwanmanee, P., Doolittle, R., & Suvachitanont, W. (2005). Molecular cloning of a new cDNA and expression of 3-hydroxy-3-methylglutaryl-CoA synthase gene from *Hevea brasiliensis*. *Planta, 221*, 502–512.

Skrukrud, C. L. (1987). *Terpenoid biosynthesis in* Euphorbia lathyris *and* Copaifera *ssp*. California: Lawrence Berkeley Laboratory University of California Berkeley.

Stockstill, B. L., & Nessler, C. L. (1986). Ultrastructural observations on the nonar ticulated, branched laticifers in *Nerium oleander* L. (Apocynaceae). *Phytomorphology, 36*, 347–355.

Suwanmanee, P., Sirinupong, N., & Suvachittanont, W. (2004). Regulation of the expression of 3-hydroxy-3-methylglutaryl-CoA synthase gene in *Hevea brasiliensis* (B.H.K.) Mull. Arg. *Plant Science, 166*, 531–537.

Su, X. H., Yin, L., & Chen, S. Q. (2016). Cloning and expression of cDNA encoding key enzymes (isopentenyl diphosphate isomerase) in diterpenoids biosynthesis pathway from *Rabdosia rubescens* Henrsl. Hara. *Northern Horticulture, 12*, 80–85.

Takaya, A., Zhang, Y. W., Asawatreratanakul, K., Wititsuwannakul, D., Wititsuwannakuld, R., Takahashi, S., et al. (2003). Cloning, expression and characterization of a functional cDNA clone encoding geranylgeranyl diphosphate synthase of *Hevea brasiliensis*. *Biochimica et Biophysica Acta (BBA) - Gene Structure and Expression, 1625*, 214–220.

Toth, M. J., & Huwyler, L. (1996). Molecular cloning and expression of the cDNAs encoding human and yeast mevalonate pyrophosphate decarboxylase. *Journal of Biological Chemistry, 271*, 7895–7898 (CrossRef exact).

Toth, M. J., Huwyler, L., & Park, J. (1996). Purification of rat liver mevalonate pyrophosphate decarboxylase. *Preparative Biochemistry & Biotechnology, 26*, 47–51 (CrossRef exact).

Wang, Y., Darnay, B. G., & Rodwell, V. W. (1990). Identification of the principal catalytically important acidic residue of 3-hydroxy-3-methylglutaryl coenzyme A reductase. *Journal of Biological Chemistry, 265*, 21216–21634.

Wang, Z. Y., Liu, H. P., Zhang, Y. C., Guo, L. Q., Li, Z. X., & Shi, X. F. (2012). Anticancer potential of *Euphorbia helioscopia* L. extracts against human cancer cells. *The Anatomical Record, 295*, 223–233.

Wang, M., Wang, D., Zhang, Q., Chai, J., Peng, Y., & Cai, X. (2017). Identifcation and cytochemical immunolocalization of acetyl-CoA acetyltransferase involved in the terpenoid mevalonate pathway in *Euphorbia helioscopia* laticifers. *Botanical Studies, 58*, 62.

Wegener, A., Gimbel, W., Werner, T., Hani, J., Ernst, D., & Sandermann, H. (1997). Molecular cloning of ozone-inducible protein from *Pinus sylvestris* L. with high sequence similarity to vertebrate 3-hydroxy-3-methylglutaryl-CoA-synthase. *Biochimica et Biophysica Acta (BBA) - General Subjects, 1350*, 247–252.

Xing, Z. B., Long, Y. L., He, S., Zhou, M., & Xiu, L. S. (2012). Cloning and expression analysis of mevalonate diphosphate decarboxylasa gene in *Eleutherococcus senticosus*. *Acta Botanica Boreali-Occidentalia Sinica, 32*, 1950–1956.

Xu, B., Gao, Y., Zhan, S., & Ge, W. (2017). Quantitative proteomic profiling for clarification of the crucial roles of lysosomes in microbial infections. *Molecular Immunology, 87*, 122–131.

Yang, X., Wei, J. H., Liu, J., & Xu, Y. H. (2013). Cloning and expression analysis of farnesyl diphosphate synthase gene AsFPS1 from *Aquilaria sinensis*. *Chinese Materia Medica, 38*, 3251–3255.

Yang, L. L., Yang, L. M., Ma, X. J., & Han, M. (2017). Correlation between *Panax ginseng* farnesyl diphosphate synthase gene expression and ginsenoside Content. *Journal of Jilin Agricultural University, 39*, 695–702. +708.

Zhang, Q., Wang, D., Zhang, H., Wang, M., Li, P., Fang, X., et al. (2018). Detection of autophagy processes during the development of nonarticulated laticifers in *Euphorbia kansui* Liou. *Planta, 247*, 845–861.

Zhang, L., Yan, X. M., Wang, J., Li, S. S., Liao, P., & Kai, G. Y. (2011). Molecular cloning and expression analysis of a new putative gene encoding 3-hydroxy-3-methylglutaryl-CoA synthase from *Salvia miltiorrhiza*. *Acta Physiologiae Plantarum, 33*, 953–961.

Zheng, W. F. (2004). Study on in vivo antiviral activity of four diterpenoids from ethanol extracts of *Euphorbia kansui*. *Chinese Traditional and Herbal Drugs, 35*, 65–68.

Zhou, R. H. (1993). *Traditional Chinese resources*. Beijing: China medical science press.

Zhou, C. F., Li, Z., Wiberley-Bradford, A. E., Weise, S. E., & Sharkey, T. D. (2013). Isopentenyl diphosphate and dimethylallyl diphosphate/isopentenyl diphosphate ratio measured with recombinant isopentenyl diphosphate isomerase and isoprene synthase. *Analytical Biochemistry, 440*, 130–136. CrossRef partial au1.

Zou, Z., Xie, G. S., Mo, Y. Y., An, F., & Yang, L. F. (2014). Genome-wide identification and analysis of HMGS gene family in four Euphorbiaceae plants. *Southwest China of Agicultural Sciences, 6*, 2259–2268.

CHAPTER SEVEN

Biochemical composition of *Hevea brasiliensis* latex: A focus on the protein, lipid, carbohydrate and mineral contents

Céline Bottier[a,b,*]
[a]CIRAD (French Agricultural Research Center for International Development), UMR IATE (Agropolymer Engineering and Emerging Technologies Research Unit), Montpellier, France
[b]IATE, Univ Montpellier, CIRAD, INRA, Montpellier SupAgro, Montpellier, France
*Corresponding author: E-mail: celine.bottier@cirad.fr

Contents

Abstract

The latex collected from the *Hevea brasiliensis* tree is today the only commercial source of natural rubber (NR), the *cis*-1,4-polyisoprene polymer, a strategic raw material. The *Hevea* latex is a very complex material both in its structure and composition. In terms of structure, it is a colloidal dispersion where various micrometric objects, mainly rubber

Advances in Botanical Research, Volume 93
ISSN 0065-2296
https://doi.org/10.1016/bs.abr.2019.11.003

particles and lutoids, are dispersed in the cytoplasmic serum (C-serum). The rubber fraction is the most abundant, followed by the C-serum and the lutoids. In terms of composition, the fresh latex contains about 60% of water, 35% of *cis*-1,4-polyisoprene and 5% of non-isoprene molecules. These non-isoprenes are biochemical compounds mostly including proteins, lipids, carbohydrates and minerals, and their distribution in the fractions of latex is not homogeneous. Although the non-isoprenes represent a minor part of the latex, some of them are retained in NR after latex processing and recognized to play a crucial role on the NR properties. Actually, the non-isoprene molecules are likely behind the better mechanical properties of NR over its synthetic counterpart, but they are also responsible for the high variability of NR quality. This variability of NR quality is a major drawback in NR industry and is directly linked to the latex composition, which is influenced by various physical and physiological parameters. The biochemical composition of latex matters, and this chapter is thus an overview of the protein, lipid, carbohydrate and mineral contents in latex, as well as their distribution in the three main fractions of latex.

1. *Hevea* latex in brief

Hevea brasiliensis (Willd. ex A. Juss) Müll. Arg (*para* rubber tree) is a tropical plant belonging to the Euphorbiaceae family, indigenous to the tropical forests of the Amazon basin. *Hevea brasiliensis* is today the only commercial source of natural rubber (NR), the *cis*-1,4-polyisoprene polymer. Global NR production reached about 13.9 million tons in 2018 (IRSG, 2018), mostly in Asia (>90%), with Thailand and Indonesia being the two first world producers supplying 35% and 30% of NR production, respectively (Trade Map, 2018). NR is extensively used in many industries, in particular for aviation and car tires, while latex can also be used as such, for example in the production of medical gloves and catheters (Vaysse, Bonfils, Sainte-Beuve, & Cartault, 2012). Latex is the cytoplasm of laticiferous cells (Andrews & Dickenson, 1960) which are anastomosed to form vessels arranged in concentric rings in the phloem (Dickenson, 1969) and specialized in the synthesis of *cis*-1,4-polyisoprene. Due to the high pressure of turgor in vessels, latex spontaneously exudes from *Hevea* tree after tapping, an operation which consists in doing a small incision in the bark to sever laticiferous cells (Fig. 1). This feature of laticifers is exploited commercially to collect latex and manufacture NR. Latex can flow for few hours after incision until it flows so slowly that it coagulates by evaporation (Blackley, 1997). The flow of latex can be reactivated few days later by doing a new incision in the cut bark. *Hevea* rubber plantations are managed according to well-defined

Fig. 1 Collection of latex from a tapped *Hevea* tree, Thailand.

systems including tapping-panel management strategies (frequency of tapping, length of the tapping cut) (Lacote, Obouayeba, Dian, Gnagne, & Gohet, 2004) and cultural practices (stimulation and/or fertilization) (Chotiphan et al., 2019).

Freshly tapped *Hevea* latex is a white opaque fluid of density between 0.97 and 0.98 depending on the rubber content. It is almost neutral having a pH in the range of 6.5–7.0. Latex usually contains 25%–50% dry matter (D'Auzac & Jacob, 1989). It mostly comprises *cis*-1,4-polyisoprene which represents about 35% of its fresh weight or 87% of its dry weight. The remaining 5%–6% w/w fresh latex or 13% w/w dry latex are naturally occurring non-isoprene molecules, *i.e.* lipids, proteins, carbohydrates and minerals (Vaysse et al., 2012). Latex is a complex colloidal dispersion which means it contains particles of colloidal size, *i.e.* several nm to several microns. As schematically described in Fig. 2B, latex is made of rubber particles, lutoids and Frey-Wyssling particles dispersed in the cytoplasmic serum (C-serum) (Cockbain & Philpott, 1963; Homans, Dalfsen, & van Gils, 1948; Homans & van Gils, 1950; Southorn, 1960). Rubber particles are the major constituent of latex and represent ~50%–70% w/w of fresh latex, according to the clone of the tree (Table 1). They are spherical and described as a core of *cis*-1,4-polyisoprene surrounded by a monolayer made of lipids and proteins (Cornish, Wood, & Windle, 1999; Nawamawat et al., 2011; Siler, Goodrich-Tanrikulu, Cornish, Stafford, & Mckeon, 1997; Wood & Cornish, 2000) (Fig. 2C). Interestingly, the rubber particles of *Hevea* latex exhibit a bimodal size distribution with the presence of large (diameter ~0.4–1.0 μm) and small particles (diameter ~0.1–0.4 μm) (Singh, Wi,

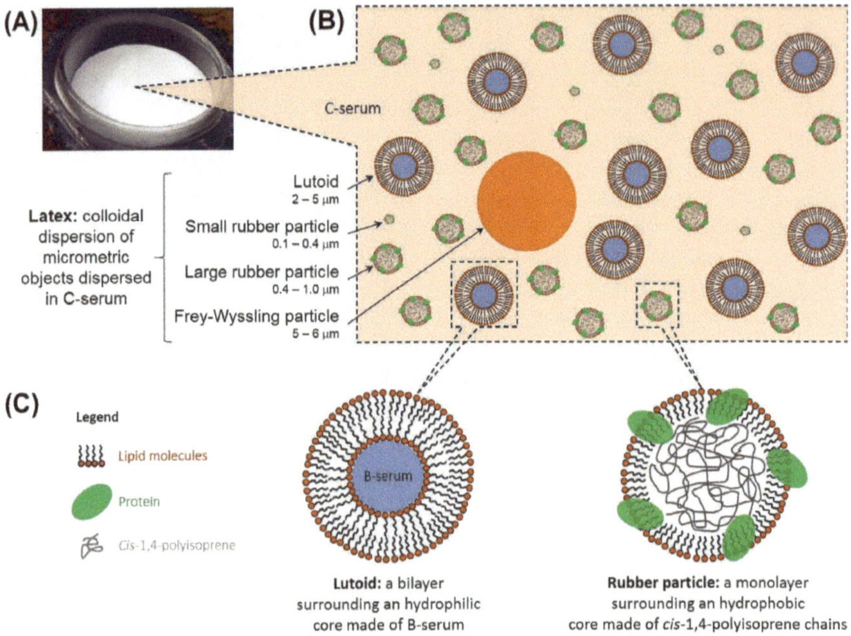

Fig. 2 Latex of *H. brasiliensis* seen at various scales. (A) Picture of a cup of latex in a rubber tree field at the macroscale. (B) Schematic view of the latex at the mesoscale showing lutoids, rubber particles and Frey-Wyssling particles dispersed in the aqueous C-serum. (C) Schematic views of the major constituents of latex at the microscale: a lutoid and a rubber particle. The diagrams in (B) and (C) are not drawn to scale.

Chung, Kim, & Kang, 2003; Wood & Cornish, 2000), named cream and skim fractions, respectively. Lutoids are complex lysosomal vacuoles of ~2–5 μm in diameter (D'Auzac, Crétin, Marin, & Lioret, 1982; Gomez, 1948) (Fig. 2C) and represent ~12%–22% w/w of fresh latex (Table 1). Their hydrophilic content, named B-serum, is rich in various enzymes (Subroto, Koningsveld, Schreuder, & Soedjanaatmadja, 1996; van Parijs, Broekaert, Goldstein, & Peumans, 1991). Frey-Wyssling particles (~2%–3% w/w fresh latex, ~5–6 μm in diameter) are less abundant in latex. They are yellowish colored non-rubber particles (Frey-Wyssling, 1929) with a complex structure suggesting that they might have a functional role (Dickenson, 1969).

As delineated in Fig. 3A, when fresh latex is subjected to centrifugation, it has the capacity to fractionate in its constituting fractions, *i.e.* cream (large rubber particles), skim (small rubber particles), C-serum and lutoids (Cook & Sekhar, 1954; Moir, 1959). The centrifugal speed applied during the first

Table 1 Comparison of bibliographic data on the mass balance (% w/w fresh latex) of the three main fractions of latex: rubber (cream + skim), C-serum and lutoids.

Reference		Clone & latex fractionnation protocol				Mass balance of fractions in latex (%w/w fresh latex)		
Author	**Year**	**Clone of the tree**	**Centrifugal force (g)**	**Time (minutes)**	**T°C**	**Rubber**	**C-serum**	**Lutoids**
Cook & Sekhar	1954	Mixed	59,000	45	5	60	25	12 to 15
Liengprayoon et al.	2017	RRIM600	16,000	45	4	49	29	22
		PB235	16,000	45	4	67	18	15

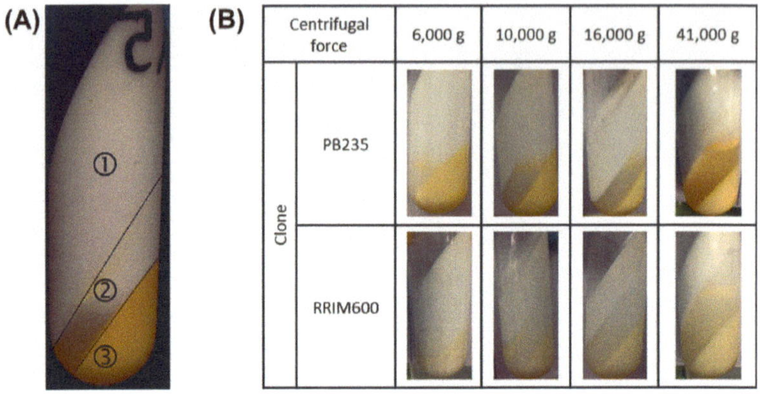

Fig. 3 (A) Picture of a 45 mL clear tube filled with fresh latex of PB235 clone after 45 min of centrifugation at 16,000 g (4 °C). Zone 1: large rubber particles (cream), zone 2: small rubber particles (skim) and Frey-Wyssling particles dispersed in C-serum, zone 3: lutoids. (B) Pictures of 45 mL clear tubes filled with fresh latex of PB235 and RRIM600 clones after 45 min of centrifugation (4 °C) at different centrifugal forces ranging from 6000 to 41,000 g.

centrifugation cycle of the fractionation protocol visually impacts the separation of the fractions, as shown in Fig. 3B. Although external parameters, such as the clone of the tree, can influence the mass balance of the fractions in fresh latex (Liengprayoon et al., 2017), the centrifugal force applied during the fractionation protocol also impacts the separation of the fractions as shown by the dispersed values reported in Table 1. In particular, the integrity of lutoids might be damaged if the centrifugal force becomes too strong resulting in lutoid bursting. The variation in the determination of the mass balance of the fractions inevitably results in dispersed data regarding the distribution of biochemical components in the latex fractions. For instance, the disparity obtained from various authors in the distribution of magnesium, phosphorus and potassium in the lutoid fraction might arise from varying fractionation protocols (Table 2). This discrepancy in the distribution of biochemical components between the latex fractions was previously mentioned by Pakianathan, Tata, Chon, and Sethuraj (1992).

Without addition of a preservative, freshly tapped latex is subjected to an inevitable process of spontaneous coagulation occurring within few hours after tapping (Blackley, 1997). Coagulation corresponds to the aggregation of rubber particles (Bauer et al., 2014) and is an essential parameter of rubber production. Indeed, production is limited if coagulation occurs too early while, in contrast, a prolonged latex flow can exhaust rubber trees. Many

Table 2 Comparison of bibliographic data on the distribution of nitrogen, magnesium, phosphorus and potassium among the three main fractions of latex: rubber (cream + skim), C-serum and lutoids.

Reference		Clone & latex fractionnation protocol				Nitrogen (%)			Magnesium (%)			Phosphorus (%)			Potassium (%)		
Author	Year	Clone of the tree	Centrifugal force (g)	Time (min)	T°C	Rubber	C-serum	Lutoids	Rubber	C-serum	Lutoids	Rubber	C-serum	Lutoids	Rubber	C-serum	Lutoids
Cook & Sekhar	1954	Mixed	59,000	45	5	34	35	31	16	30	54	34	54	12	7	74	19
Liengprayoon et al.	2017	RRIM600	16,000	45	4	38	22	40	8	10	82	33	15	52	29	40	31
Liengprayoon et al.	2017	PB235	16,000	45	4	50	20	30	12	7	81	34	12	54	38	33	29

hypotheses have been formulated to explain the coagulation mechanism which corresponds to a destabilization of the latex colloidal dispersion (Hanower, Brzozowska, & Lioret, 1976). The colloidal stability of the latex results from the negative charges mostly carried by rubber particles and lutoids (Chen & Ng, 1984; Cockbain & Philpott, 1963; Ho, Kondo, Muramatsu, & Oshima, 1996; Kumarn et al., 2018). These negative charges arise from a mixed layer of proteins and phospholipids (Ho et al., 1996). When it exudes from the tree, latex is susceptible to bacterial and enzymatic attacks, initiating biochemical reactions that generate acidic ions, reducing the pH of the fresh latex to the acidic range and leading to spontaneous coagulation of the latex. Moreover, lutoids have been identified as promotors of latex coagulation. This is due to their strong osmotic sensitivity making them fragile objects susceptible to easily burst and release their enzyme content (Gidrol, Chrestin, Tan, & Kush, 1994; Southorn, 1968; Southorn & Edwin, 1968; Southorn & Yip, 1968; Wang et al., 2013; Wititsuwannakul, Pasitkul, Jewtragoon, et al., 2008; Wititsuwannakul, Pasitkul, Kanokwiroon, et al., 2008; Wititsuwannakul, Rukseree, et al., 2008). In particular, hevein, the most abundant protein of lutoids, would behave as a lectin-like protein by creating multivalent bridges between rubber particles through its binding to the N-acetylglucosamine moiety of the glycosylated 22 kDa receptor, a protein located at the surface of the rubber particles. Preservation methods are widely spread in rubber plantations to stabilize latex at liquid state including ammonia, sodium sulfite, formalin and tetra methyl thiuram disulphide (TMTD), or zinc oxide (ZnO) (Jawjit, Pavasant, & Kroeze, 2015).

This chapter presents the review of the biochemical composition of *Hevea* latex in terms of proteins, lipids, carbohydrates and minerals. These non-isoprene components play a significant role in the re-synthesis of latex to replace the part which is drained off trough tapping. The non-isoprenes were shown to be non-homogeneously distributed within the latex fractions (Archer, Audley, McSweeney, & Hong, 1969; Homans et al., 1948; Liengprayoon et al., 2017; Tata, 1980; Wang et al., 2010; Yeang, Arif, Yusof, & Sunderasan, 2002). Hence, the four following paragraphs will describe the composition and the location of each group of molecules within the latex fractions.

The review focuses exclusively on works dealing with fresh non-ammoniated latex. Data on concentrated latex, *i.e.* ammonia-stabilized latex, will not be mentioned, as this preservation method was shown to affect the biochemical composition of latex (Archer & Sekhar, 1955; Bottier et al., 2019; Chen & Ng, 1984; Cook & Sekhar, 1954; Kekwick et al., 1996).

In addition, it must be kept in mind that latex cannot be defined by a single biochemical composition due to its strong variability (Le Roux et al., 2000). Indeed, many parameters are known to influence the biochemical composition of latex such as the clone or the age of the tree (Arreguín, Lara, & Rodríguez, 1988; Bellacicco et al., 2018; Liengprayoon et al., 2013), the season (Alam et al., 2003; Galiani et al., 2011; Trinidad, Gallego, Lucía, & Láinez, 2019) or even the farmer's practices (Wisunthorn, Chambon, Sainte-Beuve, & Vaysse, 2015). Finally, regarding the localization of proteins, lipids, carbohydrates and minerals within the latex fractions, caution should be taken when comparing data from several studies as the fractionation process itself can influence the biochemical content of the fractions (Table 2).

2. Protein content of *Hevea* latex

Latex from *Hevea brasiliensis* is rich in various proteins (Archer, Barnard, et al., 1963). Recently, 1839 unique proteins were identified from the latex of *H. brasiliensis* RRIM600 clone by *de novo* sequencing from mass spectral data (Habib et al., 2016). Latex proteins have been extensively studied because of their allergenicity issue, especially in medical gloves produced from latex (Baur, Chen, Raulf-Heimsoth, & Degens, 1997; D'Amato et al., 2010; Peixinho, Tavares, Tomáz, Taborda-Barata, & Tomáz, 2006, 2014; Raulf & Rihs, 2017; Sussman, Beezhold, & Kurup, 2002; Tomazic, Withrow, & Hamilton, 1995; Yeang et al., 2002). The

Table 3 Comparison of bibliographic data on the protein content of fresh latex (% w/w fresh latex) and the distribution (%) of proteins in the rubber, C-serum and lutoid fractions.

Reference			Proteins (%w/w fresh latex)	Distribution of proteins in fractions (%)		
Author	**Year**	**Clone of the tree**		**Rubber**	**C-serum**	**Lutoids**
Cook & Sekhar	1954	Mixed clones	–	34.0	35.0	31.0
Tata	1980	Mixed: RRIM600, RRIM501, Tjir1	1.57	26.0	45.8	28.2
Yeang et al.	2002	Unknown	1.40	25.0	43.0	32.0
Liengprayoon et al.	2017	RRIM600	1.95	41.5	15.4	43.2
		PB235	1.71	51.1	15.2	33.8

protein content in latex varies between 1% and 2% of its fresh weight (Table 3). Proteins of *Hevea* latex are not homogeneously distributed within latex fractions (Table 3). A part of proteins is adsorbed at the surface of rubber particles (26%–51%), while the remainder is dispersed in the cytoplasmic serum (C-serum) (15%–46%) and the serum contained in the core of lutoids (B-serum) (28%–43%). As mentioned above, dispersed values of protein distribution within fractions can be explained by the centrifugation protocol and/or by physiological parameters.

2.1 Proteins in C-serum fraction

C-serum is rich in proteins: a typical sample of C-serum contains about 12 mg proteins per milliliter (Yeang et al., 2002). Proteins of C-serum are numerous, most of them being water-soluble and acidic. One of the most abundant protein of C-serum is an α-globuline discovered by Archer and Cockbain (Archer & Cockbain, 1955). This water-soluble protein displays the same isoelectric pH as latex (4.55) and its capacity to adsorb on rubber particles suggests that it might contribute to the colloidal stability of the latex. Another protein of C-serum specifically recognized by its strong interactions with *Hevea* latex lectin was detected by Wititsuwannakul, Pasitkul, Jewtragoon, et al. (2008). This 40 kDa C-serum lectin-binding protein was proposed to play a crucial role in a new model of rubber latex coagulation.

Nearly half of the enzymes of *Hevea* latex are found in C-serum, where they are involved in the glycolytic pathway (Bealing, 1969; D'Auzac & Jacob, 1969) as well as in the rubber biosynthesis (Archer & Audley, 1967; Archer, Audley, et al., 1963). About 100 enzymes have been studied and characterized in the C-serum so far. A very detailed review about enzymes and enzymatic activities found in *Hevea* latex was provided by Jacob, Prevôt, and Kekwick (1989). From their work, here is a brief list of the enzymes found in C-serum (the original references corresponding to each enzyme can be found in (Jacob, Prevôt, & Kekwick, 1989)):

- Oxidoreductases: alcohol dehydrogenase, lactate dehydrogenase, malate dehydrogenase, isocitrate dehydrogenase, 6-phosphogluconate dehydrogenase, galactose dehydrogenase, glucose-6-phosphate dehydrogenase, aldehyde dehydrogenase, glyceraldehyde phosphate dehydrogenase, glutathione reductase, phenol oxidase, catalase, peroxidase.
- Transferases: acetyl-CoA-acetyl transferase (thiolase), UDPG fructose glucosyl transferase (sucrose synthetase), aspartate amino transferase (glutamate oxaloacetate transaminase), alanine amino transferase,

hexokinase, glucokinase, fructokinase, phosphofructokinase, mevalonate kinase, pyruvate kinase, phosphoglycerate kinase, pyrophosphate fructose-6-phosphate phosphotransferase, adenylate kinase, phosphoglucomutase, phosphoglyceromutase, sulfate adenyltransferase (ATP sulphurylase), UDPG pyrophosphorylase, thiosulfate sulfurtransferase (rhodanese).

- Hydrolases: carboxylesterase, neutral phosphatase, NADP phosphatase, phosphodiesterase, phospholipase C, phospholipase D, leucine, aminopeptidase, inorganic pyrophosphatase.
- Lyases: dopa decarboxylase, phosphoenolpyruvate carboxylase, aldolase.
- Isomerases: triose phosphate isomerase, glucose phosphate isomerase, isopentenyl pyrophosphate isomerase, 1-L myo-inositol-1-phosphate synthase.

C-serum contains four allergenic proteins including Hev b 5, Hev b 7, Hev b 8 and Hev b 9 (Yeang et al., 2002). Hev b 5 is classified as a major allergen while Hev b 7, Hev b 8 and Hev b 9 are considered as minor ones (Raulf & Rihs, 2017). Hev b 5 is an acidic (pH 3.5), heat-stable 16-kDa protein, rich in glutamic acid, as well as in proline residues and has multiple isoforms (Akasawa, Hsieh, Martin, Liu, & Lin, 1996). Hev b 7 is a 42-kDa patatin-like protein, which isoforms with post-translational modifications are present in C-serum (Jekel, Hofsteenge, & Beintema, 2003; Seppälä et al., 2000). Hev b 8 is a 15-kDa latex profilin belonging to a group of panallergens that are widespread in plants (Valuer, Balland, Harf, Valenta, & Deviller, 1995). Finally, Hev b 9 is a 51-kDa enolase displaying cross-reactivity (Wagner et al., 2000).

One-dimensional gel electrophoresis studies of C-serum proteins indicate a large number of different protein bands. Over the years and with improving and more powerful methods, the number of C-serum protein bands detected by electrophoresis has constantly enlarged. 7 protein bands were highlighted by Roe and Ewart (1942) and Archer and Sekhar (1955), 22 bands by Tata and Moir (1964), 24 by Tata and Edwin (1970) and 27 by Jacob, Nouvel, and Prévôt (1978). In 1957, Yeang et al. highlighted 26 bands in the electrophoretic patterns of C-serum proteins without noting any difference between various *Hevea* clones (Yeang, Ghandimathi, & Paranjothy, 1957). Later, Arreguin et al. identified 30—40 electrophoretic patterns in C-serum proteins (Arreguín et al., 1988). Among them, the 15 most intense bands showed molecular weights ranging from 126 to 18 kDa. More recently, Hasma (1992), Wang et al. (2010, 2015), Havanapan, Bourchookarn, Ketterman, and Krittanai (2016), and Liengprayoon et al. (2017) obtained comparable patterns for C-serum proteins with numerous protein bands in the high-molecular weight area. Two-dimensional electrophoresis (2-DE) was applied as well to study C-serum proteins. Posch et al. observed 200 spots in 1997 (Posch et al., 1997) while Li et al. could detect 447 spots in 2005 (Li et al., 2009). This number was strongly increased by Wang et al. to 1248 spots in 2010 (Wang et al., 2010) and 1824 in 2018 (Wang et al., 2018). In the later study, the authors could identify 1837 unique proteins by shotgun proteomics (Wang et al., 2018).

2.2 Proteins in lutoid fraction

Among the proteins of the lutoid fraction, 80% are comprised in the B-serum while the remainder 20% are part of lutoid membranes (Tata, 1980). B-serum mostly contains water-soluble proteins that can be acidic and basic (Moir & Tata, 1960). B-serum typically comprises about 24 mg proteins per milliliter (Yeang et al., 2002). Although B-serum is twice richer in proteins than C-serum, its protein population is much less diversified than the one of C-serum. Indeed, there are less than 20 major proteins in B-serum with a single protein, hevein, representing 50%–70% of the total B-serum soluble proteins (Archer et al., 1969). The concentration of hevein in latex is 1–2 g/L. Hevein was the first protein isolated from latex by Archer (1960). A recent and very complete review dedicated to hevein (Berthelot, Peruch, et al., 2016) rightly reminds that hevein should not be mistaken with hevain, a protein behaving as a protease (Lynn & Clevette-Radford, 1984), which is found in both C-serum and lutoid fractions (Lynn & Clevette-Radford, 1986). Hevein is a small (43 amino acids, 4.7 kDa), anionic and cysteine-rich protein (Walujono, Scholma, Beintema, Mariono, & Hahn, 1975) having anti-microbial properties (van Parijs et al., 1991). Due to its chitin-binding capacity, hevein was proposed to be a lectin-like protein able to interact with a 22 kDa protein of the rubber particle via a sugar linkage thus resulting in the formation of aggregates ultimately leading to the coagulation of latex (Gidrol et al., 1994). Hevein is the amino-terminal posttranslational cleavage product of its precursor - prohevein (Lee, Broekaert, & Raikhelsii, 1991; Soedjanaatmadja, Subroto, & Beintema, 1995a). Indeed, prohevein (187 amino acids, 18.5 kDa) is able to cleave into two fragments: the N-terminal domain hevein (43 amino acids, 4.7 kDa) and the C-terminal domain (144 amino acids, 13.3 kDa). Both prohevein and the C-terminal domain are present in B-serum. Moreover, the C-terminal domain was shown to possess amyloid properties (Berthelot, Lecomte, et al., 2016). These amyloid properties might result in the formation of microfibrils (Audley, 1966; Dickenson, 1969; Tian, Han, Wu, & Hu, 2003) and microhelices (Tata & Gomez, 1980) that were previously observed in B-serum of lutoids. The proteins prohevein, hevein and prohevein C-terminal domain all possess allergenic properties and were named Hev b 6.01, Hev b 6.02 and Hev b 6.03, respectively (Archer, 1960; Rozynek, Posch, & Baur, 1998; Soedjanaatmadja et al., 1995a).

Another protein abundant in B-serum of lutoids is hevamine existing under two forms (A and B) (Archer, 1976). Hevamines are 29.5 kDa cationic proteins having a lysozyme/chitinase activity (Jekel, Hartmann, & Beintema, 1991; Tata, Boyce, Archer, & Audley, 1976). The crystal structure of hevamine was determined with a resolution of 2.2 Å showing a completely new lysozyme/chitinase fold (Terwisscha van Scheltinga, Kalk, Beintema, & Dijkstra, 1994). β-1,3-glucanase (36 kDa) is also an important protein of B-serum (Subroto et al., 1996). It was shown to be one of the most allergenic among latex proteins and is known as Hev b 2 (Sunderasan et al., 1995). Two other allergens found in B-serum include Hev b 4, a component of the microhelix (Sunderasan et al., 1995), and Hev b 10, a 45 kDa manganese-superoxide dismutase (Posch et al., 1997; Rihs, Chen, Rozynek, & Cremer, 2001).

Several enzymes were found in the membrane of lutoids including: an ATPase functioning as an influx proton pump (Cretin, 1982), a NADH-cytochrome c reductase (Moreau, Jacob, Dupont, & Lance, 1975) and a pyrophosphatase (Marin, 1993). Recently, Wang et al. studied the membranes of lutoids by one- (SDS-PAGE) and two-dimensional (2-DE) electrophoresis combined to proteomics (Wang et al., 2013). Among the 19 bands detected in the SDS-PAGE profile of lutoid membrane, 3 bands were prominent. Moreover, 902 spots were highlighted in the 2-DE gel. Bands and spots were excised and later studied by mass spectrometry allowing to identify 57 proteins. Among them, the five most abundant proteins of lutoids membrane were: hydroxynitrile lyase, transposon protein, resistance protein RGC2, NBS-LRR resistance protein and hevein.

2.3 Proteins in rubber fraction

In this paragraph, the rubber fraction is defined as the mixture of large rubber particles (cream fraction) and small rubber particles (skim fraction). Although the protein populations of cream and skim fractions share many similarities, one has to keep in mind that differences also exist. Proteome analysis of the large and the small rubber particles by 2D-DIGE indicated that some proteins are differentially expressed in both populations of particles (Xiang et al., 2012). The visual comparison of the SDS-PAGE electrophoretic profiles of proteins extracted from large and small rubber particles shows different patterns as well (Liengprayoon et al., 2017). Many proteins are present in the rubber fraction. In 2013, Dai et al. identified 186 proteins in the rubber fraction by using shotgun tandem mass spectrometry (Dai et al., 2013). The identified proteins have molecular mass ranging from 3.9 to 194.2 kDa and many of them were observed for the first time in the rubber fraction, including for instance cyclophilin, phospholipase D, cytochrome P450, small GTP-binding protein, clathrin, eukaryotic translation initiation factor, annexin, ABC transporter, translationally controlled tumor protein or ubiquitin-conjugating enzymes. Recently, Wang et al. used 2-DE combined to shotgun proteomics experiments and identified 123 proteins that were specific to the rubber fraction (Wang et al., 2018).

The two most abundant transcripts in latex are two proteins, REF1 and SRPP1 (Chow et al., 2007), localized at the surface of rubber particles (Berthelot, Lecomte, Estevez, & Peruch, 2014). Whereas it is agreed that the rubber particle is made of a hydrophobic core of *cis*-1,4-polyisoprene, the structure of its lipid-protein surface membrane is still controversial. Indeed, three models of membrane have been proposed: a lipid monolayer biomembrane surrounded by proteins (or a film of proteins) (Cornish et al., 1999), a mixed monolayer model (patchy) (Nawamawat et al., 2011; Siler et al., 1997) and a shell interfacial model (Rochette, Crassous, Drechsler, Gaboriaud, & Eloy, 2013). In the first mentioned study, the polymer core interacts with a contiguous monolayer of lipid fatty acid chains, whereas the lipid polar headgroups interact with proteins, which are in contact with the C-serum medium (Cornish et al., 1999). The second model proposes a polymer core surrounded by a mixed lipid and protein monolayer (Nawamawat et al., 2011). In the third model the polymer core faces a protein layer interacting with lipids where the fatty acid chains of lipids are in contact with the protein layer and hydrophilic lipid headgroups face the C— serum medium (Rochette et al., 2013).

REF1 (Rubber Elongation Factor, 138 amino acids, 14.6 kDa) was identified in 1989 by Dennis & Light and was given its name for its role in the increase of polyisoprene chain length (Dennis & Light, 1989). SRPP1 (Small Rubber Particle Protein, 204 amino acids, 22.4 kDa) was named by Oh et al. in 1999 as it was detected at the surface of small rubber particles (Oh et al., 1999). At first, REF1 was detected on large rubber particles, but it is now established that REF1 can be found on both types of particles (Archer, Audley, et al., 1963; Bahri & Hamzah, 1996; Singh et al., 2003;

Yeang et al., 1996). REF1 makes up 10%–60% of the total proteins in the whole latex and it has been calculated that the stoichiometry of REF1 molecules to rubber molecules is 1:1 (Dennis & Light, 1989). The amino acid sequences for REF1 and SRPP1 share 72% similarity (Oh et al., 1999). Many REF and SRPP isoforms have been reported in *Hevea* rubber tree. Rahman et al. found 10 REF genes and 12 SRPP genes in the genome of rubber tree (Rahman et al., 2013), Tang et al. (2016) recently characterized 8 and 10, and Lau et al. (2016) found 9 and 8, respectively. However, REF1 and SRPP1 proteins still prevail in latex as compared to other isoforms (Tang et al., 2016). REF1 and SRPP1 are two strong allergens of *Hevea* latex and are known as Hev b 1 and Hev b 3, respectively (Yeang et al., 2002). Both proteins are highly hydrophobic (Berthelot, Lecomte, Estevez, & Peruch, 2014) and REF1 was shown to possess amyloid properties (Berthelot et al., 2012).

Although the function of REF1 and SRPP1 still remains elusive (Berthelot, Lecomte, Estevez, & Peruch, 2014), both have been suggested to be part of the rubber biosynthesis machinery occurring at the surface of rubber particles. In particular, REF1 would be involved in the rubber transferase complex as an architectural protein. Indeed, the REF1 protein, in association with another protein called bridging protein (CPT-REF1 bridging protein), would participate in the support of the enzyme cis-prenyl transferase (CPT) at the surface of the rubber particle. This would allow an appropriate arrangement and an adequate folding of the CPT to perform properly its enzymatic activity of rubber chain elongation (Yamashita et al., 2016). Interestingly, this heteromeric protein complex involved in rubber biosynthesis that was demonstrated in latex of *Hevea brasiliensis* (Yamashita et al., 2016), was established as well in the latex of lettuce (*Lactuca sativa*) (Qu et al., 2015), dandelion (*Taraxacum brevicorniculatum*) (Epping et al., 2015) and guayule (*Parthenium argentatum*) (Lakusta et al., 2019). Very recently, a different model of rubber transferase complex was proposed by Cornish et al. (2018). In this work, the authors showed in multiple plant species (*H. brasiliensis*, *P. argentatum* and *F. elastica*) the presence of an enzymatically-active complex made of three proteins of low and high molecular weights (for *H. brasiliensis*: 241 kDa, 3.65 kDa and 1.6 kDa). The question of rubber biosynthesis still raises questions and is discussed in two recent reviews (Cherian, Ryu, & Cornish, 2019; Men, Wang, Chen, Zhang, & Xian, 2019).

As the two major proteins of the rubber fraction and in light of the similar models recently proposed for Hevea, lettuce, dandelion and guayule,

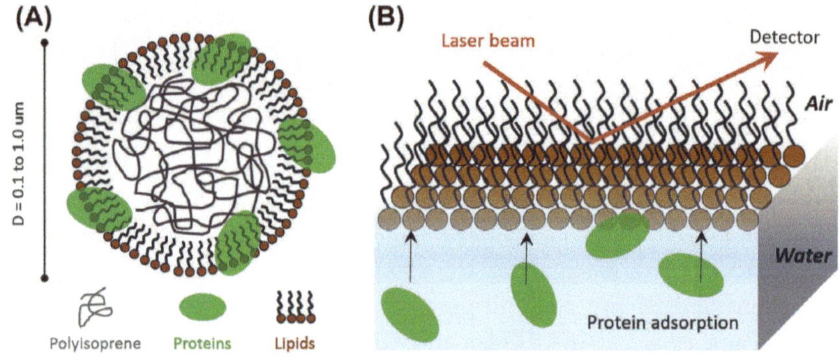

Fig. 4 (A) Schematic view of the rubber particle with the hydrophobic core of cis-1,4-polyisoprene chains surrounded by lipids and proteins organized in monolayer. (B) Schematic view of a Langmuir film of lipids formed at the air/water interface where proteins adsorb after injection in water. Interactions developed between lipids and proteins are monitored *in situ* at the air/water interface using biophysical tools. The diagrams in (A) and (B) are not drawn to scale.

the arrangement of REF1 and SRPP1 proteins in the lipid membrane of rubber particles undoubtedly plays a crucial role in the rubber biosynthesis machinery. An original approach in Langmuir monolayers was proposed to investigate these lipid-protein interactions. Indeed, such Langmuir monomolecular films formed at the air/water interface can simply and easily mimic the structure of the lipid-protein monolayer surrounding the core of *cis*-1,4-polyisoprene (Fig. 4). Langmuir films can then be characterized *in situ* at the air/water interface by various biophysical tools (*e.g.* Brewster angle microscopy, ellipsometry and polarization modulated-infrared reflection adsorption spectroscopy) to determine the type of interactions occurring between lipids and proteins. This strategy was applied to recombinant REF1 and SRPP1 proteins interacting with synthetic lipids (Berthelot, Lecomte, Estevez, Zhendre, et al., 2014). This work carried out mainly with DMPC or asolectin have brought to light the different behaviors of REF1 and SRPP1, namely lipid monolayer-inserting and particle surface-covering, respectively. Later, the interactions of recombinant REF1 and SRPP1 proteins with native lipids of *Hevea* latex were studied (Wadeesirisak et al., 2017). It was shown that both REF1 and SRPP1 proteins behave differently depending on the lipid family they interact with, *i.e.* phospholipids, glycolipids or neutral lipids. In particular, while the secondary structure of REF1 in α-helices was maintained when interacting with phospholipids and glycolipids, in the presence of neutral lipids, it switched to β-sheets as in its amyloid form. This conformational change might play a crucial role in the coagulation and/or biosynthesis mechanism.

3. Lipid content of *Hevea* latex

Lipids are hydrophobic compounds that are soluble in organic solvents (chloroform, methanol) but sparingly in aqueous solution. Due to their hydrophobicity, lipids are the major non-isoprene molecules found in dry rubber. As proteins, the lipids are non-homogeneously distributed within latex fractions and are almost absent from the aqueous C-serum fraction (Liengprayoon et al., 2017). Thus only the lipid composition of the lutoid and rubber fractions will be detailed below. The lipid content in *Hevea* latex varies from 1% to 2% of its fresh weight (Hasma & Subramaniam, 1986; Liengprayoon et al., 2013, 2017) and comprises various lipid classes including neutral lipids, phospholipids and glycolipids (Table 4). Among total lipids of *Hevea* latex, neutral lipids are the major species (49%–68%) while the contents of phospholipids and glycolipids are roughly similar, varying from 14% to 26% and 18%–32%, respectively.

Liengprayoon et al. extensively studied the lipids of latex and natural rubber (Liengprayoon et al., 2008, 2013, 2011) and confirmed the results previously obtained by Hasma and Subramaniam (Hasma & Subramaniam, 1986). The fatty acid composition of *Hevea* lipid extracts shows 9 different species including myristic acid (C14:0), palmitic acid (C16:0), palmitoleic acid (C16:1), stearic acid (C18:0), oleic acid (C18:1), linoleic acid (C18:2), linolenic acid (C18:3), arachidic acid (C20:0) and furanoid fatty acid

Table 4 Comparison of bibliographic data on the lipid content in fresh latex (% w/w fresh latex) and share of the three lipid classes (phospholipids, glycolipids, neutral lipids) among the total lipids.

Reference			Lipids (% w/w fresh latex)	Lipid class (% of total lipids)		
Author	**Year**	**Clone of the tree**		**Phospholipids**	**Glycolipids**	**Neutral lipids**
Hasma & Subramaniam	1986	RRIM501	1.06	14	32	54
Liengprayoon et al.	2013	PB235	–	19	21	60
		RRIM600	–	26	25	49
		BPM24	–	23	23	54
Liengprayoon et al.	2017	PB235	1.79	14	18	68
		RRIM600	1.01	21	24	55

(FuFA) (Liengprayoon et al., 2013). The clonal origin was shown to impact the share of these fatty acid species. For both RRIM600 and BPM24 clones, the C18:2 species is the main fatty acid (representing 45.8% and 35.5%, respectively), followed by the FuFA species (16.1% and 27.8%, respectively). For PB235, FuFA is the major fatty acid accounting for 72.9% of the total fatty acids, followed by the C18:2 species (11.4%).

The neutral lipids of *Hevea* latex were shown to contain several species including sterols (β-sitosterol, stigmasterol and Δ^5-avenasterol), tocotrienols (γ-tocotrienol and α-tocotrienol which is a strong anti-oxidant) and fatty alcohols (octadecanol C18−OH and eicosanol C20−OH) (Liengprayoon et al., 2013). For RRIM600, BPM24 and PB235 clones, the three most abundant species of the neutral lipid class were β-sitosterol (44%, 43% and 37%, respectively), Δ^5-avenasterol (26%, 19% and 29%, respectively) and stigmasterol (9%, 7% and 7%, respectively).

Six phospholipids were detected in the phospholipids of *Hevea* latex, *i.e.* phosphatidylcholine (PC), lysophosphatidylcholine (LPC), phosphatidyl ethanolamine (PE), phosphatidylinositol (PI), lysophosphatidylinositol (LPI) and phosphatidic acid (PA) (Liengprayoon et al., 2013). Among them, PC, is the most abundant species for all clones (62%-55%) followed, in decreasing order, by LPC (28%-19%), PA (11%-7%), PE (4%-5%), LPI (5%-2%) and PI (3%-1%). The major fatty acid of the phospholipid class was the species C18:2 representing 59%, 53% and 36% of the phospholipid fatty acids in clones RRIM600, BPM24 and PB235, respectively (Liengprayoon et al., 2013, 2011).

Four main species were identified in the glycolipid class of *Hevea* latex including digalactosyl diacylglycerol (DGDG), steryl glucoside (SG), esterified steryl glucoside (ESG) and monogalactosyl diacylglycerol (DGDG) (Liengprayoon, Sriroth, Dubreucq, & Vaysse, 2011). For the three clones RRIM600, BPM24 and PB235, the contents of glycolipid species were roughly similar. DGDG was the most abundant species (47%−51%), followed, in decreasing order, by SG (30%−34%), ESG (7%−12%) and MGDG (8%). The fatty acid composition of glycolipids reflected that of total lipids, with linoleic acid (C18:2) being the main fatty acid in the clones RRIM600 (47%) and BPM24 (44%), and FuFA the main one in the clone PB235 (42%).

3.1 Lipids in lutoid fraction

As a vacuole, the lutoid comprises a 80 Å thick bilayer membrane (Gomez & Southorn, 1969), named tonoplast, allowing to compartmentalize the lutoid aqueous content (B-serum) (Chrestin, Jacob, & d'Auzac, 1984). While lipids are absent from the B-serum, they are elementary components of membranes, and thus represent a significant amount of the lutoid fraction representing up to 10% w/w of the lutoid dry matter (Liengprayoon et al., 2017). Dupont et al. carried out the first research on the chemical composition of the lutoid membrane which were separated on a discontinuous sucrose gradient (Dupont, Moreau, Lance, & Jacob, 1976). They characterized the phospholipid composition of the lutoid membrane and showed that it contains almost exclusively PA which represents 82% of the total phospholipid fraction. Phospholipase D was demonstrated not to be responsible for this high content in PA. Two unidentified components formed the remaining 18% of the total phospholipid fraction. PC and PE, two major constituents of most biological membranes, were absent from lutoid membrane. Both saturated and unsaturated species were detected among fatty acids of lutoids. They include C18:2 (38.8%), C18:0 (25.3%), C16:0 (20.5%), C18:1 (13.8%) and C14.0 (1.6%). Interestingly, the linolenic acid (C18:3) was not detectable.

Liengprayoon et al. measured the content of each lipid class in the lutoid fraction of two *Hevea* clones. The lipids of the lutoid fraction of latex from PB235 clone contains roughly similar amounts of neutral lipids, glycolipids and phospholipids (31%, 39% and 30%, respectively). However, the lipids of the lutoid fraction from RRIM600 clone contains 50% neutral lipids, 36% of glycolipids and 14% of phospholipids (Liengprayoon et al., 2017).

3.2 Lipids in rubber fraction

The lipids of the rubber fraction are localized in the monolayer membrane surrounding the rubber particles. On a dry weight basis, the lipids represent 2%–3% of large rubber particles (cream) and 4%–6% of small rubber particles (skim), respectively (Liengprayoon et al., 2017). In other words, skim is twice more concentrated in lipids than cream which was explained by the specific surface that is higher in small particles than in large ones. Hasma provided a detailed composition of the lipids from rubber particles of six *Hevea* clones (Hasma, 1991). For five of the six studied clones, neutral lipids were the major class varying from 37% to 70% of the total lipids. The glycolipid and phospholipid contents were in the same range, *i.e.* 13%–36% and 15%–30%, respectively. Only the clone RRIM600 displayed a content of phospholipids (44%) that was higher than neutral lipids (34%) and glycolipids (22%). For all the *Hevea* clones, the neutral lipids comprised triglycerides, diglycerides, free fatty acids, free and esterified sterols, free and esterified tocotrienols and fatty alcohols and their acetates. The glycolipids consisted mainly of free and esterified SG and both species MGDG and DGDG, while the phospholipids consisted of PE, PC and PI.

Siler et al. later studied the composition of rubber particles of four plants including *Hevea brasiliensis*, *P. argentatum*, *F. elastica*, and *Euphorbia lactiflua* (Siler et al., 1997). For *Hevea brasiliensis* (clone PB260), the fatty acid composition of rubber particles comprises eight species including C16:0, C16:1, C18:0, C18:1, C18:2, C18:3 C20:0 and FuFA. All the fatty acid species were found in the three lipid class (neutral lipid, phospholipids, glycolipids) except the C16:1 species which was detected in phospholipid class only. In contrast, FuFA was absent from the phospholipids while it is the major fatty acid in the neutral lipids and glycolipids of rubber particles. In the mentioned study, FuFA was not detected in the rubber particles of other latex plants.

In an attempt to compare the molecular species of acylglycerols (AG) in the lipid extracts of rubber particles from guayule (*P. argentatum*) and *Hevea brasiliensis*, Hathwaik et al. identified 56 molecular species of AG in the lipid extract of *Hevea* rubber particle (Hathwaik, Lin, & McMahan, 2018). Among the total fatty acids of *Hevea*, FuFA represents about 81%, normal fatty acids about 17% and hydroxy fatty acids about 1.6%.

4. Carbohydrate content of *Hevea* latex

The carbohydrate content of *Hevea* latex is about 1.5% of its fresh weight (Wititsuwannakul & Wititsuwannakul, 2001). The major soluble carbohydrates in latex are total cyclitols, sucrose and glucose in that order (Low, 1978). Bellacicco et al. studied the sugar and polyol composition of latex from five *Hevea* clones (Bellacicco et al., 2018). They found that quebrachitol and sucrose contents were both in the range of 0.5% w/w fresh latex. Other carbohydrates were found to be 10 times lower than sucrose and quebrachitol for myo-inositol and glycerol, or even 100 times lower for fructose, mannose, galactose and sorbitol. The clonal origin impacted the carbohydrate contents. For instance, the clone PB235 showed significantly lower content of sucrose and higher content of galactose as compared to other clones. Carbohydrates are mainly confined to the C-serum fraction where they serve as substrate for the numerous enzymes present in C-serum. Only glucose was mostly found in the lutoid fraction (Bealing, 1969; D'Auzac & Jacob, 1969). As compared to C-serum, the content in carbohydrates of the lutoid fraction is much less documented in the literature. Blackley wrote that the lutoids may or may not have associated with them small amounts of carbohydrates (Blackley, 1997). The carbohydrates are nearly absent from the rubber fraction. Thus, only the C-serum fraction is detailed below.

The two most abundant carbohydrates in C-serum are sucrose and quebrachitol (2-*O*-methyl-L-inositol). Quebrachitol is the major carbon compound in fresh latex representing up to 1.2% (w/v), in comparison with 0.4% (w/v) for sucrose (Bealing, 1969, 1981). Sucrose is the predominant soluble sugar (Bealing, 1969) and its intracellular concentration is used to evaluate the metabolic activity of laticifer cells (Jacob, Prevôt, Clement-Vidal et al., 1989). The biosynthesis of *cis*-1,4-polyisoprene derives from isopentenyl pyrophosphate (IPP) which itself originates from the pyruvate obtained by sucrose degradation (glycolysis). Sucrose is thus the unique precursor of rubber biosynthesis (D'Auzac, 1964). Quebrachitol, a cyclic polyol, is the most important single component among the non-isoprene molecules of latex (Vaysse et al., 2012). Amazingly, the physiological function of quebrachitol remains unclear, although it was shown to contribute significantly to the formation and maintenance of turgor pressure and thus to the latex flow after tapping (Sheldrake, 1979). As quebrachitol was found at high concentration in a clone displaying a rapid coagulation (Tjir 1) and at lower concentration in a clone having a slow coagulation (RRIM501), it was suggested that quebrachitol negatively impacts the colloidal stability of the latex (Low, 1978).

In addition to sucrose and quebrachitol, the C-serum also contains myo-inositol, glycerol, sorbitol, mannose, glucose, galactose and fructose (D'Auzac & Pujarniscle, 1959; Smith, 1954). A low amount of raffinose was also detected by Tupy & Resing (Tupy & Resing, 1968). These authors explained the low amount of fructose which would immediately serve in metabolism, thus leading to glucose accumulation.

5. Mineral content of *Hevea* latex

About 0.5% of fresh *Hevea* latex weight (5000 ppm) is composed of inorganic ions (Jacob, D'Auzac, & Prevôt, 1993). The overall mineral content is known as the ash content and it can be influenced by both seasonal and clonal parameters (Le Roux et al., 2000; Moreno, Ferreira, Gonçalves, & Mattoso, 2005; Yip, 1990). For instance, the ash contents of both clones

GT1 and PB235 showed similar trends with a gradual increase from the major rainy season (+0.30% in July) to the major dry season (+0.55% in February) and then a progressive decrease to May–June (Le Roux et al., 2000). Other authors observed that, regarding the ash content, the clone RRIM600 was the least susceptible to the climate variation in contrast to the clone GT1 which was the most susceptible (Moreno et al., 2005).

Rolere et al. recently gathered values from the literature on the detailed mineral composition of fresh latex (Rolere, Char, et al., 2016). Potassium is the most abundant mineral of latex accounting for 1000–5000 ppm, followed by sodium (70–1000 ppm), magnesium (10–1200 ppm), phosphorus (100–700 ppm), calcium (0.5–300 ppm), iron (10–120 ppm), rubidium (7–40 ppm) and cooper (2–5 ppm). Manganese, zinc and lead were detected at the level of traces. These figures are in accordance with a recent study assessing the long term effects of the fertilization on the yield and the quality of latex (Chotiphan et al., 2019). It was indeed found that the mean content of minerals in fresh latex harvested from non-fertilized trees over three years (2014–7) was: 1512 ppm of potassium, 522 ppm of phosphorus, 402 ppm of magnesium and 14 ppm of calcium.

As observed for the overall mineral composition of latex, significant differences in the concentrations of specific components were also noticed among the clones and the seasons (Gopalakrishnan, Thomas, Philip, Krishnakumar, & Jacob, 2010). In particular, potassium and magnesium contents in the latex were influenced significantly by seasonal effect and varied differently among clones. These mineral elements, in addition with phosphorus, are known to influence the colloidal stability of latex (Philpott & Westgarth, 1953). Low potassium and phosphorus and high magnesium correlate with a low stability of latex. In addition, Gomez observed that high latex stability was associated with low Mg/P and (Mg + Ca)/P ratios, and vice versa (Gomez, 1980).

Liengprayoon et al. studied the mineral composition of the latex and its fractions for the clones RRIM600 and PB235 (Liengprayoon et al., 2017). In latex, the sum of the contents of sulfur, phosphorus, potassium, calcium and magnesium in RRIM600 and PB235 clones accounts for 0.30% and 0.25% w/w fresh latex, respectively. For the clone RRIM600, C-serum, lutoid and rubber fractions contain 26%, 50% and 24% of the total minerals, respectively. These figures slightly differ for the clone PB235 where C-serum, lutoid and rubber fractions comprise 22%, 45% and 33% of the total minerals, respectively.

Table 5 Distribution (%) on a fresh latex basis of potassium, phosphorus, magnesium, sulfur and calcium in the rubber, C-serum and lutoid fractions.

	Potassium (%)	Phosphorus (%)	Magnesium (%)	Sulfur (%)	Calcium (%)
C-serum	33–40	12–15	7–10	17–20	5–18
Lutoids	29–31	52–54	81–82	48–55	41–56
Rubber	29–38	33–34	8–12	25–35	39–41

From Liengprayoon et al. (2017).

5.1 Minerals in C-serum fraction

The most abundant mineral element of C-serum is potassium, followed by phosphorus and magnesium (Jacob et al., 1993; Liengprayoon et al., 2017; Yip & Chin, 1977). Sulfur is also present and traces of calcium as well. On a fresh latex basis and for both clones RRIM600 and PB235, the C-serum contains 17–20% of the total sulfur, 12%–15% of the total phosphorus, 33%–40% of the total potassium, 5%–18% of the total calcium and 7%–10% of the total magnesium (Liengprayoon et al., 2017) (Table 5). Traces of iron, manganese and zinc were also detected in the C-serum fraction (Jacob et al., 1993).

5.2 Minerals in lutoid fraction

As compared to C-serum and rubber fractions, the lutoid fraction is the richest in minerals (Jacob et al., 1993; Liengprayoon et al., 2017). The two most abundant mineral elements in lutoids are magnesium and phosphorus, followed by potassium, sulfur and calcium (Liengprayoon et al., 2017). On a fresh latex basis and for both clones RRIM600 and PB235, the lutoid fraction contains 48%–55% of the total sulfur, 52%–54% of the total phosphorus, 29%–31% of the total potassium, 41%–56% of the total calcium and 81%–82% of the total magnesium (Table 5). According to Jacob et al. the ratios lutoids/C-serum for the potassium, magnesium, phosphorus and calcium contents were 1.0, 8.0, 8.7 and 6.0, respectively (Jacob et al., 1993); whereas these ratios for the clones RRIM600 and PB235 from Liengprayoon et al. are 0.8–0.9, 8.2–11.2, 3.6–4.5 and 2.2–10.5, respectively (Liengprayoon et al., 2017).

5.3 Minerals in rubber fraction

On a dry matter basis, the concentration in minerals in the rubber fractions is extremely low as compared to C-serum and lutoids. While the mineral concentration in the rubber fraction is about 0.1%–0.2% w/w dry matter, it is around 7%–8% w/w dry matter for C-serum and 7%–9% w/w dry matter for lutoids (Liengprayoon et al., 2017). The two major minerals of the rubber fraction are phosphorus and potassium, followed by sulfur, magnesium and calcium. On a fresh latex basis and for both clones RRIM600 and PB235, the rubber fraction contains 25%–35% of the total sulfur, 33%–34% of the total phosphorus, 29%–38% of the total potassium, 39%–41% of the total calcium and 8%–12% of the total magnesium (Liengprayoon et al., 2017) (Table 5).

6. Concluding remarks

The latex of *Hevea brasiliensis* is a very complex material both in its structure and composition. In terms of structure, the *Hevea* latex is a colloidal dispersion where various micrometric objects, including rubber particles, lutoids and Frey-Wyssling particles, are dispersed in the cytoplasmic serum (C-serum). The rubber fraction is the most abundant in latex (37–67% w/w fresh latex), followed by the C-serum (18–29% w/w fresh latex), the lutoids (12%–22% w/w fresh latex) and the Frey-Wyssling particles (2%–3% w/w fresh latex). In terms of composition, fresh latex comprises: 35.0% *cis*-1,4-polyisoprene, 1.5% proteins, 1.3% lipids, 1.5% carbohydrates, 0.5% minerals, 0.5% organic solutes and 59.7% water (Fig. 5A and B). The three main fractions of latex, *i.e.* C-serum, lutoids and rubber, can be easily

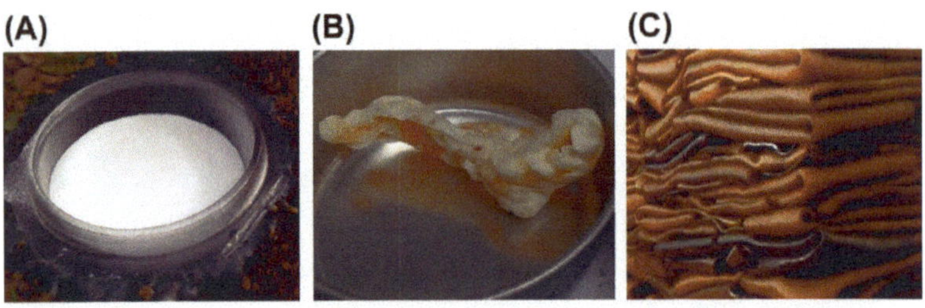

Fresh latex composition (% w/v fresh latex)	Dry latex composition (% w/w dry latex)	Dry rubber composition (% w/w dry rubber)
Polyisoprene: 35.0 %	Polyisoprene: 87.0 %	Polyisoprene: 94.0 %
Proteins: 1.5 %	Proteins: 3.7 %	Proteins: 2.2 %
Lipids: 1.3 %	Lipids: 3.2 %	Lipids: 3.4 %
Carbohydrates: 1.5 %	Carbohydrates: 3.7 %	Carbohydrates: 0.4 %
Minerals: 0.5 %	Minerals: 1.1 %	Minerals: 0.2 %
Organic solutes: 0.5 %	Organic solutes: 1.2 %	Organic solutes: 0.1 %
Water: 59.7 %		

Fig. 5 Pictures and corresponding compositions of: (A) fresh latex, (B) dry latex and (C) dry rubber (in the form of unsmoked sheet). *Figures from Vaysse et al. (2012).*

separated by centrifugation and the non-isoprene components, *i.e.* proteins, lipids, carbohydrates and minerals, are not homogeneously distributed in these fractions. Each class of non-isoprene displays a diverse and varied composition. In very brief, the proteins comprise numerous water-soluble enzymes and hydrophobic proteins as well. Three lipid classes are present in the latex including neutral lipids, phospholipids and glycolipids and each lipid class contains many species. The carbohydrates mostly contain sucrose and quebrachitol while the major minerals are potassium, magnesium and phosphorus. Finally, the composition of *Hevea* latex is subjected to a strong variability under the influence of various parameters such as the clone, the age of the tree, the season, the location, etc.

Natural rubber (NR) is derived from the *Hevea* latex, and thus the complexity and the variability of its composition have heavy consequences on NR properties. Indeed, a part of the biochemical components originally contained in the latex is retained in dry rubber after processing (Fig. 5C). Due to their hydrophobic nature, lipids are the most retained in NR (3.4% w/w dry rubber), followed by proteins (2.2% w/w dry rubber), carbohydrates (0.4% w/w dry rubber) and minerals (0.2% w/w dry rubber). Although the overall part of non-isoprenes in NR is rather low (about 6% w/w dry rubber), the non-isoprene molecules are suspected of being responsible for the superior properties of NR over its synthetic counterpart. However, they would also be involved in the inconsistent quality of NR. Many studies have demonstrated the great importance of those components on the NR quality and properties (Hasma & Othman, 1990; Karino, Ikeda, Yasuda, Kohjiya, & Shibayama, 2007; Lotti, Moreno, Gonçalves, & Bhattacharya, 2012; Nun-anan, Wisunthorn, Pichaiyut, Nathaworn, & Nakason, 2019; Thuong, Yamamoto, Nghia, Cornish, & Kawahara, 2017; Wei, Liu, Zhang, Zhao, & Luo, 2019). For instance, phospholipids and proteins take part in the peculiar structure of NR (Rolere, Bottier, et al., 2016; Tarachiwin, Sakdapipanich, Ute, Kitayama, Bamba, et al., 2005; Tarachiwin, Sakdapipanich, Ute, Kitayama, & Tanaka, 2005). The saturated fatty acids play an important role in the crystallization behavior of NR as nucleating agents (Kawahara, Kakubo, Sakdapipanich, Isono, & Tanaka, 2000), while the unsaturated fatty acids act as plasticizer (Kawahara & Tanaka, 1995). The minerals are involved in the structure of rubber particles and NR material (Rippel, Alberto, Leite, & Galembeck, 2002; Rolere, Char, et al., 2016), while quebrachitol crystals impact the mechanical properties of NR (Toki et al., 2008). The proportion of serum and/or lutoids, and consequently their biochemical components, has an important effect on the change of structure and properties during storage of NR (Chaiyut et al., 2017). The crucial role of non-isoprenes on the properties of NR is now well recognized and documented. The growing interest in the characterization of such compounds consequently led to the development of new methods allowing to determine their presence and amounts in NR (Nun-anan, Wisunthorn, Pichaiyut, Vennemann, & Nakason, 2018; Rolere, Liengprayoon, Vaysse, Sainte-Beuve, & Bonfils, 2015; Wu et al., 2017).

Although the rubber industry is facing challenges to deal with the variability of NR quality, it could also take advantage of the great richness and diversity of the biochemical composition of *Hevea* latex. This would be a tremendous opportunity to move forward in the era of the circular economy and sustainable development. Indeed, a way to make the rubber industry more sustainable is to valorize the co-products found in the wastewater of factories where latex is processed into NR. While most of the hydrophobic components are retained in NR, the water-soluble compounds are eliminated in the wastewater of factories. For instance, hevein is found in the effluent of a rubber factory at a concentration of 0.7 g/L effluent (Soedjanaatmadja, Subroto, & Beintema, 1995b), and quebrachitol is one of the major organic pollutants in the processes involving latex coagulation and release of serum (Danwanichakul, Pohom, & Yingsampancharoen, 2019; Jiang, Zhang, Wu, Meng, & Xue, 2014; Xue et al., 2014). These biomolecules display interesting properties with potential value, *e.g.* hevein is antifungal, quebrachitol has medicinal properties, thiols and ascorbate are strong antioxidants (Zhang, Leclercq, & Montoro, 2017) and proteins of C-serum as well (Kerche-Silva et al., 2016). Therefore, the effluents of rubber factories could be an infinite source of biomolecules to be valorized. This is precisely the objective of an international patent filed by a Malaysian-Japanese team claiming that 9 kg of quebrachitol can be purified from 1 ton of produced NR (Udagawa, Machida, & Ogawa, 1989). With a global NR production reaching 14 million tons in 2018, this would mean a potential production of 126,000 tons of quebrachitol (Vaysse et al., 2012). Another way to recycle the factory wastewater is to use it to irrigate crops allowing to reach higher yields (Chaiprapat & Sdoodee, 2007). This could be explained by the high content of minerals found in the wastewater. As an example, Vaysse et al. has recently demonstrated that about one-third of the amount of phosphorus and potassium brought by the fertilization of a rubber plantation is lost in the wastewater (Vaysse et al., 2017). The topics of recycling and valorization of co-products in the rubber industry are essential to make it more sustainable, and efforts in the future should definitively be directed toward these research questions.

Acknowledgments

I am very grateful to my colleagues from CIRAD working in the field of natural rubber and I thank them very much, especially to Dr. Jérôme Sainte-Beuve, Dr. Frédéric Bonfils and Dr. Laurent Vaysse for constant support, valuable advices and sharing of their large knowledge on latex and natural rubber. Many thanks as well to my colleagues Dr. Siriluck Liengprayoon and Dr. Natedao Musigamart from KAPI (Kasetsart University, Bangkok) for those very enriching eight past years in Thailand.

References

Akasawa, A., Hsieh, L.-S., Martin, B. M., Liu, T., & Lin, Y. (1996). A novel acidic allergen, Hev b 5, in latex. Purification, cloning and characterization. *The Journal of Biological Chemistry, 271*, 25389–25393.

Alam, B., Das, G., Raj, S., Roy, S., Pal, T. K., & Dey, S. K. (2003). Studies on yield and biochemical sub-components of latex of rubber trees (*Hevea brasiliensis*) with a special reference to the impact of low temperature in a non-optimal environment. *Journal of Rubber Research, 6*, 241–257.

Andrews, E. H., & Dickenson, P. B. (1960). Preliminary electron microscope observations on the ultrastructure of the latex vessels and its contents in young tissue of *Hevea brasiliensis*. *Proceedings of the Natural Rubber Research Conference*, 756–760 (Kuala Lumpur).

Archer, B. L. (1960). The proteins of *Hevea brasiliensis* latex. 4. Isolation and characterization of crystalline hevein. *Biochemical Journal, 75*, 236–240.

Archer, B. L. (1976). Hevamine: A crystalline basic protein from *Hevea brasiliensis* latex. *Phytochemistry, 15*, 297–300.

Archer, B. L., & Audley, B. G. (1967). Biosynthesis of rubber. *Advances in Enzymology and Related Areas of Molecular Biology, 29*, 221–257.

Archer, B. L., Audley, B. G., Cockbain, E., & McSweeney, G. P. (1963). The biosynthesis of rubber. Incorporation of mevalonate and isopentenyl pyrophosphate into rubber by *Hevea brasiliensis*-latex fractions. *Biochemical Journal, 89*, 560–565.

Archer, B. L., Audley, B. G., McSweeney, G. P., & Hong, T. C. (1969). Studies on composition of latex serum and "bottom fraction" particles. *Journal of the Rubber Research Institute of Malaysia, 21*, 560–569.

Archer, B. L., Barnard, D., Cockbain, E. G. C., Dickenson, P. B., & McMullen, A. I. (1963). Structure, composition and biochemistry of *Hevea* latex. In L. Bateman (Ed.), *The chemistry and physics of rubber-like substances: Studies of the natural rubber producers' research association*. Applied Science Publishers.

Archer, B. L., & Cockbain, E. G. (1955). The proteins of *Hevea brasiliensis* Latex 2. Isolation of the alpha-globulin of fresh latex serum. *Biochemical Journal, 61*, 508–512.

Archer, B. L., & Sekhar, B. C. (1955). The proteins of *Hevea brasiliensis* 1. Protein constituents of fresh latex serum. *Biochemical Journal, 61*, 413–416.

Arreguín, B., Lara, P., & Rodríguez, R. (1988). Comparative study of electrophoretic patterns of latex proteins from clones of *Hevea brasiliensis*. *Electrophoresis, 9*, 323–326.

Audley, B. G. (1966). The isolation and composition of helical protein microfibrils from *Hevea brasiliensis* latex. *Biochemical Journal, 98*, 335–341.

Bahri, A. R. S., & Hamzah, S. (1996). Immunocytochemical localisation of rubber membrane protein in *Hevea* latex. *Journal of Natural Rubber Research, 11*, 88–95.

Bauer, G., Gorb, S. N., Klein, M.-C., Nellesen, A., von Tapavicza, M., & Speck, T. (2014). Comparative study on plant latex particles and latex coagulation in *Ficus benjamina, Campanula glomerata* and three *Euphorbia species. PLoS One, 9*, e113336.

Baur, X., Chen, Z., Raulf-Heimsoth, M., & Degens, P. (1997). Protein and allergen content of various natural latex articles. *Allergy, 52*, 661–664.

Bealing, F. J. (1969). Carbohydrate metabolism in *Hevea* latex— availability and utilization of substrates. *Journal of the Rubber Research Institute of Malaysia, 21*, 445–455.

Bealing, F. J. (1981). Quebrachitol synthesis in *Hevea brasiliensis. Journal of the Rubber Research Institute of Malaysia, 29*, 111–122.

Bellacicco, S., Prades, A., Char, C., Vaysse, L., Granet, F., Lacote, R., et al. (2018). The sugar and polyol composition of *Hevea brasiliensis* latex depends on the clonal origin of the tree. *Journal of Rubber Research, 21*, 224–235.

Berthelot, K., Lecomte, S., Coulary-Salin, B., Bentaleb, A., & Peruch, F. (2016). *Hevea brasiliensis* prohevein possesses a conserved C-terminal domain with amyloid-like properties in vitro. *BBA - Proteins and Proteomics, 1864*, 388–399.

Berthelot, K., Lecomte, S., Estevez, Y., Coulary-Salin, B., Bentaleb, A., Cullin, C., et al. (2012). Rubber elongation factor (REF), a major allergen component in *Hevea brasiliensis* latex has amyloid properties. *PLoS One, 7*, e48065.

Berthelot, K., Lecomte, S., Estevez, Y., & Peruch, F. (2014). *Hevea brasiliensis* REF (Hevb1) and SRPP (Hevb3): An overview of possible functions of rubber particle proteins. *Biochimie, 106*, 1–9.

Berthelot, K., Lecomte, S., Estevez, Y., Zhendre, V., Henry, S., Thévenot, J., et al. (2014). Rubber particle proteins, HbREF and HbSRPP, show different interactions with model membranes. *BBA - Biomembranes, 1838*, 287–299.

Berthelot, K., Peruch, F., & Lecomte, S. (2016). Highlights on *Hevea brasiliensis* (pro)hevein proteins. *Biochimie, 127*, 258–270.

Blackley, D. C. (1997). Volume 2: Types of latices. In *Polymer latices - science and technology*.

Bottier, C., Gross, B., Wadeesirisak, K., Srisomboon, S., Jantarasunthorn, S., Musigamart, N., et al. (2019). Rapid evolution of biochemical and physicochemical indicators of ammonia- stabilized *Hevea* latex during the first twelve days of storage. *Colloids and Surfaces A, 570*, 487–498.

Chaiprapat, S., & Sdoodee, S. (2007). Effects of wastewater recycling from natural rubber smoked sheet production on economic crops in southern Thailand. *Resources, Conservation and Recycling, 51*, 577–590.

Chaiyut, J., Liengprayoon, S., Bottier, C., Lecomte, S., Peruch, F., Granet, F., et al. (2017). Fractionation of *Hevea brasiliensis* latex by centrifugation: (ii) a mean to locate the drivers of natural rubber unique structure and properties?. In *Conference of the International Rubber Research and Development Board* (pp. 661–678). Jakarta.

Chen, S.-F., & Ng, C.-S. (1984). The natural higher fatty acid soaps in natural rubber latex and their effect on the mechanical stability of the latex. *Rubber Chemistry and Technology, 57*, 243–253.

Cherian, S., Ryu, S. B., & Cornish, K. (2019). Natural rubber biosynthesis in plants, the rubber transferase complex, and metabolic engineering progress and prospects. *Plant Biotechnology Journal, 17*, 2041–2061.

Chotiphan, R., Vaysse, L., Lacote, R., Gohet, E., Thaler, P., Sajjaphan, K., et al. (2019). Can fertilization be a driver of rubber plantation intensification? *Industrial Crops and Products, 141*, 111813.

Chow, K., Wan, K., Noor, M., Isa, M., Bahari, A., & Tan, S. (2007). Insights into rubber biosynthesis from transcriptome analysis of *Hevea brasiliensis* latex. *Journal of Experimental Botany, 58*, 2429–2440.

Chrestin, H., Jacob, J.-L., & d'Auzac, J. (1984). Role of the lutoidic tonoplast in the control of the cytosolic homeostasis within the laticiferous cells of *Hevea. Zeitschrift für Pflanzenphysiologie, 114*, 269–277.

Cockbain, E. G., & Philpott, M. W. (1963). Colloidal properties of latex. In L. Bateman (Ed.), *The chemistry and physics of rubber and rubber-like substances*. London: Maclaren and Sons.

Cook, A. S., & Sekhar, B. C. (1954). Fractions from *Hevea brasiliensis* latex centrifuged at 59,000g. *Rubber Chemistry and Technology, 27*, 297–301.

Cornish, K., Scott, D. J., Xie, W., Mau, C. J. D., Feng, Y., Liu, X.-H., et al. (2018). Unusual subunits are directly involved in binding substrates for natural rubber biosynthesis in multiple plant species. *Phytochemistry, 156*, 55–72.

Cornish, K., Wood, D. F., & Windle, J. J. (1999). Rubber particles from four different species, examined by transmission electron microscopy and electron-paramagnetic-resonance spin labeling , are found to consist of a homogeneous rubber core enclosed by a contiguous, monolayer biomembrane. *Planta, 210*, 85–96.

Cretin, H. (1982). The proton gradient across the vacuo-lysosomal membrane of lutoids from the latex of *Hevea brasiliensis*. I. Further evidence for a proton-translocating ATPase on the vacuo-lysosomal membrane of intact lutoids. *The Journal of Membrane Biology, 65*, 175–184.

Dai, L., Kang, G., Li, Y., Nie, Z., Duan, C., & Zeng, R. (2013). In-depth proteome analysis of the rubber particle of *Hevea brasiliensis* (para rubber tree). *Plant Molecular Biology*, 155–168.

Danwanichakul, P., Pohom, W., & Yingsampancharoen, J. (2019). L-Quebrachitol from acidic serum obtained after rubber coagulation of skim natural rubber latex L -Quebrachitol. *Industrial Crops and Products, 137*, 157–161.

Dennis, M. S., & Light, D. R. (1989). Rubber elongation factor from *H. brasiliensis. The Journal of Biological Chemistry, 264*, 18608–18617.

Dickenson, P. B. (1969). Latex vessel system of *Hevea brasiliensis. Journal of the Rubber Research Institute of Malaysia, 21*, 535–559.

Dupont, J., Moreau, F., Lance, C., & Jacob, J.-L. (1976). Phospholipid composition of the membrane of lutoids from *Hevea brasiliensis* latex. *Phytochemistry, 15*, 1215–1217.

D'Amato, A., Bachi, A., Fasoli, E., Boschetti, E., Peltre, G., Sénéchal, H., et al. (2010). In-depth exploration of *Hevea brasiliensis* latex proteome and "hidden allergens" via combinatorial peptide ligand libraries. *Journal of Proteomics, 73*, 1368–1380.

D'Auzac, J. (1964). Mise en evidence de la glycolyse et de ses relations avec la biosynthèse du caoutchouc au sein du latex d'*Hevea brasiliensis. Revue Generale des Caoutchoucs & Plastiques, 41*, 1831–1834.

D'Auzac, J., Crétin, H., Marin, B., & Lioret, C. (1982). A plant vacuolar system: The lutoids from *Hevea brasiliensis* latex. *Physiologie Vegetale, 20*, 311–331.

D'Auzac, J., & Jacob, J.-L. (1969). Regulation of glycolysis in latex of *Hevea brasiliensis. Rubber Research Institute of Malaya, Journal, 21*, 417–444.

D'Auzac, J., & Jacob, J.-L. (1989). Section 2. The composition of latex from *Hevea brasiliensis* as A laticiferous cytoplasm. In J. D'Auzac, J.-L. Jacob, & H. Chrestin (Eds.), *Physiology of rubber tree latex* (pp. 56–96). Boca Raton, Florida: CRC press.

D'Auzac, J., & Pujarniscle, S. (1959). Les glucides de l'*Hevea brasiliensis. Revue Generale des Caoutchoucs & Plastiques, 36*, 1687–1692.

Epping, J., van Deenen, N., Niephaus, E., Stolze, A., Fricke, J., Huber, C., et al. (2015). A rubber transferase activator is necessary for natural rubber biosynthesis in dandelion. *Nature Plants, 1*, 15048.

Frey-Wyssling, A. (1929). Microscopisch onderzock haar het woorkomen van harsen in latex van *Hevea. Archief voor de Rubbercultuur, 13*, 394–434.

Galiani, P. D., Martins, M. A., Gonc, P. D. S., McMahan, C. M., Henrique, L., & Mattoso, C. (2011). Seasonal and clonal variations in technological and thermal properties of raw *Hevea* natural rubber. *Journal of Applied Polymer Science, 122*, 2749—2755.

Gidrol, X., Chrestin, H., Tan, H.-L., & Kush, A. (1994). Hevein, a lectin-like protein from *Hevea brasiliensis* (Rubber Tree) is involved in the coagulation of latex. *The Journal of Biological Chemistry, 269*, 9278—9283.

Gomez, J. B. (1948). Lutoids of *Hevea* latex : Morphological considerations. *Journal of Natural Rubber Research, 5*, 231—240.

Gomez, J. B. (1980). Factors influencing the colloidal stability of fresh clonal *Hevea* latices as determined by the aerosol OT test. *Journal of the Rubber Research Institute of Malaysia, 28*, 86—106.

Gomez, J. B., & Southorn, W. A. (1969). Studies on lutoid membrane ultrastructure. *Journal of the Rubber Research Institute of Malaysia, 21*, 513—523.

Gopalakrishnan, J., Thomas, M., Philip, A., Krishnakumar, R., & Jacob, J.-L. (2010). Contribution of latex cations to the water relations and latex yield in *Hevea brasiliensis*. *Journal of Natural Rubber Research, 23*, 93—97.

Habib, M. A. H., Yuen, G. C., Othman, F., Zainudin, N. N., Latiff, A. A., & Ismail, M. N. (2016). Proteomics analysis of latex from *Hevea brasiliensis* (clone RRIM 600). *Biochemistry and Cell Biology, 1—33.*

Hanower, P., Brzozowska, J., & Lioret, C. (1976). Étude du mécanisme de la coagulation du latex d'*Hevea brasiliensis* (Kunth) Müll. Arg. 1. Facteurs agissant sur la coagulation. *Physiologie Vegetale, 14*, 677—693.

Hasma, H. (1991). Lipids associated with rubber particles and their possible role in mechanical stability of latex concentrates. *Journal of Natural Rubber Research, 6*, 105—114.

Hasma, H. (1992). Proteins of natural rubber latex concentrate. *Journal of Natural Rubber Research, 7*, 102—112.

Hasma, H., & Othman, A. B. (1990). Role of some non rubber constituents on thermal oxidative ageing of natural rubber. *Journal of Natural Rubber Research, 5*, 1—8.

Hasma, H., & Subramaniam, A. (1986). Composition of lipids in latex of *Hevea brasiliensis* clone RRIM501. *Journal of Natural Rubber Research, 1*, 30—40.

Hathwaik, U., Lin, J.-T., & McMahan, C. (2018). Molecular species of triacylglycerols in the rubber particles of *Parthenium argentatum* and *Hevea brasiliensis*. *Biocatalysis and Agricultural Biotechnology, 16*, 107—114.

Havanapan, P., Bourchookarn, A., Ketterman, A. J., & Krittanai, C. (2016). Comparative proteome analysis of rubber latex serum from pathogenic fungi tolerant and susceptible rubber tree (*Hevea brasiliensis*). *Journal of Proteomics, 131*, 82—92.

Ho, C., Kondo, T., Muramatsu, N., & Oshima, H. (1996). Surface structure of natural rubber latex particles from electrophoretic mobility data. *Journal of Colloid and Interface Science, 445*, 442—445.

Homans, L. N. S., Dalfsen, J. W., & van Gils, G. E. (1948). Complexity of fresh *Hevea* latex. *Nature, 161*, 177—178.

Homans, L. N. S., & van Gils, G. E. (1950). Fresh *Hevea* Latex. A complex colloidal system. *Rubber Chemistry and Technology, 23*, 644—651.

IRSG. (2018). *Quarterly Statistics by the International Rubber Study group.*

Jacob, J.-L., D'Auzac, J., & Prevôt, J.-C. (1993). The composition of natural latex from *Hevea brasiliensis*. *Clinical Reviews in Allergy, 11*, 325—337.

Jacob, J.-L., Nouvel, A., & Prévôt, J.-C. (1978). Electrophorèse et mise en évidence d'activités enzymatiques dans le latex d'*Hevea brasiliensis*. *Revue Generale des Caoutchoucs & Plastiques, 55*, 87—90.

Jacob, J.-L., Prevôt, J.-C., Clement-Vidal, A., & D'Aauzac, J. (1989). Inorganic pyrophosphate metabolism in *Hevea brasiliensis* latex - Characteristics of cytosolic alkaline inorganic pyrophosphatase. *Plant Physiology and Biochemistry, 27*, 355—364.

Jacob, J.-L., Prevôt, J.-C., & Kekwick, R. G. O. (1989). Section 3. The metabolism of the laticiferous cells of *Hevea brasiliensis*. In J. D'Auzac, J.-L. Jacob, & H. Chrestin (Eds.), *Physiology of rubber tree latex* (pp. 99–144). Boca Raton, Florida: CRC Press.

Jawjit, W., Pavasant, P., & Kroeze, C. (2015). Evaluating environmental performance of concentrated latex production in Thailand. *Journal of Cleaner Production, 98*, 84–91.

Jekel, P. A., Hartmann, B. H., & Beintema, J. J. (1991). The primary structure of hevamine , an enzyme with lysozyme/chitinase activity from *Hevea brasiliensis* latex. *European Journal of Biochemistry, 200*, 123–130.

Jekel, P. A., Hofsteenge, J., & Beintema, J. J. (2003). The patatin-like protein from the latex of *Hevea brasiliensis* (Hev b 7) is not a vacuolar protein. *Phytochemistry, 63*, 517–522.

Jiang, S.-K., Zhang, G.-M., Wu, Y., Meng, Z.-H., & Xue, M. (2014). Isolation and characterisation of L-quebrachitol from rubber factory wastewater. *Journal of Rubber Research, 17*, 23–33.

Karino, T., Ikeda, Y., Yasuda, Y., Kohjiya, S., & Shibayama, M. (2007). Nonuniformity in natural rubber as revealed by small-angle neutron scattering, small-angle X-ray scattering, and atomic force microscopy. *Biomacromolecules, 8*, 693–699.

Kawahara, S., Kakubo, T., Sakdapipanich, J. T., Isono, Y., & Tanaka, Y. (2000). Characterization of fatty acids linked to natural rubber - role of linked fatty acids on crystallization of the rubber. *Polymer, 41*, 7483–7488.

Kawahara, S., & Tanaka, Y. (1995). Plasticization and crystallization of cis- 1,4 polyisoprene mixed with methyl linolate. *Journal of Polymer Science, 33*, 753–758.

Kekwick, R., Bhambri, S., Chabane, M. H., Autegarden, J. E., Levy, D. a, & Leynadier, F. (1996). The allergenic properties of fresh and preserved *Hevea brasiliensis* latex protein preparations. *Clinical and Experimental Immunology, 104*, 337–342.

Kerche-Silva, L. E., Cavalcante, D. G. S. M., Danna, C. S., Gomes, A. S., Carrara, I. M., Cecchini, A. L., et al. (2016). Free-radical scavenging properties and cytotoxic activity evaluation of latex C-serum from *Hevea brasiliensis* RRIM 600. *Free Radicals and Antioxidants,* 7, 107–114.

Kumarn, S., Churinthorn, N., Nimpaiboon, A., Sriring, M., Ho, C.-C., Takahara, A., et al. (2018). Investigating the mechanistic and structural role of lipid hydrolysis in the stabilization of ammonia-preserved *Hevea* rubber latex. *Langmuir, 34*, 12730–12738.

Lacote, R., Obouayeba, S., Dian, K., Gnagne, M., & Gohet, E. (2004). Panel management in rubber (*Hevea brasiliensis*) tapping and impact on yield, growth, and latex diagnosis. *Journal of Rubber Research,* 7, 199–217.

Lakusta, A. M., Kwon, M., Kwon, E. G., Stonebloom, S., Scheller, H. V., & Ro, D. (2019). Molecular studies of the protein complexes involving cis-Prenyltransferase in guayule (*Parthenium argentatum*), an alternative rubber-producing plant. *Frontiers in Plant Science, 10*, 1–14.

Lau, N., Makita, Y., Kawashima, M., Taylor, T. D., Kondo, S., Othman, A. S., et al. (2016). The rubber tree genome shows expansion of gene family associated with rubber biosynthesis. *Scientific Reports, 6*, 28594.

Le Roux, Y., Ehabe, E., Sainte-Beuve, J., Nkengafac, J., Nkeng, J., Ngolemasango, F., et al. (2000). Seasonal and clonal variations in the latex and raw rubber of *Hevea brasiliensis*. *Journal of Rubber Research, 3*, 142–156.

Lee, H., Broekaert, W. F., & Raikhelsii, N. V. (1991). Co- and post-translational processing of the hevein preproprotein of latex of the rubber tree (*Hevea brasiliensis*). *The Journal of Biological Chemistry, 266*, 15944–15948.

Liengprayoon, S., Bonfils, F., Sainte-Beuve, J., Sriroth, K., Dubreucq, E., & Vaysse, L. (2008). Development of a new procedure for lipid extraction from *Hevea brasiliensis* natural rubber. *European Journal of Lipid Science and Technology, 110*, 563–569.

Liengprayoon, S., Chaiyut, J., Sriroth, K., Bonfils, F., Sainte-Beuve, J., Dubreucq, E., et al. (2013). Lipid compositions of latex and sheet rubber from *Hevea brasiliensis* depend on clonal origin. *European Journal of Lipid Science and Technology, 115*, 1021–1031.

Liengprayoon, S., Sriroth, K., Dubreucq, E., & Vaysse, L. (2011). Glycolipid composition of *Hevea brasiliensis* latex. *Phytochemistry, 72*, 1902–1913.

Liengprayoon, S., Vaysse, L., Jantarasunthorn, S., Wadeesirisak, K., Chaiyut, J., Srisomboon, S., et al. (2017). Fractionation of *Hevea brasiliensis* latex by centrifugation :(i) a comprehensive description of the biochemical composition of the 4 centrifugation fractions. In *Conference of the International rubber research and Development Board* (pp. 645–660). Jakarta.

Li, D. J., Yan, J., Deng, Z., Chen, C. L., He, P., Wu, M., et al. (2009). Evaluation of three methods for protein extraction suitable for 2-DE of *Hevea brasiliensis* C-serum. *Chinese Agricultural Science Bulletin, 25*, 273–279.

Lotti, C., Moreno, R. M. B., Gonçalves, P. S., & Bhattacharya, S. (2012). Extensional rheology of raw natural rubber from new clones of *Hevea brasiliensis*. *Polymer Engineering and Science, 52*, 139–148.

Low, F. C. (1978). Distribution and concentration of major soluble carbohydrates in *Hevea* latex, the effect of ethephon stimulation and the possible role of these carbohydrates in latex flow. *Journal of the Rubber Research Institute of Malaysia, 26*, 21–32.

Lynn, K. R., & Clevette-Radford, N. A. (1984). Purification and characterization of hevain, a serine protease from *Hevea brasiliensis*. *Phytochemistry, 23*, 963–964.

Lynn, K. R., & Clevette-Radford, N. A. (1986). Hevains: Serine-centred proteases from the latex of *Hevea brasiliensis*. *Phytochemistry, 25*, 2279–2282.

Marin, B. P. (1993). The traffic of molecules across the tonoplast of plant cells. In D. J. Morré, K. E. Howell, & J. J. M. Bergeron (Eds.), *Molecular mechanisms of membrane traffic* (pp. 109–110). Berlin, Heidelberg: Springer.

Men, X., Wang, F., Chen, G., Zhang, H., & Xian, M. (2019). Biosynthesis of natural Rubber : Current state and perspectives. *International Journal of Molecular Sciences, 20*, 50.

Moir, G. F. J. (1959). Ultracentrifugation and staining of *Hevea* latex. *Nature, 184*, 1626–1628.

Moir, J., & Tata, S. J. (1960). The proteins of *Hevea brasiliensis* latex. Part 3 the soluble proteins of bottom fraction. *Journal of the Rubber Research Institute of Malaysia, 16*, 155–165.

Moreau, F., Jacob, J.-L., Dupont, J., & Lance, C. (1975). Electron transport in the membrane of lutoids from the latex of *Hevea brasiliensis*. *BBA - Bioenergetics, 396*, 116–124.

Moreno, R. M. B., Ferreira, M., Gonçalves, P. D. S., & Mattoso, L. H. C. (2005). Technological properties of latex and natural rubber of *Hevea brasiliensis* clones. *Scientia Agricola, 62*, 122–126.

Nawamawat, K., Sakdapipanich, J. T., Ho, C. C., Ma, Y., Song, J., & Vancso, J. G. (2011). Surface nanostructure of *Hevea brasiliensis* natural rubber latex particles. *Colloids and Surfaces A: Physicochemical and Engineering Aspects, 390*, 157–166.

Nun-anan, P., Wisunthorn, S., Pichaiyut, S., Nathaworn, C., & Nakason, C. (2019). Influence of nonrubber components on properties of unvulcanized natural rubber. *Polymers for Advanced Technologies*, 1–16.

Nun-anan, P., Wisunthorn, S., Pichaiyut, S., Vennemann, N., & Nakason, C. (2018). Novel approach to determine non-rubber content in *Hevea brasiliensis* : Influence of clone variation on properties of un-vulcanized natural rubber. *Industrial Crops and Products, 118*, 38–47.

Oh, S. K., Kang, H., Shin, D. H., Yang, J., Chow, K. S., Yeang, H. Y., et al. (1999). Isolation, characterization, and functional analysis of a novel cDNA clone encoding a small rubber particle protein from *Hevea brasiliensis*. *The Journal of Biological Chemistry, 274*, 17132–17138.

Pakianathan, S. W., Tata, S. J., Chon, L. F., & Sethuraj, M. R. (1992). Chapter 14 - Certain aspects of physiology and biochemistry of latex production. In M. R. Sethuraj, & N. M. Mathew (Eds.), *Natural rubber: Biology, cultivation and technology* (pp. 298–323). Elsevier.

van Parijs, J., Broekaert, W. F., Goldstein, I. J., & Peumans, W. J. (1991). Hevein: An antifungal protein from rubber-tree latex. *Planta*, 258–264.

Peixinho, C. M., Gabriel, M. F., Caramelo-Nunes, C., Tavares-Ratado, P., & Tomaz, C. T. (2014). Separation of the four most important latex allergens from latex gloves : A potential tool for diagnosis and immunotherapy purposes. *Allergologia et Immunopathologia, 20*, 380–383.

Peixinho, C. M., Tavares, P., Tomáz, M. R., Taborda-Barata, L., & Tomáz, C. (2006). Differential expression of allergens on the internal and external surfaces of latex surgical gloves. *Allergologia et Immunopathologia, 34*, 206–211.

Philpott, M. W., & Westgarth, D. R. (1953). Stability and mineral composition of *Hevea* latex. *Journal of Rubber Research, 14*, 133–148.

Posch, A., Chen, Z., Wheeler, C., Dunn, M. J., Raulf-Heimsoth, M., & Baur, X. (1997). Characterization and identification of latex allergens by two-dimensional electrophoresis and protein microsequencing. *The Journal of Allergy and Clinical Immunology, 99*, 385–395.

Qu, Y., Chakrabarty, R., Tran, H. T., Kwon, E.-J. G., Kwon, M., Nguyen, T.-D., et al. (2015). A lettuce (Lactuca sativa) homolog of human Nogo-B receptor interacts with cis-prenyltransferase and is necessary for natural rubber biosynthesis. *The Journal of Biological Chemistry, 290*, 1898–1914.

Rahman, A. Y. A., Usharraj, A. O., Misra, B. B., Thottathil, G. P., Jayasekaran, K., Feng, Y., et al. (2013). Draft genome sequence of the rubber tree *Hevea brasiliensis*. *BMC Genomics, 14*, 75.

Raulf, M., & Rihs, H.-P. (2017). Latex allergens: Source of sensitization and single allergens. In J. Kleine-Tebbe, & T. Jakob (Eds.), *Molecular allergy diagnostics* (pp. 459–470).

Rihs, H.-P., Chen, Z., Rozynek, P., & Cremer, R. (2001). Allergenicity of rHev b 10 (manganese-superoxide dismutase). *Allergy, 56*, 85–86.

Rippel, M., Alberto, C., Leite, P., & Galembeck, F. (2002). Elemental mapping in natural rubber latex films by electron energy loss spectroscopy associated with transmission electron microscopy. *Analytical Chemistry, 74*, 2541–2546.

Rochette, C. N., Crassous, J., Drechsler, M., Gaboriaud, F., & Eloy, M. (2013). Shell structure of natural rubber particles : Evidence of chemical stratification by electrokinetics and cryo-TEM. *Langmuir, 29*, 14655–14665.

Roe, C. P., & Ewart, R. H. (1942). An electrophoretic study of the proteins of milk serum. *Journal of the American Chemical Society, 64*, 177–183.

Rolere, S., Bottier, C., Vaysse, L., Sainte-Beuve, J., & Bonfils, F. (2016). Characterisation of macrogel composition from industrial natural rubber samples: Influence of proteins on the macrogel crosslink density. *Express Polymer Letters, 10*, 408–419.

Rolere, S., Char, C., Taulemesse, J.-M., Bergeret, A., Sainte-Beuve, J., & Bonfils, F. (2016). The majority of minerals present in natural rubber are associated with the macrogel: An ICP-MS and SEM/EDX investigation. *Journal of Applied Polymer Science, 133*, 1–11.

Rolere, S., Liengprayoon, S., Vaysse, L., Sainte-Beuve, J., & Bonfils, F. (2015). Investigating natural rubber composition with Fourier transform infrared (FT-IR) spectroscopy: A rapid and non-destructive method to determine both protein and lipid contents simultaneously. *Polymer Testing, 43*, 83–93.

Rozynek, P., Posch, A., & Baur, X. (1998). Cloning, expression and characterization of the major latex allergen prohevein. *Clinical and Experimental Allergy, 28*, 1418–1426.

Seppälä, U., Palosuo, T., Kalkkinen, N., Seppälä, U., Ylitalo, L., Reunala, T., et al. (2000). IgE reactivity to patatin-like latex allergen, Hev b 7, and to patatin of potato tuber, Sol t 1, in adults and children allergic to natural rubber latex. *Allergy, 55*, 266–273.

Sheldrake, A. R. (1979). Effects of osmotic stress on polar auxin transport in arena mesocotyl sections A. *Planta, 145*, 113–117.

Siler, D. J., Goodrich-Tanrikulu, M., Cornish, K., Stafford, A. E., & Mckeon, T. A. (1997). Composition of rubber particles of *Hevea brasiliensis, Parthenium argentatum, Ficus elastica,* and *Euphorbia lactiflua* indicates unconventional surface structure. *Plant Physiology and Biochemistry, 35*, 881–889.

Singh, A. P., Wi, S. G., Chung, G. C., Kim, Y. S., & Kang, H. (2003). The micromorphology and protein characterization of rubber particles in *Ficus carica, Ficus benghalensis* and *Hevea brasiliensis. Journal of Experimental Botany, 54*, 985–992.

Smith, R. H. (1954). The phosphatides of the latex of *Hevea brasiliensis*. 3. Carbohydrate and polyhydroxy constituents. *Biochemical Journal, 57*, 140–144.

Soedjanaatmadja, U. M. S., Subroto, T., & Beintema, J. J. (1995a). Processed products of the hevein precursor in the latex of the rubber tree (*Hevea brasiliensis*). *FEBS Letters, 363*, 211–213.

Soedjanaatmadja, U. M. S., Subroto, T., & Beintema, J. J. (1995b). The effluent of natural rubber factories is enriched in the antifungal protein hevein. *Bioresource Technology, 53*, 39–41.

Southorn, W. A. (1960). Complex particles in *Hevea* latex. *Nature, 187*, 1048–1049.

Southorn, W. A. (1968). Latex flow studies I. Electron microscopy of *Hevea brasiliensis* in the region of the tapping cut. *Journal of the Rubber Research Institute of Malaysia, 20*, 176–186.

Southorn, W. A., & Edwin, E. E. (1968). Latex flow studies II. Influence of lutoids on the stability and flow of *Hevea* latex. *Journal of the Rubber Research Institute of Malaysia, 20*, 187–200.

Southorn, W. A., & Yip, E. (1968). Latex flow studies III. Electrostatic considerations in the colloidal stability of fresh *Hevea* latex. *Journal of the Rubber Research Institute of Malaysia, 20*, 201–215.

Subroto, T., Koningsveld, G. A., Schreuder, H. a, & Soedjanaatmadja, U. M. S. (1996). Chitinase and beta-1,3 glucanase in the lutoid-body fraction of *Hevea* Latex. *Phytochemistry, 43*, 29–37.

Sunderasan, E., Hamzah, S., Hamid, S., Ward, M. A., Yeang, H. Y., & Cardosa, M. J. (1995). Latex B-Serum b-1,3-glucanase (Hev b II) and a component of the microhelix (Hev b IV) are major latex allergens. *Journal of Rubber Research, 10*, 82–99.

Sussman, G. L., Beezhold, D. H., & Kurup, V. P. (2002). Allergens and natural rubber proteins. *The Journal of Allergy and Clinical Immunology, 110*, 33–39.

Tang, C., Yang, M., Fang, Y., Luo, Y., Gao, S., Xiao, X., et al. (2016). The rubber tree genome reveals new insights into rubber production and species adaptation. *Nature Plants, 2*, 16073.

Tarachiwin, L., Sakdapipanich, J., Ute, K., Kitayama, T., Bamba, T., Fukusaki, E.-I., et al. (2005). Structural characterization of alpha-terminal group of natural rubber. 1. Decomposition of branch-points by lipase and phosphatase treatments. *Biomacromolecules, 6*, 1851–1857.

Tarachiwin, L., Sakdapipanich, J., Ute, K., Kitayama, T., & Tanaka, Y. (2005). Structural characterization of alpha-terminal group of natural rubber. 2. Decomposition of branch-points by phospholipase and chemical treatments. *Biomacromolecules, 6*, 1858–1863.

Tata, S. J. (1980). Distribution of proteins between the fractions of *Hevea* latex separated by ultracentrifugation. *Journal of the Rubber Research Institute of Malaysia, 28*, 77–85.

Tata, S. J., Boyce, A. N., Archer, B. L., & Audley, B. G. (1976). Lysozymes: Major components of the sedimentable phase of *Hevea brasiliensis* latex. *Journal of the Rubber Research Institute of Malaysia, 24*, 233–236.

Tata, S. J., & Edwin, E. E. (1970). *Hevea* latex enzymes detected by zymogram technique after starch gel electrophoresis. *Journal of the Rubber Research Institute of Malaysia, 23*, 1–12.

Tata, S. J., & Gomez, J. B. (1980). Isolation and characterisation of microhelices from lutoids of *Hevea* latex. *Journal of the Rubber Research Institute of Malaysia, 28*, 67–76.

Tata, S. J., & Moir, G. F. J. (1964). Part 5. Starch gel electrophoresis of C-serum proteins. *Journal of the Rubber Research Institute of Malaysia, 18*, 97–108.

Terwisscha van Scheltinga, A. C., Kalk, K. H., Beintema, J. J., & Dijkstra, B. W. (1994). Crystal structures of hevamine, a plant defence protein with chitinase and lysozyme activity, and its complex with an inhibitor. *Structure, 2*, 1181–1189.

Thuong, N. T., Yamamoto, O., Nghia, P. T., Cornish, K., & Kawahara, S. (2017). Effect of naturally occurring crosslinking junctions on green strength of natural rubber. *Polymers for Advanced Technologies, 28*, 303–311.

Tian, W.-M., Han, Y.-Q., Wu, J.-L., & Hu, Z.-H. (2003). Fluctuation of microfibrillar protein level in lutoids of primary laticifers in relation to the 67 kDa storage protein in *Hevea brasiliensis*. *Acta Botanica Sinica, 45*, 127–130.

Toki, S., Burger, C., Hsiao, B. S., Amnuaypornsri, S., Sakdapipanich, J., & Tanaka, Y. (2008). Multi-scaled microstructures in natural rubber characterized by synchrotron X-Ray scattering and optical microscopy. *Journal of Polymer Science Part B*, 2456–2464.

Tomazic, V. J., Withrow, T. J., & Hamilton, R. G. (1995). Characterization of the allergen(s) in latex protein extracts. *The Journal of Allergy and Clinical Immunology, 96*, 635–642.

Trade Map. (2018). *Monthly, quarterly and yearly trade data.*

Trinidad, N., Gallego, Z., Lucía, M., & Láinez, Á. (2019). Effect of the phenological stage in the natural rubber latex properties. *Journal of Polymers and the Environment, 27*, 364–371.

Tupy, J., & Resing, W. L. (1968). Anaerobic respiration in latex of *Hevea brasiliensis* substrate and limiting factors. *Biologica Plantarum (Praha), 10*, 72–80.

Udagawa, Y., Machida, M., & Ogawa, S. (1989). *Method of recovering L-quebrachitol from rubber latex serums*. US Patent 5,041,689. The Yokohama Rubber Co., Ltd., Tokyo, Japan; The Board of the Rubber Research Institute of Malaysia, Kuala Lumpur, Malaysia.

Valuer, P., Balland, S., Harf, R., Valenta, R., & Deviller, P. (1995). Identification of profilin as an IgE-binding component in latex from *Hevea brasiliensis*: Clinical implications. *Clinical and Experimental Allergy, 25*, 332–339.

Vaysse, L., Bonfils, F., Sainte-Beuve, J., & Cartault, M. (2012). Natural rubber. In *Polymer science: A comprehensive reference* (pp. 281–293).

Vaysse, L., Sathornluck, S., Bottier, C., Liengprayoon, S., Char, C., Bonfils, F., et al. (2017). Effect of fertilization and stimulation of *Hevea brasiliensis* trees on mineral compositions and properties of produced latex and rubber. In *Conference of the International Rubber Research and Development Board* (pp. 358–369). Jakarta.

Wadeesirisak, K., Castano, S., Berthelot, K., Vaysse, L., Bonfils, F., Peruch, F., et al. (2017). Rubber particle proteins REF1 and SRPP1 interact differently with native lipids extracted from *Hevea brasiliensis* latex. *BBA - Biomembranes, 1859*, 201–210.

Wagner, S., Breiteneder, H., Simon-Nobbe, B., Susani, M., Krebitz, M., Niggemann, B., et al. (2000). Hev b 9, an enolase and a new cross-reactive allergen from *Hevea* latex and molds. *European Journal of Biochemistry, 267*, 7006–7014.

Walujono, K., Scholma, R. A., Beintema, J. J., Mariono, A., & Hahn, A. M. (1975). Amino acid sequence of hevein. *Proceedings of the International Rubber Conference*, 518–531. Kuala Lumpur.

Wang, X., Shi, M., Lu, X., Ma, R., Wu, C., Guo, A., et al. (2010). A method for protein extraction from different subcellular fractions of laticifer latex in *Hevea brasiliensis* compatible with 2-DE and MS. *Proteome Science, 8*, 35–45.

Wang, X., Shi, M., Wang, D., Chen, Y., Cai, F., Zhang, S., et al. (2013). Comparative proteomics of primary and secondary lutoids reveals that chitinase and glucanase play a crucial combined role in rubber particle aggregation in *Hevea brasiliensis*. *Journal of Proteome Research, 12*, 5146–5159.

Wang, D., Sun, Y., Chang, L., Tong, Z., Xie, Q., & Jin, X. (2018). Subcellular proteome profiles of different latex fractions revealed washed solutions from rubber particles contain crucial enzymes for natural rubber biosynthesis. *Journal of Proteomics, 182*, 53–64.

Wang, X., Wang, D., Sun, Y., Yang, Q., Chang, L., Wang, L., et al. (2015). Comprehensive proteomics analysis of laticifer latex reveals new insights into ethylene stimulation of natural rubber production. *Scientific Reports, 5*, 13778.

Wei, Y.-C., Liu, G.-X., Zhang, H.-F., Zhao, F., & Luo, M.-C. (2019). Non-rubber components tuning mechanical properties of natural rubber from vulcanization kinetics. *Polymer, 183*, 121911.

Wisunthorn, S., Chambon, B., Sainte-Beuve, J., & Vaysse, L. (2015). Natural rubber quality starts at the smallholdings : Farmers' cup coagulum production in southern Thailand. *Journal of Rubber Research, 18*, 87–98.

Wititsuwannakul, R., Pasitkul, P., Jewtragoon, P., & Wititsuwannakul, D. (2008). *Hevea* latex lectin binding protein in C-serum as an anti-latex coagulating factor and its role in a proposed new model for latex coagulation. *Phytochemistry, 69*, 656–662.

Wititsuwannakul, R., Pasitkul, P., Kanokwiroon, K., & Wititsuwannakul, D. (2008). A role for a *Hevea* latex lectin-like protein in mediating rubber particle aggregation and latex coagulation. *Phytochemistry, 69*, 339–347.

Wititsuwannakul, R., Rukseree, K., Kanokwiroon, K., & Wititsuwannakul, D. (2008). A rubber particle protein specific for *Hevea* latex lectin binding involved in latex coagulation. *Phytochemistry, 69*, 1111–1118.

Wititsuwannakul, R., & Wititsuwannakul, D. (2001). Biochemistry of natural rubber and structure of latex. In T. Koyama, & A. Steinbüchel (Eds.), *Polyisoprenoids: Vol. 2. Biopolymers* (pp. 151–201).

Wood, D. F., & Cornish, K. (2000). Microstructure of purified rubber particles. *International Journal of Plant Sciences, 161*, 435–445.

Wu, J., Qu, W., Huang, G., Wang, S., Huang, C., & Liu, H. (2017). Super-resolution fluorescence imaging of spatial organization of proteins and lipids in natural rubber. *Biomacromolecules, 18*, 1705–1712.

Xiang, Q., Xia, K., Dai, L., Kang, G., Li, Y., Nie, Z., et al. (2012). Proteome analysis of the large and the small rubber particles of *Hevea brasiliensis* using 2D-DIGE. *Plant Physiology and Biochemistry, 60*, 207–213.

Xue, M., Lv, Z., Chen, J., Dong, X., Meng, Z., Dong, X., et al. (2014). Analysis of L-quebrachitol from the wastewater of rubber latex serum using hydrophilic interaction chromatography and evaporative light scattering detector method. *Journal of Liquid Chromatography & Related Technologies, 38*, 92–96.

Yamashita, S., Yamaguchi, H., Waki, T., Aoki, Y., Mizuno, M., Yanbe, F., et al. (2016). Identification and reconstitution of the rubber biosynthetic machinery on rubber particles from *Hevea brasiliensis*. *eLife, 5*, 1–28.

Yeang, H. Y., Arif, S. A. M., Yusof, F., & Sunderasan, E. (2002). Allergenic proteins of natural rubber latex. *Methods, 27*, 32–45.

Yeang, H. Y., Cheong, K. F., Sunderasan, E., Hamzah, S., Chew, N. P., Hamid, S., et al. (1996). The 14.6 kd rubber elongation factor (Hev b 1) and 24 kd (Hev b 3) rubber particle proteins are recognized by IgE from patients with spina bifida and latex allergy. *The Journal of Allergy and Clinical Immunology, 98*, 628–639.

Yeang, H. Y., Ghandimathi, H., & Paranjothy, K. (1957). Protein and enzyme variation in some *Hevea* cultivars. *Journal of the Rubber Research Institute of Malaysia, 25*, 9–18.

Yip, E. (1990). Clonal characterization of latex and rubber properties. *Journal of Natural Rubber Research*, 1–4.

Yip, E., & Chin, H. C. (1977). Latex flow studies, X. distribution of metallic ions between phases of *Hevea* [rubber] latex and the effects of yield stimulation on this distribution. *Journal of the Rubber Research Institute of Malaysia, 25*, 31–49.

Zhang, Y., Leclercq, J., & Montoro, P. (2017). Reactive oxygen species in *Hevea brasiliensis* latex and relevance to Tapping Panel Dryness. *Tree Physiological Reviews, 37*, 261–269.